BIOLOGICAL HUSBANDRY
A scientific approach to organic farming

Biological Husbandry
A scientific approach to organic farming

B. STONEHOUSE
University of Bradford

Butterworths
London - Boston - Sydney - Durban - Toronto - Wellington

First published 1981

© The several contributors named in the list of contents 1981

British Library Cataloguing in Publication Data

Biological husbandry.
  1. Organic farming   2. Plants, Cultivated
  I. Stonehouse, Bernard
631.5'8   S605.5

ISBN 0–408–10726–X

Reproduced from copy supplied
Printed and bound in Great Britain
by Billing and Sons Limited and Kemp Hall Bindery
Guildford, London, Oxford, Worcester

FOREWORD

The International Institute of Biological Husbandry was founded in 1975 to promote the scientific development of biological or organic agriculture, as a viable alternative to orthodox modern agriculture. Its first international symposium was held at Wye College, Ashford, Kent, on August 26–30, 1980, attracting over 100 delegates—mainly scientists or practical farmers—from more than 20 countries. This volume includes most of the papers presented at that symposium.

Biological husbandry is a system of agriculture that seeks to maintain and improve productivity of land as far as possible by encouraging and enhancing natural biological processes, minimizing the use of chemical fertilizers or pesticides. It is based on traditional methods and has been developed—mostly by practical farmers—to incorporate much that is new in agricultural research and technology. Rejecting the use of resource-extravagant and energy-demanding chemicals, it stresses the need for further understanding and development of natural processes which help the farmer—processes that can be managed or manipulated in nondestructive ways to help him further. Supporters of biological husbandry see particular applications for it in poor countries, where unimproved traditional methods are no longer sufficient but where the insensitive introduction of orthodox, energy-consuming methods of farming leads rapidly to economic and agricultural disaster.

As the proceedings published in this book show, there is little doctrinal orthodoxy in biological husbandry—only a determination to try, in whatever ways may be appropriate to the circumstances, to produce good yields of high-quality food through stable, self-perpetuating and self-contained agriculture. The discussions that followed these papers, continuing well beyond the programmed hours, emphasized this point clearly. There was, however, an awareness of a need for far more research and systematic enquiry into the methods used in biological husbandry, and measurement of quantities and quality of yields obtained. If biological husbandry is to progress—to establish a level of credibility equal to that of orthodox farming—its supporters must themselves graduate beyond the levels of enthusiasm and mystique that have brought them so far, and develop a firm foundation of scientific theory, research and practice.

This volume shows that research is already in progress in many organizations throughout the world, and indicates the important future role for the International Institute of Biological Husbandry in coordinating and disseminating research information.

Bernard Stonehouse
University of Bradford

CONTENTS

INTRODUCTION: ORIGINS OF THE INTERNATIONAL INSTITUTE OF BIOLOGICAL HUSBANDRY

D. R. STICKLAND

International Institute of Biological Husbandry, Ipswich

I have been interested in the organic or biological approach to farming since about 1949 and, while trying to make a living in agriculture in various parts of the world, I have pursued this interest subsequently with a gradually growing determination to succeed.

Back in those days, organic farming was represented in the UK mainly by Friend Sykes, Newman Turner, Professor Lindsay Robb, Lady Eve Balfour and a few others. They were battling against far more difficult circumstances than we have now. There were, for example, Government subsidies on artificial fertilizers and lime, but none on rock phosphate and other so-called organic fertilizers. Any farmer using organic methods was therefore saving the taxpayer money, but also considerably increasing his costs. One cannot at all blame the conventional farmer of that time: after having had all sorts of weed and growth problems he was suddenly presented with magical products that killed the weeds and left the crop nice and clean. Such weeds as poppies and charlock ceased to be a nuisance almost overnight. As far as growing heavier crops was concerned, a few bags of the correct fertilizer were all that was needed, instead of handling (often literally, by handfork) tons of manure.

All over Great Britain, wherever corn is growing, can be seen evidence of those times—broken-down and empty cattle sheds and cattle courts. Many cattle were kept solely for their manure value, and very often were unprofitable as far as a farm enterprise was concerned.

A growing market for agricultural chemicals developed, and new products frequently appeared, but things started to go wrong. For instance, in the Okanagan Valley in British Columbia where I lived for some time, a great many apples and pears are grown. In those days lead arsenate was used as a spray for various orchard pests and there was no mite problem at all, presumably because lead arsenate did not kill off the predator of the mite. Then came the wonder drug DDT, and instead of lead arsenate (which was fairly unpleasant and somewhat dangerous) growers switched to the full use of DDT, which killed off every pest imaginable. Suddenly they had a mite problem, and they have had it ever since. Of course, this was the simple effect of killing off the mite predator. Unfortunately, mites are more persistent and reproduce more rapidly than a predator, unless the predator is allowed complete freedom to live naturally. At that time not much significance was attached to this, but we know the progress of chemicals since those years.

I first became interested in organic farming because of problems with our Guernsey herd in Kent. The book by Newman Turner, *Fertility Pastures*, seemed to have the right answers. Human nutrition was of no interest to me then: I have always been more interested in the quality of food grown for stock than in that of food grown for human consumption. After all, humans in Great Britain and other developed countries can eat what they choose, but stock have to eat what they are given, or go hungry. Nevertheless, I must admit that human nutrition is of interest now, because of the different problems for humans these days.

We organic enthusiasts of those early years had to work on a basis of traditional husbandry, but with modern machinery if available. We had to follow careful rotations, whereas the conventional farmer could forget about the worry of rotations and grow the same crop for several years without too many problems. Any problem that did arise just meant finding the right cure from the right agrochemical company, and applying it properly. If it did not work it was generally thought to be because the farmer's application was wrong, the weather was wrong, or it was the wrong cure for that particular problem. The failure was never attributed to the chemical.

Conditions have not altered very much to date, but specialist machines are being developed for the organic farmer; various products, such as the seaweed range, have been marketed, and the organic farmer of today is far more than just an updated 1930s farmer.

When, eventually, I owned a farm in New Zealand and tried to run it organically, I was faced with problems to which I could not find any answers. Going to agricultural colleges was useless and, to a certain extent, still is: you would be given a chemical answer, or told that you would go bankrupt if you farmed organically. All the free advice from commercial firms was, again, chemical, because there was little money for any company in producing for organic farmers. So, like every other organic farmer I tried to find my own answers—quite often without satisfaction.

Returning to England in 1972, I worked for the Soil Association, and started the Soil Association Organic Marketing Company for setting standards of production and marketing members' crops. This was not entirely appropriate, so it was decided to form a separate Farmers' Cooperative, owned and run by farmers. With the help of colleagues I formed Organic Farmers & Growers Limited, which has progressed reasonably well. Now we have farmers all over the UK and are being recognized as a serious commercial concern by the farming community and agricultural business. However, soon after forming this company, I realized that I was going to have to answer a lot of queries from farmers, and the more successful the company became, the more farmers there would be with problems to be answered. The questions then arose, of where to find the answers, and how to develop organic farming to keep it viable with ever-worsening national economic circumstances. It was obvious that it would be necessary to interest scientists in these particular research problems and to develop better methods, but there was no institute or organization to attract scientists at that time. I therefore founded, again with the help of colleagues, the International Institute of Biological Husbandry Limited, a nonprofitmaking organization with educational and research objectives.

The purpose of the Institute would be to gather together information, sort it and disseminate it on a world-wide basis. Building up a catalogue of problems and solutions, we would at the same time be encouraging research, so that our farmers could look forward to ever-increasing crops and ever-increasing economic success. The overall aim of the Institute is to provide a sound scientific base for biological husbandry.

I have been extremely lucky in finding scientific colleagues of various disciplines and of great enthusiasm, ability and willingness to cooperate. This has already made a great difference to the future of biological husbandry, and bears out my original contention—that if the right organization could be formed, the scientists wishing to work in this particular field would certainly be found.

Looking to the future, I see biological agriculture developing along two lines. There will always be pure organic husbandry, working to the Number One standard of Organic Farmers & Growers Ltd, which does not permit the use of any artificial salts, pesticides or other agrochemicals, and which involves certain methods of cultivation, rotation, etc. There will also be the other standard—involving a complete biological approach to management and to all basic techniques, but with the use of chemicals for problem-solving or crop topping-up, which will not do any harm to the biological life or condition in the soil. Although this approach has caused some arguments, I see no problem at all in following both lines, provided that one is absolutely honest about what one is doing.

In the Institute, our first college is coming into being and we hope that this development will spread to other countries. It is essential to provide these facilities to fulfil the wishes of many students who are very disillusioned with the usual agricultural colleges and institutes. In addition, when a farmer changes to biological methods he may need some tuition, because many have forgotten (or have never known) the older traditional farming methods. Many of us consider that successful organic farming requires more skill, foresight and planning than farming which relies on chemicals.

The Conference at which the papers published in this book were given was the first conference on biological husbandry in the United Kingdom with such an emphasis on a scientific approach. It is hoped that this book may provide a basis for a textbook on biological principles that will be renewed continually as the years go by and as more developments are perfected.

My colleagues put in a tremendous amount of work on the Conference and on these papers; I thank them very much for their efforts. Dr David Hodges and his colleagues at Wye College are particularly to be thanked for running the conference so well. I hope they see it as I do—a great advance towards making biological farming the farming of the future—which it undoubtedly is.

1

AN AGRICULTURE FOR THE FUTURE

R. D. HODGES
Wye College (University of London), Ashford, Kent, UK

Introduction

The title of this chapter—also the title of the conference of which this volume is the Proceedings—poses a number of questions. What is meant by an agriculture for the future? Will the present system of agriculture serve humanity into the future? If so, why is an alternative agriculture needed? What potential alternatives are there to the present system? In order to ensure that most important of needs, a continuing and adequate food supply, mankind has to possess or to develop an agricultural system which can provide enough food to supply the needs, not only of the present world population but also of that of future population growth, until this can be brought under control. This agricultural system must also be one which is adaptable to various levels of development and industrialization, and which does not rely too heavily on nonrenewable resources, or on fuels whose depletion could endanger its long-term continuity. Another important factor is that an agriculture for the future must have a minimal impact on the natural environment—the biosphere—on the overall integrity of which we rely ultimately for our well-being and even survival.

The present system of modern Western agriculture (called here conventional agriculture), which is practised to the exclusion of almost any other system in the industrialized countries, and which in recent years has been introduced more and more into the developing countries, does not easily fulfil these, or other, criteria for 'an agriculture for the future'. The first part of this chapter outlines the reasons why I think that conventional agriculture does not fulfil these criteria: the second part considers potential alternative systems, biological agriculture in particular.

An assessment of conventional agriculture

While recognizing that the development of conventional agricultural techniques has resulted in great increases in productivity, the negative aspects of the system must be considered as well, in order to make a true assessment of its impact and potential. In summarizing the problems associated with conventional agriculture, it should be recognized that some are specifically the result of the system, and others are caused by the interaction between agriculture and various political, economic and social factors which are part of our way of life.

1

BIOLOGICAL AND ECOLOGICAL FACTORS

Effects on soil

Man's increasing impact on the environment is resulting in a world-wide tendency towards the degradation and erosion of soils (Eckholm, 1976; Hare, 1980). Although many of the factors causing this are nonagricultural, nevertheless failure to practise an environmentally sound agriculture does result in unacceptably high rates of soil erosion or degradation (*see for example* Pimental and Krummel, 1977; Rennie, 1979).

Effects of pesticides

Widespread use of pesticides has been followed by a number of detrimental effects. Pesticide resistance, particularly of arthropods, has now become a serious world-wide problem. By the late 1970s, well over 200 pest species had developed resistance to one or more of the major pesticide types (Watson and Brown, 1977; Brown, 1978). Unfortunately, the development of resistance among arthropod pests has not been paralleled by a similar development among the predators and parasites which normally control the pests (Croft, 1977). As pests develop resistance, so pesticides lose effectiveness—but the problem still remains, necessitating the development of new pesticides. However, the substantial costs and risks involved in developing a commercially viable product (Goring, 1977) seriously reduce the incentive to discover new pesticides. The extensive application of pesticides to monocultures of crops over a number of seasons can give rise to new pest species from insects which previously had no significant impact upon the crop. This phenomenon arises from the ecological imbalance caused by the destruction of predator insects and the consequent removal of natural controls on potential pest species. Examples of secondary pests developing on cotton intensively sprayed with pesticides have been quoted by Van Den Bosch (1971).

By definition, pesticides are highly toxic chemicals. Very few have a high degree of specificity in relation to their target organisms: rather, they are normally broad-spectrum compounds which have a relatively indiscriminate effect on wildlife as well as pests. Many cases of the effects of pesticides on nontarget insects, as well as on fish, birds and mammals, have been described by authors such as Carson (1962), Rudd (1964, 1975) and Gillett (1970). Pesticides have frequently been reported as having only transient effects upon soil microorganisms, although soil processes such as nitrification are very sensitive and may be sharply inhibited by some pesticides (Vlassak, Govindaraju and Verstaeten, 1977). However, although much is known about acute effects of pesticides upon the soil microflora, there are few data on long-term chronic effects of repeated pesticide exposures (Parr, 1974).

Pesticides are also toxic to humans. In spite of all the precautions taken in the production and application of pesticides, their high toxicity can and does result in poisoning and death. Although the organochlorine pesticides do not have a very high acute toxicity, others such as some organophosphorus compounds are highly toxic, with a lethal dose of less than 5 mg/kg body weight (Hodges, 1977a). Accurate data on cases of acute pesticide poisoning or death are not easily obtained, while the possible long-term chronic effects of pesticide exposure are even more difficult to determine. Nevertheless, the World Health Organization (WHO, 1975) has estimated that approximately 500 000 cases of acute poisoning occur annually, with a fatality rate of about 1 per cent, and both Rudd (1975) and Davies (1977) consider that pesticide exposure is a serious and increasing world-wide health hazard. Long-term effects may occur in humans as a result of chronic exposure to low pesticide levels over periods of time, particularly because some pesticides, such as toxaphene, have been shown to be mutagenic (Hooper *et al.*, 1979) or such as BHC, DDT, dieldrin, mirex or kepone, to be carcinogenic in rodents (Davies, 1977; Ames, 1979). Long-term effects could result, as much from the presence of many pesticides as ubiquitous contaminants throughout the environment, as from continuing new applications. For example, a wide range of chlorinated hydrocarbons have been detected as residues in human fat (Ames, 1979) and also in human milk (Aubert, 1977; Ames, 1979).

Use of fertilizers

The almost complete reliance of conventional agriculture upon the use of soluble fertilizers for the renewal of soil fertility may result in a number of problems which can arise from the excessive application of fertilizer nutrients. Because large inputs of fertilizer are often required for maximum production, there may be an overlap between those amounts of fertilizer leading to high productivity and those causing ecological problems. The potential detrimental effects of fertilizers on plants have been summarized by Phillips (1972) as: reduction in germination; retardation of seedling growth; scorching; increase in soil acidity, and development of nutrient imbalances. Other effects can be development of root damage (Imai, 1977) and increased susceptibility to disease, as well as reduction of legume root nodulation and plant mycorrhizal associations. Day, Doner and McLaren (1978) have summarized the environmental effects as follows: '. . . too much nitrate fertilizer can lead to water pollution, can result in soil humus reduction and can disturb the atmosphere composition'. In spite of the need for a high fertilizer input for productivity and the problems that this may engender, the efficiency of fertilizer uptake apparently can be quite low. For example, the National Academy of Sciences of the United States of America (NAS, 1975) stated that, in the USA, only 50 per cent of nitrogen fertilizer and only 35 per cent of phosphorus and potash are recovered by crops. In the tropics, losses of fertilizer nitrogen may be even greater, with recovery averaging only 25–35 per cent for rice crops.

ENERGY AND RESOURCE FACTORS

During the past decade many reviews have stressed the energy and resource intensiveness of conventional agriculture, and have commented on the relatively poor energy efficiency, the ratio of energy produced by crops over the energy needed to produce them, when compared with many traditional agricultural systems (*see* Pimentel *et al.*, 1973; Steinhart and Steinhart, 1974; Leach, 1975). During times of cheap, abundant energy such inefficiency did not matter but, since the oil crisis of 1973, increasing prices and shortages of oil, natural gas and minerals mean that the main tools of conventional agriculture, fertilizers and pesticides, have become increasingly more expensive. Such escalating costs can cause problems for agriculture in the developed countries, but far more so for many developing countries which need to increase food production but may not have the resources to produce or buy the necessary compounds. Pimentel and Krummel (1977) have commented on the impossibility, in energy terms, of ever being able to provide the world population with an American-style diet produced by high-technology conventional agriculture. Conventional agriculture is very wasteful in energy and mineral resources in that it is almost entirely a linear process. Industrially produced plant nutrients are used to grow crops, and the farm products pass out into the community, normally to be lost to agriculture, each crop needing fresh nutrients and energy inputs. Although this is largely a sociopolitical problem in which the community must decide whether or not to recycle all its wastes, nevertheless conventional agriculture often has not seriously attempted to recycle those wastes (straw, slurries, etc.) which remain under its control. The energy and resource constraints facing agriculture seem unlikely to improve in the foreseeable future.

ECONOMIC AND SOCIAL FACTORS

Mechanization

One of the major factors associated with the development of conventional agriculture has been the continuing trend towards mechanization and the consequent loss of manpower from the land. This trend has usually been considered to be an increase in agricultural efficiency. In the West, where this process has been taking place over many years, the workers leaving the land have been able to move to towns and cities to obtain urban and industrial jobs. Even so, considerable social problems have arisen, particularly due to rural depopulation. A recent example of this trend can be found in the EEC where, between 1960 and 1976, half of the original 16 million agricultural workers moved from the land.

The Green Revolution

The application of conventional agricultural methods to developing countries, and particularly the comparative rapidity with which those changes have taken place in recent years, has often contributed to the worsening of social conditions instead of improving them. Some of these changes are a direct result of the Green Revolution which was originally intended to improve conditions in the developing countries. Much has been written about the Green Revolution in recent years, both for and against it. In the areas where it has worked well there is no doubt that the production of the two main crops, rice and wheat, has increased greatly (Gill, 1978). Nevertheless, with its need for the specific inputs of credit, fertilizers, pesticides, seeds and irrigation in order to obtain its high yields, it often turns out to be a package of agricultural technology which is unsuited to the needs of many farmers and smallholders in developing countries. Analyses of the impact of the Green Revolution on world food production have come to differing conclusions but, in the opinion of many analysts, it has failed to meet its early expectations. Among the more pessimistic reviews, Griffin (1974) has calculated that the Green Revolution has not resulted in an acceleration of food production in any part of the Third World. Aziz (1977) has gone further in stating that annual increases in food production in developing countries have dropped from 3.1 per cent in 1952–62 to 2 per cent in 1972–75, in spite of the Green Revolution. The complex and capital-intensive nature of Green Revolution technology has often resulted in benefits to large farmers, to the detriment of the smaller ones, in the loss of the jobs of many agricultural labourers, and in increases in social and class divisions (Havens and Flynn, 1973; Castillo, 1977; ILO, 1977; Ledesma, 1977). Chang, one of those associated with the development of the new strains of high-yielding rice, admits that the intensive techniques required for this rice-growing technology have resulted in 'devastating epidemics of diseases, insect pests, or both', and considers that the high-yielding varieties do not produce as well on farms as in the research institutes (Chang, 1979). He believes that the full potential of these varieties can be achieved only by increasing inputs of fertilizer, irrigation, pest controls, credit and all the industrial and governmental support normally provided to farmers in the developed countries. Many developing countries are unable to provide these inputs at the required levels.

Productivity

The increasing mechanization and capital-intensiveness of conventional agriculture has been considered to have resulted in increased efficiency and productivity. Although, when viewed in certain terms, its efficiency has increased, when considered basically in terms of productivity per man or per hectare, improvement in productivity is much less certain. First, in the case of productivity per man, although only a few farm workers 'in the field' actually produce the food, they are supported by many workers in the agricultural or associated industries. Thus in the US the whole food production and processing complex employs about 20 per cent of all workers (Pimentel *et al.*, 1973). Secondly, a number of recent studies have suggested that smaller, labour-intensive operations are more productive per hectare than large, highly mechanized farms (*see for example* Best and Ward, 1956; Britton and Hill, 1975; Perelman, 1975; Vale, 1977).

CONCLUSION

All the above factors, together with the probability that many crops grown under conventional practices are close to their maximum productivity, suggest that conventional agriculture may now be coming up against considerable environmental, social, energy and resource constraints, and further development along the same lines will be counterproductive. Indeed, it could be said that, during the past 30 or 40 years, conventional agriculture has increasingly followed a path of 'chemical development' which has now reached its maximum levels of productivity and is turning out to be a cul-de-sac. Even though the developed countries have plenty of food, and are even overproducing many commodities, on a global basis humanity is short of food and population is increasing. The overall world situation can be summed up by three short quotations. Boerma (1975), the then Director-General of FAO, considers that to meet the world's food needs between now and 2000 AD '. . . agricultural production will have to double every 18 years—a rate never before achieved over a sustained period . . .'. Olembo (1977) states that in order to match food supplies to growing population '. . . we must urgently and dramatically increase food production . . .'. Finally, Manocha (1975) considers that 'at the present time the land under cultivation has been under great agricultural stress, which in turn has put strains on the biological ecosystems of this planet. The "Green Revolution" in the poorer countries of the world cannot be maintained for an indefinite period and it is doubtful that the high yield per acre achieved in the developed countries can be expanded or even maintained for a long stretch of time'.

If this analysis is correct, and conventional agriculture does not have the potential to qualify as an agriculture for the future, what alternative system or systems are there? The alternatives to be considered here are biological agriculture and integrated agriculture.

Biological agriculture

Biological agriculture is a system of farming based upon the principle that agriculture is, first and foremost, a biological science. The term biological agriculture is one which, in recent years, has tended to supersede the original term, organic farming; however, *the two terms are virtually synonymous.* Biological husbandry has a wider meaning in that it includes biological systems of agriculture, horticulture, forestry, gardening, etc.

Biological agriculture can be defined as a system that attempts to provide a balanced environment, in which the maintenance of soil fertility and the control of pests and diseases are achieved by the enhancement of natural processes and cycles, with only moderate inputs of energy and resources, while maintaining an optimum productivity. The introduction, into a biological system, of chemicals such as fertilizers tends to short-circuit these natural processes and thus a proper interpretation of this definition does not allow the use of soluble fertilizers or synthetic pesticides in the system. A biological system in which some fertilizers and pesticides are used on a regular basis is more properly called integrated agriculture (*see below* (p. 9)). The aims of biological agriculture are to develop:

1. A sustainable agriculture, i.e. a system which maintains and improves soil fertility such as to guarantee adequate food production into the foreseeable future

2. A self-sufficient or, more realistically, a self-sustaining agriculture; i.e. a system which relies as much as possible upon resources from within its own area and which is not reliant upon large quantities of imported resources. This applies particularly to the developing countries and fertilizers, etc.
3. An agriculture which takes as its guide the working of biological processes in natural ecosystems. It must always be remembered that agriculture is primarily applied biology and is most likely to be successful when it accepts and follows biological principles.

Aim 3 is most important—it is the basic principle of biological agriculture. All agricultural systems interfere with natural processes and are frequently a gross simplification of natural ecosystems. Nevertheless, it is possible to devise a system of agriculture which is founded on basic biological principles and which does not attempt to ignore, or to contravene, these principles. Biological agriculture is such a system. From aim 3 one can derive a number of points which can be considered as the basic principles of biological agriculture. They are:

1. The health of soil, plant, animal and man are linked by a common nutritional cycle
2. The health of the whole cycle will be diminished by a loss of soil fertility or by any imbalance introduced into the soil by improper husbandry practices
3. All living materials and waste products must be returned to the soil for the maintenance and improvement of its fertility
4. This return is necessary for the purification of waste materials, which would otherwise cause pollution, and for the recycling of essential elements
5. The soil should retain an ordered structure, with decomposing material on the surface and humus-enriched soil below. This implies a minimum of soil disturbance
6. As in natural ecosystems, plants and animals should coexist, each as mixed communities; crop rotations and mixed stocking constitute a practical expression of this principle
7. As far as possible the soil should always be covered by living and decaying material
8. The resources of an area are usually adequate for sustained growth in that area.

The necessity for taking 'natural conditions' as the basic model for agricultural practice is most important because all living things have evolved to fit precisely into such conditions and any significant deviation from them is likely to put stress upon the crops or animals being farmed. For example, the complex biological system represented by plant roots, associated microflora and soil organisms in general (Matile, 1973), which has developed into its present complexity over millions of years of evolution, cannot be short-circuited by chemical means without risk of damage to the health of the soil and of the plant.

Other basic concepts behind biological agriculture are:

1. It concentrates on building up the biological fertility of the soil so that the crops take the nutrients they need from the steady turnover within the soil. Nutrients produced in this way are released in harmony with the needs of the plants, and are not presented to the plants in excessive amounts at any time. Biological soil fertility is directly related to the development and maintenance of a high level of organic matter. Organic matter is stressed here rather than humus: the former is a highly complex mixture of organic materials in varying stages of decomposition and in a state of dynamic interaction with the soil life; the latter is well-decomposed organic residues which do not support the full spectrum of biological activity. Organic matter performs a wide range of very important functions in a fertile soil, indeed it can be said to be one of the most important factors in the development of a biologically fertile soil. When applied to the soil in the right way, organic matter has a wide range of effects—physical, chemical and biological—which not only develop the soil structure and fertility but also can help to control pests and diseases within the soil and on the plants (Baker and Cook, 1974). The combination of all the properties and effects of organic matter is what helps to produce the fertile soils which are the basis of biological agriculture. Conversely, it is the reduction of soil organic matter levels in conventional farming which can cause many problems

2. Control of pests, diseases and weeds is maintained largely by the development of an ecological balance within the system and by the use of various cultural techniques such as rotations and cultivations. Practical experience has shown that plants grown in a biological system tend to be less susceptible to disease

3. Biological agriculture attempts at all times to work on a cyclical basis, which is the only way a sustainable and self-sufficient agriculture can be maintained. Biological farmers recycle all wastes and manures within the farm, but the export of products from the farm results in a steady drain of nutrients. In a situation where conservation of energy and resources is considered to be important, a community or country would make every effort to recycle all urban and industrial organic wastes back to agriculture and thus the system would require only a very small input of new resources to 'top up' soil fertility. The potential availability of enough organic wastes to maintain soil fertility has been discussed by, among others, Hodges (1977b), Nager (1977) and El Bassam and Thorman (1978). The production of fertilizers, particularly nitrogenous fertilizer, is a very energy-intensive process. Because biological agriculture uses few, if any, of these compounds in maintaining soil fertility, its energy intensity often can be shown to be less than that of conventional agriculture. Lockeretz *et al.* (1976) have shown that, largely because of the use of fertilizers, conventional agriculture can be about 2.3 times as energy-intensive as biological agriculture, while Crouau (1977) calculates the figure at 2.5 times as energy-intensive.

In the long run, and particularly in the context of this discussion, the value of biological agriculture as a potential agriculture for the future depends largely upon its ability to produce enough food to feed the world. In recent years numerous experimental results have shown that inputs of manures or other organic fertilizer materials are capable of producing crop levels as great as, and sometimes greater than, the equivalent chemical fertilizer inputs (*see for example* Ionescu-Sisesti *et al.*, 1975; Mathers, Stewart and Thomas, 1975; Cooke, 1976, 1977; Nilsson, 1979). In summarizing the results of the long-term comparative trials carried out at experimental stations such as Rothamsted, Thorne and Thorne (1978) state: '. . . in recent years, with improved tillage and weed-control practices, plots treated with farmyard manure now yield somewhat more than those receiving equivalent amounts of nutrients in inorganic fertilizers.' Hodges (1977b) has reviewed a number of reports which examined the productivity of biological agriculture in comparison with its conventional counterpart. The general conclusions drawn from these reports are that, for most crops and products, biological agriculture can be as productive as conventional agriculture. A very recent report from the United States Department of Agriculture (USDA, 1980) supports this conclusion. Of much greater significance for increased food production, particularly in Third World conditions, is the development in recent years of what is called the Biodynamic–French Intensive method (Jeavons, 1974, 1976; Kaffka, 1976). This is a small-scale, labour-intensive, minimal energy-input technique for growing vegetables which, it is claimed, can produce significantly higher levels of output than its conventional horticultural equivalent. Other biologically based, intensive and highly productive systems of agriculture/horticulture from the Third World have been reported by King (1926), Clayton (1964) and Gleave and White (1967).

Integrated agriculture

Integrated agriculture is a system which attempts to combine aspects of both the biological and the conventional systems. In general, this system maintains soil fertility as much as possible by biological means (such as the recycling of all manures and wastes) but also adds moderate amounts of chemical fertilizers in order to attain maximum yields. Some experimental studies have suggested that the combination of manures and fertilizers can be more highly productive than either separately. Control of pests and diseases in this system may be by integrated control—primarily the use of cultural and biological techniques, backed up where necessary by pesticides and herbicides.

Conclusions

After nearly 40 years of intensive research and development, conventional agriculture appears to have reached a position where, although it is highly productive under Western conditions, it is becoming subject to a number of environmental and other constraints. Significant further increases in productivity seem unlikely as a result of extensions of conventional technology but, even if they are, the problems associated with these constraints will almost certainly increase. The widespread application of conventional technology to developing countries is possible only if enormous economic resources are allocated to such a project. Many developing countries are unlikely to possess these resources, even if the technology is applicable to their needs. The evidence available at present suggests that biological agriculture is capable of competing with conventional agriculture, and that it has attained this position of apparent parity without the backing of extensive research support. It seems likely that the present position of biological agriculture is similar to that of conventional agriculture 30 or 40 years ago and that, if an extensive research effort is applied to its development, it has an enormous potential for food production while remaining an environmentally benign system with relatively low energy and resource requirements. The need for extensive research has been high-lighted by the recent report from America (USDA, 1980). In particular, the development of biological techniques along the lines of the Biodynamic–French Intensive method are urgently required to enable the poorer parts of the Third World to begin working towards self-sufficiency in food produc-tion. Biological agriculture is an ideal system in these circumstances because it is potentially adaptable to all levels of agriculture, from the smallest peasant holding to large-scale mechanized agriculture.

The widespread development and application of efficient biological tech-niques requires not only research, but also the political decision within individual countries or areas to develop satisfactory means of recycling organic wastes and thus to reuse the nutrients removed from the land as food. The present method of disposing of wastes so that they are per-manently lost to agriculture is an irresponsible misuse of finite resources. However, once a decision has been made to develop recycling techniques, then efficient biological agriculture becomes possible and adequate food production can be assured for the future. During any period of trans-formation from conventional to biological agriculture, particularly if it occurred on a large scale, it would probably be advisable to introduce a phase of integrated agriculture because complete withdrawal of the use of fertilizers often results in a drop of productivity before biological soil fertility can be established.

All that has been said in this paper can be summed up by quoting a few words from Dr Sicco Mansholt, the former EEC Commissioner for Agri-culture and Netherlands Minister of Agriculture: 'If we utilize biological means we shall be able to farm for ever. There are no limits set here. As long as there is sun, water, soil and microorganisms, we shall—in as much as we maintain an ecological balance—be able to maintain an assured production' (Mansholt, 1979).

References

AMES, B. N. (1979). Identifying environmental chemicals causing mutations and cancer. *Science*, **204**, 587–589

AUBERT, C. (1977). *L'Agriculture Biologique*. 3rd edn. Paris; Le Courrier du Livre

AZIZ, S. (1977). The world food situation—today and in the year 2000. In *Proceedings of the World Food Conference of 1976*, pp. 15–31. Ames, Iowa; Iowa State University Press

BAKER, K. F. and COOK, R. J. (1974). *Biological Control of Plant Pathogens.* San Francisco; Freeman

BEST, R. H. and WARD, J. T. (1956). *The Garden Controversy.* Studies in Rural Land Use, Report No. 2. Wye College, University of London

BOERMA, A. H. (1975). Foreword. In *The Man/Food Equation.* Eds F. Steele and A. Bourne. London; Academic Press

BRITTON, D. K. and HILL, B. (1975). *Size and Efficiency in Farming.* Farnborough; Saxon House

BROWN, A. W. A. (1978). *Ecology of Pesticides.* Chichester; Wiley

CARSON, R. (1962). *Silent Spring.* Boston; Houghton Mifflin

CASTILLO, G. T. (1977). The farmer revisited: Toward a return to the food problem. In *Proceedings of the World Food Conference of 1976*, pp. 33–53. Ames, Iowa; Iowa State University Press

CHANG, T. T. (1979). The Green Revolution's second decade. *Span*, **22**, (1), 2–3

CLAYTON, E. S. (1964). *Agrarian Development in Peasant Economies.* Oxford; Pergamon Press

✦ COOKE, G. W. (1976). Long-term fertilizer experiments in England: The significance of their results for agricultural science and for practical farming. *Ann. Agron.*, **27**, 503–536

COOKE, G. W. (1977). The roles of manures and organic matter in managing soils for higher crop yields—a review of the experimental evidence. In *Proceedings of the International Seminar on Soil Environment and Fertility Management in Intensive Agriculture*, pp. 53–64. Japan; Soc. Science of Soil and Manure

CROFT, B. A. (1977). Resistance in arthropod predators and parasites. In *Pesticide Management and Insecticide Resistance*, pp. 377–393. Eds D. L. Watson and A. W. A. Brown. New York; Academic Press

CROUAU, M. (1977). Comparative study of energy consumption in biological and conventional agriculture. In *IFOAM Bulletin*, No. 20, p. 4

DAVIES, J. E. (1977). Pesticide management safety—From a medical point of view. In *Pesticide Management and Insecticide Resistance*, pp. 157–167. Eds D. L. Watson and A. W. A. Brown. New York; Academic Press

DAY, P. R., DONER, H. E. and McLAREN, A. D. (1978). Relationships among microbial populations and rates of nitrification and denitrification in a Hanford Soil. In *Nitrogen in the Environment*, vol. 2, pp. 305–363. Eds D. R. Nielsen and J. G. MacDonald. New York; Academic Press

ECKHOLM, E. P. (1976). *Losing Ground: Environmental Stress and World Food Prospects.* New York; W. W. Norton

EL BASSAM, N. and THORMAN, A. (1978). Potentials and limits of organic wastes in crop production. *Compost Science/Land Utilisation*, Nov./Dec., 30–35

GILL, M. S. (1978). Success in the Indian Punjab. In *Conservation and Agriculture*, pp. 194–206. Ed. J. G. Hawkes. London; Duckworth

GILLETT, J. W. (1970). *The Biological Impact of Pesticides in the Environment.* Portland Oreg., Oregon State University

GLEAVE, M. B. and WHITE, H. P. (1967). Population density and agricultural systems in West Africa. In *Environmental Land Use in Africa.* Eds M. F. Thomas and G. W. Whittington. London; Methuen

GORING, C. A. I. (1977). The costs of commercializing pesticides. In *Pesticide*

Management and Insecticide Resistance, pp. 1–33. Eds D. L. Watson and A. W. A. Brown. New York; Academic Press

GRIFFIN, K. (1974). *The Political Economy of Change*. London; Macmillan

HARE, F. K. (1980). First 150th Anniversary Lecture of the Royal Geographical Society. *Times Higher Educ. Suppl.*, 4 April, pp. 10–11

HAVENS, A. E. and FLINN, W. (1973). *Green Revolution Technology and Community Development. The Limits of Action Programs.* L.T.C. No. 93. Madison; The Land Tenure Center, University of Wisconsin

HODGES, L. (1977a). *Environmental Pollution*, 2nd edn. New York; Holt, Rinehart and Winston

HODGES, R.D. (1977b). Who needs inorganic fertilizers anyway? The case for biological agriculture. In *Proceedings of the International Conference on Granular Fertilizers and their Production, 1977*. London; British Sulphur Corporation

HOOPER, N. K., AMES, B. N., SALEH, M. A. and CASIDA, J. E. (1979). Texaphene, a complex mixture of polychloroterpenes and a major insecticide, is mutagenic. *Science*, **205**, 591–593

ILO (1977). *Poverty and Landlessness in Rural Asia*. Geneva; International Labour Office

IMAI, H. (1977). The harmful effects of ammonium nitrogen on crop roots. In *Proceedings of the International Symposium on Soil Environment and Fertility Management in Intensive Agriculture*, pp. 634–640. Japan; Soc. Science of Soil and Manure

IONESCU-SISESTI, V., JINGA, I., ROMAN, G. and PROCOP, G. (1975). The efficiency of using sludge from pig growing complexes as organic fertilizer. In *Managing Livestock Wastes. Proceedings of the 3rd International Symposium on Livestock Wastes*, pp. 271–273. St. Joseph, Michigan; Amer. Soc. Agr. Eng.

JEAVONS, J. (1974). *How to Grow More Vegetables*. Palo Alto, California; Ecology Action of the Midpeninsula

JEAVONS, J. (1976). Quantitative research on the French Intensive/Biodynamic method. In *Small Scale Intensive Food Production*, pp. 32–38. Santa Barbara, California; League for International Food Education

KAFFKA, S. (1976). The French Intensive system of gardening. In *Small Scale Intensive Food Production*, pp. 24–31. Santa Barbara, California; League for International Food Education

KING, F. H. (1926). *Farmers of Forty Centuries*. London; Jonathan Cape

LEACH, G. (1975). *Energy and Food Production*. London; International Institute for Environment and Development

LEDESMA, A. Y. (1977). The potentials of the human work force to produce food. In *Proceedings of the World Food Conference of 1976*, pp. 405–413. Ames, Iowa; Iowa State University Press

LOCKERETZ, W., KEPPLER, R., COMMONER, B., GERTLER, M., FAST, S. and O'LEARY, D. (1976). *Organic and Conventional Crop Production in the Corn Belt: a Comparison of Economic Performance and Energy Use for Selected Farms*. Report CBNS-AE-7. St. Louis, Mo; Center for the Biology of Natural Systems, Washington University

MANOCHA, S. L. (1975). *Nutrition and Our Overpopulated Planet*. Springfield, Illinois; Thomas

MANSHOLT, S. (1979). *The Common Agricultural Policy. Some New Thinking*. Haughley, Suffolk; The Soil Association

MATHERS, A. C., STEWART, B. A. and THOMAS, J. D. (1975). Residual and annual rate effects of manure on grain sorghum yields. In *Managing Livestock Wastes. Proceedings of the 3rd International Symposium on Livestock Wastes*, pp. 252–254. St. Joseph, Michigan; Amer. Soc. Agr. Eng.

MATILE, P. (1973). In *Environmental Problems and Agriculture. Wasser und Abwasser, July, 1973 Supplement.* Reported in *Soil Assoc. J.*, **1**, No. 8, 5–8

NAGER, B. R. (1977). Alternatives to energy intensive fertilizers: organic materials as fertilizers. In *Agriculture and Energy*, pp. 657–667. Ed. W. Lockeretz. New York; Academic Press

NAS (1975). *Enhancement of Food Production for the United States. World Food and Nutrition Study.* Washington, D.C.; Board on Agriculture and Renewable Resources of the Commission on Natural Resources, National Research Council

NILSSON, T.(1979). Yield, storage ability, quality and chemical composition of carrot, cabbage and leek at conventional and organic farming. *Acta Hortic.*, **93**, 209–223

OLEMBO, R. J. (1977). Environmental issues in current food production, marketing and processing practices. In *Proceedings of the World Food Conference of 1976*, pp. 145–162. Ames, Iowa; The Iowa State University Press

PARR, J. F. (1974). Effects of pesticides on microorganisms in soil and water. In *Pesticides in Soil and Water*, pp. 315–340. Ed. W. D. Guenzi. Madison, Wisconsin; Soil Science Society of America

PERELMAN, M. (1975). The real and fictitious economics of agriculture and energy. In *Energy, Agriculture and Waste Management*, pp. 133–139. Ed. W. J. Jewell. Michigan; Ann Arbor Science Publishers

PHILLIPS, J. (1972). Problems in the use of chemical fertilizers. In *The Careless Technology*, pp. 549–560. Eds M. T. Farvar and J. P. Milton. London; Tom Stacey

PIMENTEL, D. and KRUMMEL, J. (1977). America's agricultural future. *The Ecologist*, **7**, No. 7, 254–261

PIMENTEL, D., HURD, L. E., BELLOTTI, A. C., FORSTER, M. J., OKA, I. N., SHOLES, O. D. and WHITMAN, R. J. (1973). Food production and the energy crisis. *Science*, **182**, 443–449

RENNIE, D. A. (1979). Intensive cultivation. The long term effects. *Agrologist*, **8**, No. 4, 20–22

RUDD, R. L. (1964). *Pesticides and the Living Landscape.* Madison; University of Wisconsin Press

RUDD, R. L. (1975). Pesticides. In *Environment, Resources, Pollution and Society*, 2nd edn, pp. 325–353. Ed. W. W. Murdoch. Sunderland, Massachusetts; Sinauer Associates

STEINHART, J. S. and STEINHART, C. E. (1974). Energy use in the U.S. food system. *Science*, **184**, 307–316

THORNE, W. and THORNE, A. C. (1978). Land resources and quality of life. In *Renewable Resource Management for Forestry and Agriculture*, pp. 17–34. Eds J. S. Bethel and M. A. Massengale. Seattle; University of Washington Press

USDA (1980). *Report and Recommendations on Organic Farming.* 1980–0–310–944/96. Washington, D.C., US Government Printing Office

VALE, R. (1977). *Smallholding and Food Production.* Open University, Alternative Technology Group

VAN DEN BOSCH, R. (1971). The melancholy addiction of Ol' King Cotton. *Natural History*, Dec. 1971, pp. 86–92

VLASSAK, K., GOVINDARAJU, K. and VERSTAETEN, L. M. J. (1977). Nitrification in soils treated with some pesticides. In *Proceedings of the International Symposium on Soil Environment and Fertility Management in Intensive Agriculture*, pp. 750–754. Japan; Soc. Science of Soil and Manure

WATSON, D. L. and BROWN, A. W. A. (1977). *Pesticide Management and Insecticide Resistance*, section 4, pp. 237–393. New York; Academic Press

WHO (1975). Cited by Davies (1977)

I

SOIL STRUCTURE, FLORA AND FAUNA

Introduction

Soil is the prime component of any agricultural system—hydroponics excepted—and an understanding of soil processes should be fundamental to all systems. However, soil processes are notoriously complex. No two soils are alike, and there is nothing simple about the relationships between the living and non-living components of any of them.

Understanding of soils and soil processes has not come readily nor spread universally. There are still latter-day mystics whose approach to 'the living soil' is more reverential than analytical: today we may write them off 'muck-and-magic brigade', But they and their forebears flew the flag for organic husbandry when nobody else was bothering, and this symposium owes much to their vision. And there are still plenty of farmers, by no means confined to the American Midwest, who call soil 'dirt' and treat it as such—or at most as the mainly inert material that holds the crops down and (if it stays around long enough itself) keeps them from blowing away.

Somewhere between these extremes stand the soil scientists who challenge the soil's mysteries and try to unravel them. With them stand the many farmers and gardeners who use soils sensibly: supporters of organic methods might say that foremost among them in understanding are those who respect and cherish their soils, and cultivate the living components as an essential bridge between soil minerals and crops. Awareness of the importance of the soil's biological components is certainly the hallmark of a biological farmer, although not necessarily restricted to him. That a bad farmer feeds his crops, a good one his soils, is a fair aphorism for any farmer to consider seriously.

The five chapters in this section deal with various aspects of soil structure and organisms, with special emphasis on the roles and functions of the living components. Dr V. I. Stewart and Dr R. O. Salih discuss the properties of the widespread temperate soils called loams, stressing the importance of earthworms in maintaining their structure and pinpointing the dangers of concentrated fertilizers and pesticides in intensive production systems. Professor R. P. Moss emphasizes the variety to be found in tropical as well as temperate soils, and cautions against the facile imposition of temperate farming techniques on tropical soils; his paper provides a useful background to those of Section 3. Dr J. Lopez-Real draws attention to the importance of nitrogen-fixers in farming systems, and the potential for their further exploitation, especially in the tropics. Dr D. S. Madge describes the wide

17

variety of invertebrate animals in soils, and the effects on them of various systems of agricultural practice, including the use of pesticides and fertilizers, ploughing and burning. Finally Dr K. R. Gray and Dr A. J. Biddlestone discuss the process of composting, and ways in which compost can be used on the farm to return valuable humus and minerals to the soil from waste materials.

2

PRIORITIES FOR SOIL USE IN TEMPERATE CLIMATES

V. I. STEWART
Senior Lecturer in Charge, Soil Science Unit,
University College of Wales, Aberystwyth, UK

R. O. SALIH
Head of Unit of Land Reclamation,
The Institute of Applied Research on Natural Resources, Baghdad, Iraq

Where rainfall is abundant and falls uniformly throughout the year, soils are strongly leached and tend to become acid. In soils with a loamy texture this can lead to a compact structure and poor surface drainage. If soils are allowed to drift in accordance with these natural trends, the resulting poorly drained, acid soils will not favour crop production.

Essential features of a temperate climate as they affect soil

In a temperate climate, rainfall exceeds evaporation and is generally adequate for growth throughout the year. There are no extremes of temperature, but there is a winter season of biological dormancy determined by temperatures low enough to stop growth. Soil freezing is not very common or persistent, and seldom penetrates to the base of the topsoil.

Average rainfall in Britain could fairly be represented as amounting to 2–3 mm/d, whereas evapotranspiration varies between from 6 mm in mid-summer to 0.6 mm in midwinter. This means that, although the soil tends to be kept topped up to field capacity or above throughout the winter period of biological dormancy, it is liable to dry out to an extent sufficient to stop growth for a few days each summer.

The climatic characteristic which, with regard to the soil, best expresses the effect of a temperate climate on farming practice, is the number of days in an average year when the soil is warm enough to promote growth but too wet for cultivation or grazing. Even soils initially in a good structural state, if mismanaged during this very sensitive period are liable to become physiologically stressful for plant growth because of poor soil aeration. Agricultural climatic data published by MAFF (1976) suggests that this critical period for Dyfed in West Wales extends to 2 months in the spring and 3 months in the autumn, that is, two-thirds of the growing season, whereas, at the other extreme, in the climatically more continental counties of East Anglia, the growing season and the grazing season virtually coincide.

The ideal crop for a temperate climate is grass, grazed with discretion, particularly in spring and autumn, and conserved for overwintering from the excess of midseason growth. In the drier, more continental areas, summer drought can restrict midseason growth for more than a month and this shifts the emphasis on to cereals rather than grass.

The occasional cultivations required for re-seeds, cereals and root crops, must be timed to utilize those periods in late spring and early autumn when the soil is briefly in a fit state for working. The priority should be the creation and preservation of an ideal soil structure, supplemented by efficient underdrainage, so that as much as possible of the growing season can be used safely for grazing and conservation.

Soils characteristic of temperate climates

In Britain, and probably in temperate areas generally, proximity to the poles has led to a predominance of silt in the texture of the soil. Silt is a product of physical comminution and is typically a feature of glacial weathering. It was widely redistributed by wind and water after the Ice Age, and confers a degree of unity in soil texture over wide areas; in addition, it increases the water-holding capacity of the soil. Objective sampling of soil survey records suggests that, if the soils of England and Wales are typical, then 75 per cent of the soils of temperate areas are likely to be some form of loam rather than a sand or a clay. This means that either they have insufficient clay to swell and shrink in response to variation in moisture with the severity required continually to restructure themselves by fracture, or they have insufficient sand to avoid fine particle dominance of the pore space when compact. Stewart and Adams (1968) have suggested that soils will be adequately aerated when the pore space is less than 70 per cent full of water at field capacity. Loams can achieve this only when the fine particles are clustered into water-stable granules of sand size and larger. However, this type of aggregate structure, although stable to the impact of raindrops, cannot be other than vulnerable to collapse, returning to a compact, totally micropore state, prone to waterlogging, if abused physically by excessive cultivation, treading or wheel traffic. Furthermore, because much of the water stability depends on organic components liable to be inactivated by further decay, this necessary structure is only transient and requires to be continually recreated (Stewart, 1975). These are the structurally sensitive soils characteristic of much of the land that today experiences a temperate climate with its inherent requirement for effective soil drainage.

Requirements of plant roots

Inherent in what has already been discussed is the assumption that crop plants need a soil rooting environment that will allow easy root penetration through relatively large pores maintained adequately aerobic by rapid drainage. In fact, the dimensions of the underground exploring organs of temperate crop plants, root tips and root hairs, accord so well with the physical character of the soils they exploit, that they would appear to reflect the selective influence of evolutionary adaptation. Root tips, of the order of 100 μm in diameter, are of an ideal width to penetrate the macropore system which, in freely drained soils, will normally be drained rapidly of water and fully aerobic. Root hairs, of the order of 7 μm in diameter, are well able to enter the larger range of silt and clay aggregate size pores that remain water-filled at field capacity, and from which stored water can be relatively easily extracted to maintain growth. As root hair cells are liable to be at risk from the byproducts of anoxic biological activity, it is probably not without significance that they are only about 2–3 mm long, a size of the same order as estimates of the maximum water-film thickness through which aeration can be maintained by gaseous diffusion in a biologically active medium (Greenwood, 1969; 1970). Thus, for crop plants adapted to exploit freely drained soils, an ideal air/water balance could be provided by silt, or similar-sized clay aggregates, granulated into units equivalent in size to particles ranging between medium sand and coarse gravel, 200–10 000 μm in diameter. This is the same as the size range of water-stable aggregates into which a loam topsoil in good heart will readily disintegrate when gently agitated in water.

The work of Drew (1975) has identified the nitrogenous and phosphatic components of organic matter as the chemical triggers responsible for the well-known influence of organic matter on root branching. This could have adverse repercussions for those who use readily available and highly concentrated forms of these chemicals as fertilizers but, under the natural conditions that have influenced evolution, nitrogenous and phosphatic nutrients have been reliable indicators of the presence of decomposing organic matter, a wide-spectrum food source and a reservoir of water. It would be no surprise, therefore, if time had allowed the evolutionary mechanism of selection to favour the individuals that limit the expenditure of energy on root branching to the sections of their exploring, primary root system that is traversing a soil zone rich in decomposing organic matter.

Roots must continually grow to encounter fresh supplies of water and nutrients. Passive flow to the root is probably inadequate to meet all the needs of an actively growing plant. From crop modelling studies, Rowse and Barnes (1979) predict that, if only 3 per cent of the roots of a broad bean crop in a Wellesbourne soil penetrate to 50–70 cm, water uptake and growth could be increased by 30 per cent in an average year over that obtained when roots are confined to the top 50 cm. This suggests that the deeper the root-exploitable soil, the more effectively the crop should weather the stresses of seasonal variations in growing conditions.

THE ACTIVITIES OF BURROWING EARTHWORMS

In natural soils, organic matter enters the soil by two main routes: as debris deposited on the surface and as roots invading actively. In British soils, worked by burrowing earthworms, surface-deposited organic residues and root debris may be ingested along with fine particles of mineral matter less than 1 mm in diameter. As this passes through the earthworm's oesophagus it is supplemented with digestive juices, and lime is recycled back into the gut by a special gland. This compost is stored in the crop and then ground up and intimately blended in the gizzard. After completing its passage through the intestine the residual soil compost is ejected in the form of a cylindrical cast, 2–3 mm in diameter. Such a structure is water-stable and of an ideal form and composition to stimulate root branching in its vicinity and to reward, with water and a wide spectrum of nutrients, root hairs penetrating its pore space. Plant roots may not form this aggregate structure but, once stimulated to branch within it, may well extend its period of stability by mechanical reinforcement and organic replenishment.

This view of the development of soil granulation would not be news to Darwin (1881), or many other naturalists, but it has yet to find general acceptance among scientifically trained agriculturalists. Strutt (1970), in *Modern Farming and the Soil*, describes the formation of 'crumbs' and 'granules' as 'products of root action and the decomposition of fresh organic matter. They are more common in topsoils than subsoils and are produced more readily by some plant root systems than others. The root systems of grasses vary but usually excel in causing granulation.'

THE IDEAL SOIL

The requirements for the efficient use of loamy soils in temperate regions are clear but the mode of achievement is not. A water-stable, organic-enriched, granular structure is essential if crops, grown on the loamy soils that are characteristic of temperate regions, are to develop the deep rooting systems necessary to enable them to exploit fully the water and nutrient reserves available. To allow this, the granulation must be small enough to be aerated effectively by gaseous diffusion through the water-filled, biologically active, micropore structure of the granule and yet large enough to allow a macropore system between the granules that will provide for the rapid infiltration of rainfall and clearance of excess, for good aeration and for easy root penetration. It is a common experience to find just such a good soil structure associated with, and exploited by, grass roots but, under these same circumstances, we also find active populations of burrowing earthworms. The problem is to sort out cause and effect.

The development of a granular soil structure

FIELD EVIDENCE

Under natural conditions the contrast in organic matter distribution between the worm-deficient podsol profile under pine and the worm-worked, brown earth profile under ash shows clearly one easily understood consequence of burrowing earthworm activity (*Figures 2.1, 2.2*). This effect, plus the associated contrasts in root distribution, and the structure and drainage of the mineral soil, can also be seen under grass (*Figures 2.3, 2.4*). In fact the whole complex of soil/plant interactions which the ecologist may associate with the distinction between calcicole and calcifuge, lime-loving and lime-hating ecosystems, the pedologist may use to distinguish brown earths from podsols, the forester, mull from mor, and farmers the lush, fertile, well-tended pastures of the lowlands from the acid, matted

Figure 2.1 Podsol profile with mor humus form in supposed remnant of Ancient Caledonian Pinewood, Glen Tanar Estate, Nr. Aboyne, Grampian, Scotland (Stewart, 1958). Organic residues left to decompose on the surface of a typical, worm-deficient, forest soil. Roots concentrated within surface humus layer

Figure 2.2 Brown Forest Soil with mull humus form in an old birchwood on the Glen Tanar Estate, Cambus O'May, Grampian, Scotland (Stewart, 1958). Organic residues incorporated into the surface mineral layers of a worm-worked, forest soil. Roots distributed in depth

Figure 2.3 Matted pasture with mor humus form on drumlin, near Preston, Lancashire, England. Organic residues left to decompose on the surface. Roots mainly restricted to the organic, surface layer. Mineral soil compact and gleyed in tongues looping down from the organic surface

Figure 2.4 Remnant of natural, undisturbed pasture on slightly calcareous glacial drift near Preston, Lancashire, England. No organic debris left on the surface. Roots distributed in depth within mineral soil. Numerous earthworm burrows

Figure 2.5 Strong growth of ryegrass restricted to vicinity of chalk, pitch marking. Elsewhere ryegrass component of original seed mixture has failed to persist. Blaendolau playing field, UCW, Aberystwyth, Wales

Figure 2.6 Wear and desiccation less severe in vicinity of chalk, line markings on tennis court. One of a line of out-courts at Wimbledon, England

Figure 2.7 Contrasting worm populations from same site as in Figure 2.5. Each jar of worms extracted from a square yard of soil: on the right, from the immediate vicinity of the chalk line; on the left, one yard from the chalk line

pasture common on marginal land. All these well-recognized field distinctions can be linked back to the major dichotomy in soil development that is initiated directly by distinctions in soil pH and, indirectly, by the palatability of the litter determining the presence or absence of an active population of burrowing earthworms (Stewart, 1980). The well-known effect of lime markings on sports turf (*Figures 2.5, 2.6*) raises the problem of whether the persistence of the stronger-growing grasses along the line markings is a product of the higher lime status, the grasses in their turn causing the improved soil structure, or whether the presence of the lime encourages earthworms (*Figure 2.7*) and their activity allows scope for deep rooting and the survival of the more demanding, stronger-growing grasses in the original seed mixture.

In woodlands the fate of the earthworm population seems often to be determined by the palatability of the litter. Birch, for example, is able to retain a soil fauna of burrowing earthworms, whereas the litter of pine is rejected and, when dominant, can lead to worm eradication. 'Mull/mor' humus forms can be found where the litter of birch and pine are mixed. In one such soil (*Figure 2.8*) a small population of earthworms (*Lumbricus rubellus, Allolobophora caliginosa* and *Octolasium cyaneum*) was still present and at the base of the organic mat there were earthworm casts (Stewart, 1958). Litter refused by earthworms had been left to decompose

Figure 2.8 Mull/Mor humus form under lone pine within birchwood on Glen Tanar Estate, Cambus O'May, Grampian, Scotland (Stewart, 1958). Some organic debris left to decompose on the surface and some incorporated within the mineral soil. Roots in both the organic surface and the mineral soil. A small population of burrowing earthworms still active

Figure 2.9 Scrub birch thinned but not clear-felled to allow for interplanting with conifers. Near Aboyne, Grampian, Scotland

Figure 2.10   The matted surface of a fine turf lawn from which earthworms have been eradicated

on the surface (mor) but some, presumably the birch, had been consumed and incorporated into the mineral soil (mull). A sequence of tree litter palatability has been established by Bocock and Gilbert (1957): Satchell (1967) has suggested that foresters might do well to pay heed to this, whereas, for example, they are inclined to remove entirely a stand of birch scrub before planting up with conifers. An alternative that would appear to make better soil sense is illustrated in *Figure 2.9.*

On the horticulturalist's ornamental lawn, and on sports turf, particularly that used for cricket, hockey, bowls and golf, the surface-casting activities of earthworms have long since encouraged the use of chemicals for their elimination. This may help to reduce the weed problem and help in the maintenance of a smooth surface but there is no need to go far to explain why such turf care also involves coping with problems of 'thatch', surface rooting and drainage (*Figure 2.10*), and the related problems of sensitivity to nutrient and water supply, diseases and pests.

EXPERIMENTAL EVIDENCE

The experiments reported were set up to isolate the influence on soil granulation of lime, organic matter, grass roots and earthworms both separately and in various combinations (Salih, 1978).

Figure 2.11 Row of three boxes on left all granulated by screening through 3 mm sieve and treated with lime before seeding with S23 ryegrass. Those on right similarly treated but also subject to earthworm activity. Stronger, greener growth in presence of earthworms (Salih, 1978)

Figure 2.12 Growth benefits, after 3 months, arising from earthworm processing of organic matter, and from earthworm structuring of soils initially inadequately structured for satisfactory aeration. '< 3 mm' etc refers to the sieve size used for the initial screening of the soil. 'G' indicates rubbing down to completely de-structure. 'F' indicates the addition of milled, ryegrass hay. 'E' indicates the addition of burrowing earthworms (Salih, 1978)

Figure 2.13 Growth benefits from earthworm activity, after 7 months. Treatments as in Figure 2.12 (Salih, 1978)

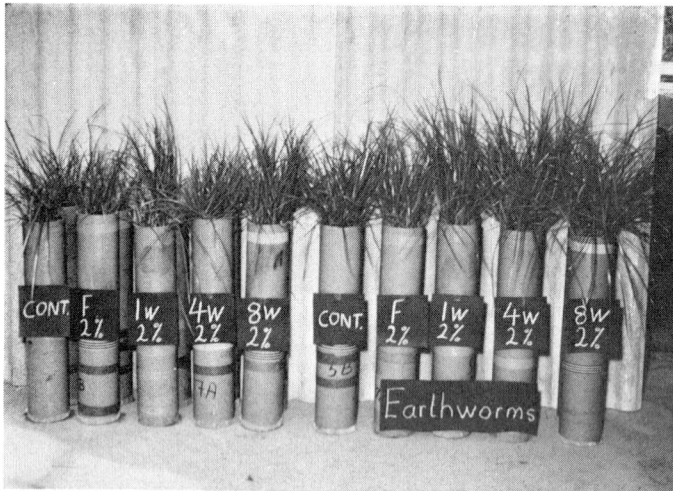

Figure 2.14 Growth benefits from 4 months of earthworm activity. Soil for all treatments initially screened to pass a 3 mm sieve. 'F', '1W', '4W' and '8W' refers to the added organic matter being either fresh milled ryegrass hay, or hay previously composted for 1, 4 or 8 weeks (Salih, 1978)

Figure 2.11 indicates that there is the possibility of a beneficial effect of earthworm activity on nutrition and yield, even in an initially well-granulated soil. Where the soil had 2 per cent fresh milled hay incorporated and was progressively less coarsely structured and, therefore, less well aerated initially, those treatments which also contained burrowing earthworms clearly benefited both in the early stages of growth (*Figure 2.12*) and for at least 7 months thereafter (*Figure 2.13*). As this effect was not entirely eliminated by adding initially well-rotted organic matter (*Figure 2.14*), the

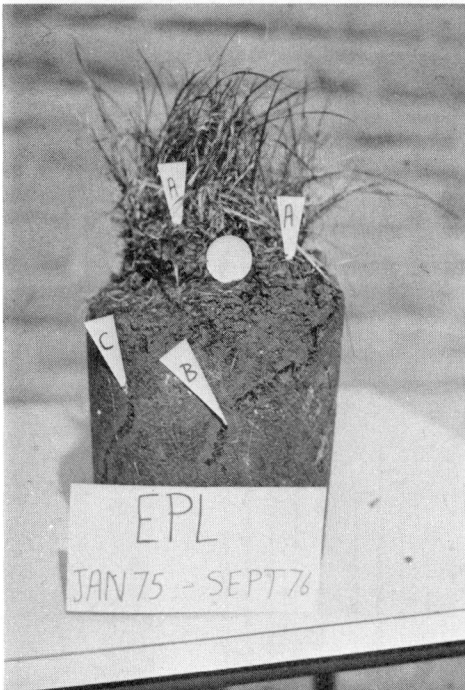

Figures 2.15, 2.16 Comparison of the effect of earthworm activity on a standard soil, initially all screened to pass a 3 mm sieve and planted with S23 ryegrass. PL = grass + lime
EPL = earthworms + grass + lime (Salih, 1978)

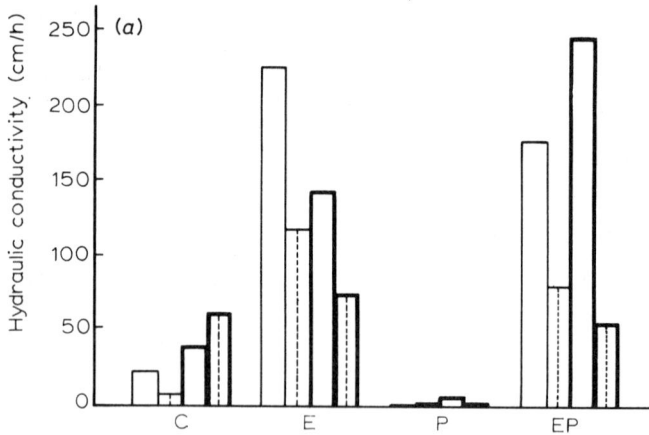

Figure 2.17(a) Hydraulic conductivity determined by 'falling-head' technique as defined by Klute (1965). Base of container removed and replaced by rigid grid so as to allow determination to be carried out on undisturbed soil

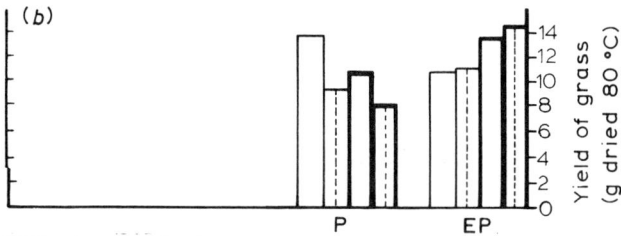

Figure 2.17(b) Average of values for 16 and 21 months. The value for 8 months is not available

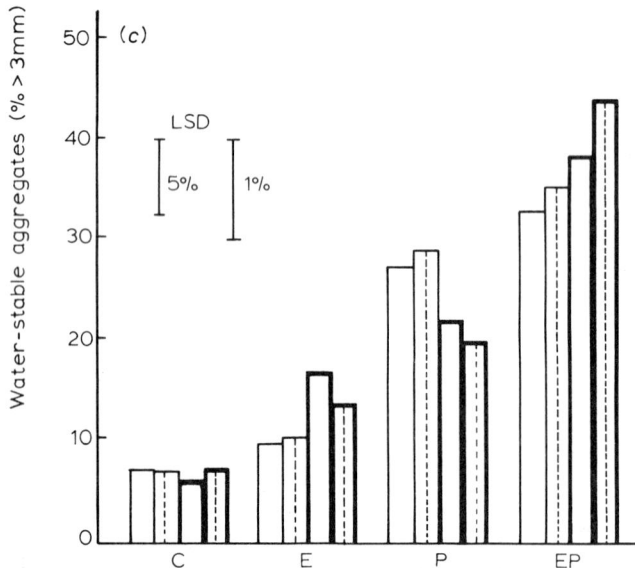

Figure 2.17(c) Water-stable aggregates assessed by wet-sieving (Williamson, Pringle and Coutts, 1956; Tinsley and Coutts, 1967)

benefit from earthworm activity must be more than merely processing the organic matter to reduce its carbon/nitrogen ratio.

Comparison of *Figures 2.15* and *2.16* indicates the obvious effects of earthworm burrowing, the walls of the burrows as well as the surface being the repository of the soil excavated from the burrows and reformed into casts. It is no surprise to find the benefit of the burrows expressed in the high values for hydraulic conductivity in the worm-worked treatments (*Figure 2.17*).

When the soil blocks were eventually air-dried and gently disintegrated to pass a 10 mm sieve, preserving as much as possible of the natural structure, it was evident that the soil was much more inclined to remain in large aggregate form in all the worm-worked treatments (*Figure 2.18*). We

Figure 2.18   Air-dried, soil cores, broken down by hand to pass a 10 mm sieve, before wet-sieving. Treatments applied for 16 months. E = earthworms; P = grass; C = incorporation of 0.7% ryegrass clippings; L = lime to adjust to pH 7. Strong granulation a feature of all earthworm treatments (Salih, 1978)

Figure 2.17   *Experimental conditions* (Salih, 1978). Standard silty clay loam, moderately well-structured soil, from 10–20 cm depth under pasture. 8% Organic matter, pH 6.0. Soil screened to pass a 3 mm sieve and placed in free-draining, opaque, plastic bins, 16 cm square and 22.5 cm deep, maintained in greenhouse under conditions favouring continuous growth.

Soil amendments: ☐ unamended control. ⊞ incorporation of lime to adjust pH to 7.0. ◼ incorporation of 0.7% Italian ryegrass clippings, plus a similar quantity applied to the surface in the second year. ⊞ incorporation of both lime and grass clippings.

Biological activity: 'C' control, no earthworms and no plants. 'E' 2 Lumbricus terrestris added initially and one each month thereafter. 'P' planted to uniform sward of S32 ryegrass, uncut for 11 months then over-seeded and maintained by cutting weekly at 7.5 cm. 'EP' earthworms plus plants assessed the water-stability of this structure by the wet-sieving technique of Williamson, Pringle and Coutts (1956), as modified by Tinsley and Coutts (1967). Our tests gave values for water-stable aggregates larger than 3 mm, indicating that the benefits occurring from a year of earthworm activity show up clearly only when plant roots are also active (*Figure 2.17*). By contrast the addition of organic matter had no effect on its own and only a slight effect when alone in the presence of earthworms. In fact, the main effect of adding organic matter in the form of hay clippings was to reduce significantly the effectiveness of plant roots. However, when well-rotted clippings were used in a second experiment this adverse effect disappeared, suggesting that it may have been only the adverse effect of the relatively high carbon/nitrogen ratio of the fresh clippings on plant growth (*Figure 2.19*). The unspectacular benefit from earthworm activity indicated by the yield data for the EP treatment in *Figure 2.17* reflects difficulties experienced in germination at the time of reseeding (11 months). The earthworms kept burying the seeds in clusters, greatly delaying re-establishment (Salih, 1978).

It is important to realize that, in the experiments reported above, much of the root benefit could have been achieved as a result of the roots reinforcing the structure created mechanically when the standard soil used was initially

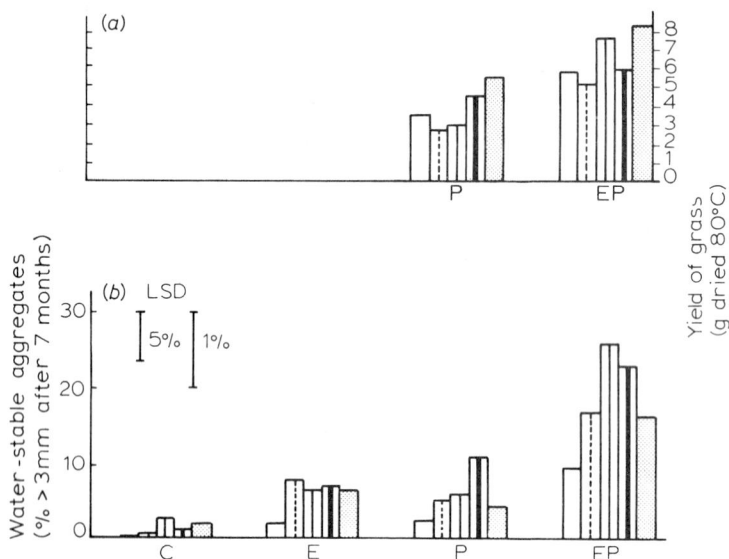

Figure 2.19 Experimental conditions (Salih, 1978). Standard silty clay loam soil of fairly poor structure from 50 cm below surface of permanent pasture. 8% Organic matter, originally pH 5.6 but adjusted to pH 7.0 by addition of lime. Soil screened to pass a 3 mm sieve and placed in 50 cm section of 7 cm diameter, opaque plastic pipe, forming a 25 cm column of soil above a 25 cm column of fine sand, soil and sand separated by a worm barrier of thin fabric. Management and biological treatments similar to those for experiment described in *Figure 2.17*. Duration 7 months. Grass cut once only at the end of the 7-month period.
(b) Water-stable aggregates assessed by wet-sieving as for *Figure 2.17*.
Soil amendments: 2% milled ryegrass hay incorporated after composting under controlled laboratory conditions to achieve various stages of decay. ☐ unamended control. ☐ fresh milled hay. ☐ 1 week composted. ☐ 4 weeks composted. ☐ 8 weeks composted.

passed through a 3 mm sieve. This initial treatment was intended to facilitate the ramification of roots and resemble what a farmer tries to achieve when creating a seed bed by mechanical cultivation. Thus roots, by merely ramifying between these mechanically created structural units, could assist in their stabilization without having had anything to do with their formation. Organic agents derived from the decay of added organic matter might also do this, by water-stabilizing the structural faces, but if this was occurring it was by no means as effective as living roots.

The cumulative effect of roots and earthworms working together on the creation and stabilization of soil aggregates may be accounted for by one of two hypotheses: either (1) roots are the effective agents both for creation and stabilization and earthworms contribute only to the extent that their activities improve plant growth or (2) earthworms actually control events, not only by their effect on plant growth, which leads to an increase in roots for stabilization, but also by their casting activities, creating new structures. In the second case the role of earthworms, incorporating organic residues into their casts, could enhance both the organic matter and rooting effect by encouraging the roots to branch into the casts that they form.

Land management unfavourable to burrowing earthworms

Whichever hypothesis explains how earthworms, in combination with organic matter and plant roots, achieve the water-stable granulation of loam soils, the activities of burrowing earthworms are essential. Therefore, any form of soil management that neglects their welfare should be avoided, including those listed below.

1. The planting of pure stands of conifers on land with a viable population of earthworms.
2. The removal of topsoil and its replacement denatured by storage.
3. The regular application of highly concentrated, readily soluble, ammonium and nitrate fertilizers (Edwards and Lofty, 1977; Salih, 1978).
4. The use of pesticides with uncontrolled, or unknown, side-effects. Some widely used, modern, systemic fungicides were first used as worm-killers in turf! (Wright, 1977; McEwen and Stephenson, 1979).

Conclusion

There seems to be cause for concern over the extent to which some farmers and horticulturalists are conditioned to systems of intensive production, that put at risk the natural mechanisms enabling plants to exploit their environment to the full. Such growers could be in a vicious circle of increasing dependence on the expensive products of the agrochemical industry unless they rethink their management procedures in the light of the benefits that can be acquired by fostering deep exploitive rooting.

Two diverging hypotheses seem possible:

1. Plant roots, especially grass roots, are responsible for the water-stable soil granulation which is so important for the loams. Use of concentrated mineral fertilizers simultaneously increases plant growth and root development and thereby may benefit soil structure.

2.   Earthworms and plant roots, working together, are responsible for water-stable soil granulation. Therefore increases in yield should be sought only by means that do not limit the effectiveness of either component and, preferably, benefit both.

The risk of the vicious circle arises when concentrated mineral fertilizers are substituted for organic manures. Increasing plant growth by the surface application of fertilizers will tend to restrict root proliferation to the more superficial soil layers. The lush leaf growth stimulated by the luxury uptake of fertilizer nitrogen may encourage insect and fungal attack and the urge to use pesticides to counterattack. The combined effect of the fertilizer-plus-pesticide programme may maintain yields but at an increasing cost for the agrochemicals and at some risk to the efficiency of the biological structuring mechanism. After extensive survey of agricultural grassland in Britain, Troughton (1957) concluded that rooting becomes progressively more superficial with time since ploughing and, in his view, no one has yet supplied a wholly satisfactory explanation. Bearing in mind the benefits to be derived from deep rooting, could it be that the use of modern fertilizers and pesticides so interferes with the natural, biological agents of soil structure development that this road leads to a dependence on these agrochemicals and, in present circumstances, leads to progressively escalating costs?

The essence of the biological approach to soil fertility is wholeness: feed the whole soil and let the soil feed the plant. This is the message of Balfour (1975) in *The Living Soil and the Haughley Experiment.* It could be good advice for farmers in difficulty on loamy soils in temperate regions.

References

BALFOUR, E. B. (1975). *The Living Soil and the Haughley Experiment.* London; Faber and Faber

BOCOCK, K. L. and GILBERT, O. J. W. (1957). The disappearance of leaf litter under different woodland conditions. *Plant and Soil,* **9,** 179–185

DARWIN, C. (1881). *The Formation of Vegetable Mould Through the Action of Worms, with Observations on Their Habits.* London; Murray

DREW, M. C. (1975). Comparison of the effects of a localized supply of phosphate, nitrate, ammonium and potassium on the growth of the seminal root system, and the shoot, in barley. *New Phytologist,* **75,** 479–490

EDWARDS, C. A. and LOFTY, J. R. (1977). *Biology of Earthworms,* 2nd edn. London; Chapman and Hall

GREENWOOD, D. J. (1969). Effect of oxygen distribution in the soil on plant growth. In *Root Growth,* pp. 202–221. Ed. W. J. Whittington. London; Butterworths

GREENWOOD, D. J. (1970). Soil aeration and plant growth. *Rep. Progr. Appl. Chem.* **55,** 423–431

KLUTE, A. (1965). Laboratory measurement of hydraulic conductivity of saturated soil. In *Methods of Soil Analysis,* Pt. 1. Ed. C. A. Black. Wisconsin, USA; American Society of Agronomy Inc.

MAFF (1976). *The Agricultural Climate of England and Wales*. Ministry of Agriculture, Fisheries and Food. London; HMSO

McEWEN, F. L. and STEPHENSON, G. R. (1979). *The Use and Significance of Pesticides and the Environment*. Chichester; Wiley

ROWSE, H. R. and BARNES, A. (1979). Weather, rooting depth and water relations of broad beans—a theoretical analysis. *Agric. Meteorol.*, **20**, 381–391

SALIH, R. O. (1978). *The Assessment of Soil Structure and the Influence of Soil Treatment on this Property*. PhD Thesis, University College of Wales, Aberystwyth

SATCHELL, J. E. (1967). Lumbricidae. In *Soil Biology*, pp. 259–322. Eds. A. Burgess and F. Raw. London and New York; Academic Press

STEWART, V. I. (1958). *A Study in Variations in the Humus Horizons of Soils in the Natural Pinewoods and Birchwoods of the Glen Tanar Estate, Aberdeenshire*. PhD Thesis, University of Aberdeen

STEWART, V. I. (1975). Soil structure. *Soil Ass. Quart. Rev.*, **1** (3)

STEWART, V. I. (1980). Soil drainage and soil moisture. In *Amenity Grassland: an Ecological Perspective*, pp. 119–124. Ed. I. H. Rorison and R. Hunt. Chichester; Wiley

STEWART, V. I. and ADAMS, W. A. (1968). The quantitative description of soil moisture states in natural habitats with special reference to moist soils. In *The Measurement of Environmental Factors in Terrestrial Ecology*, pp. 161–173. Ed. R. M. Wadsworth. Oxford; Blackwell Scientific Publications

STRUTT, N. (1970). Modern Farming and the Soil. *Report of Agricultural Advisory Council on Soil Structure and Soil Fertility*. Ministry of Agriculture, Fisheries and Food. London; HMSO

TINSLEY, J. and COUTTS, J. R. H. (1967). Aggregate stability determination to determine the true crumb structure. In *West European Methods for Soil Structure Determinations*. *International Soil*, **6**, 75

TROUGHTON, A. (1957). The underground organs of herbage grasses. *Commonwealth Bureaux Field Crops Bulletin No. 44*, 163

WILLIAMSON, W. T. II., PRINGLE, J. and COUTTS, J. R. H. (1956). Rapid method for the determination of water stable aggregates in soils. *J. Sci. Fd. Agric.*, **7**, 265–269

WRIGHT, M. A. (1977). Effects of benomyl and some other systemic fungicides on earthworms. *Ann. Appl. Biol.*, **87**, 520–524

3

ORGANIC MATTER CYCLES IN TROPICAL SOILS AND HUSBANDRY SYSTEMS (with special reference to Africa)

R. P. MOSS
Department of Geography, University of Salford, UK

Preamble

Tropical agricultural development has been bedevilled by the uncritical acceptance of two unqualified presuppositions:

1. That tropical environments and tropical soils are peculiarly fragile and intractable;
2. That tropical environments are so uniform in detailed pattern that blanket solutions to problems are possible.

In relation to the first it is often forgotten that the traditional systems of agriculture, even in their more recent modern developments, not only maintained the continuity of agricultural and biological production, but also generally met the needs of the people employing them, as well as providing a surplus and latterly a significant cash crop component. In the past, efforts at increasing production have almost always been based on temperate technologies and have made no attempt at all to use the knowledge of fundamental ecological and edaphological relationships achieved in the wisdom of indigenous cultivators. There seems to be no sustainable *a priori* reason why such technologies should necessarily be immediately applicable, let alone successful, in the very different kinds of soil characteristics and soil/plant relationships which exist in the tropics. Temperate technologies have developed by relatively gradual change in which the environment has also been changed in response to altered technology—and, be it noted, not without its fair share of environmental disaster, as in the Dust Bowl of the United States. Temperate development has been a dialogue between technology and environment. Tropical development must be the same, and it is foolish to ignore the accumulated knowledge already available and to approach low-latitude problems with high-latitude prejudices. One purpose of this chapter is to outline what seem to be some of the key elements in the relevant relationships, with particular reference to organic matter balances.

Another aim of the chapter is to attempt, in a short compass, to dispel the false assumption of tropical uniformity, and the concomitant error of seeking the blanket universal solution to tropical agricultural problems. Here again it is difficult to understand why those assumptions should have been so uncritically accepted for so long—there are numerous books on 'tropical soils', but none on 'temperate soils' as a group. It is also fascinating to examine the impact of tropical studies in soil science and ecology upon these disciplines, as has been done elsewhere (Moss, 1977).

Then, in conclusion, some implications for development will be outlined, in relation to current ecological understanding. This will in part provide an underpinning for some of the other chapters in this volume (Dalzell, page 205; Ojomo, page 187; Wilson and Kang, page 193), as well as providing a conspectus of the significance of ecological understanding to tropical agricultural development.

The variability of tropical environments

The considerable variability of the tropical environment may be focused under the three main groups of factors of most relevance to man—climatic variables, soil characteristics and ecosystematic relationships. Geological and geomorphological differences, although obvious and important in their own right, are only rarely of proximate significance to human agricultural activity: their role is important only when the focus of attention is changed from the immediately relevant contemporary and shorter-term ecological and edaphological set of questions, to the fascinating but much more academic questions of history, longer-term development and morphogenetics.

TROPICAL CLIMATES

Tropical climates vary along two main axes of change important to plants. In the first place a gradient of atmospheric water supply exists, which is represented by a decreasing mean annual rainfall, associated with an increasing length and intensity of the dry period, and a decreasing reliability of both the amount of precipitation and of the time of the onset of the principal wet periods. While very broadly related to latitude and distance from the Equator, the spatial expression of this gradient is dominated over large areas by monsoonal circulations, and by the effect of altitude, relief and distance from the oceans. The second gradient relates to temperature, not so much to mean values, but to the occurrence of extremes. Night temperatures are especially significant to a number of tropical plants (Longman and Jenik, 1974), and the frequent occurrence of night frosts in arid regions and on tropical mountains clearly emphasizes the special problems of desert and montane regions. Here again there is a very general relationship of the spatial distribution with latitude, but with considerable modifications of the pattern in response to relief, oceanity, and also moisture distributions. These climatic gradients are illustrated in *Table 3.1*.

A further environmental element of importance is photoperiod, which is, of course, directly related to latitude. Recent work has demonstrated clearly the photoperiodic sensitivity of many tropical plants (Longman and Jenik, 1974; Janzen, 1975), and the hitherto largely unsuspected importance of this factor may be critical in such decisions as species choice in agroforestry.

The various dimensions of the variability in atmospheric water supply interact with the spatial structure of particular plant communities, and with inherent properties of the solum, to produce daily, seasonal, and periodic patterns of moisture stress for the plants. These obtain the water they need largely through their root system, and therefore from the volume of the solum they are thus able to exploit. The dependence of soil moisture status upon the character of the vegetation canopy was clearly demonstrated many years ago. For example, in East Africa, under closed forest at Mbeya, even a 5 to 6-month dry season was unable to reduce the soil moisture level to wilting point in any soil layer, and there was moisture available throughout the rooting depth of most perennial components of the vegetation for the

Table 3.1 CLIMATIC VARIABILITY IN TROPICAL LATITUDES; SOME EXAMPLES
(a) CHANGE WITH LATITUDE AND OCEANITY FOR WEST AFRICA (FROM HARRISON CHURCH, 1980)

Station	Jan.	Feb.	Mar.	Apr.	May	Jun.	Jul.	Aug.	Sep.	Oct.	Nov.	Dec.	Year
Tabou:	30.3	30.7	31.1	31.4	30.2	28.7	27.8	27.3	27.8	28.8	29.7	30.1	29.5
4.25°N; 7.22°W	23.0	23.2	23.3	23.4	23.3	23.2	22.9	22.1	22.4	23.0	23.4	23.1	23.0
	38	69	99	117	417	579	175	99	198	221	221	117	2350
Calabar:	30.4	31.9	31.9	31.4	30.6	30.3	27.6	28.1	28.7	29.5	30.2	30.5	30.1
4.58°N; 8.20°E	22.8	23.0	23.6	23.4	23.6	22.9	22.6	22.4	22.6	22.5	22.8	22.9	22.9
	43	76	152	213	312	406	450	406	427	310	191	43	3030
Lagos:	32.5	32.8	33.1	32.5	31.1	29.0	27.9	27.6	28.8	29.8	31.8	31.8	30.7
6.27°N; 3.24°E	21.0	22.6	22.8	22.1	22.3	21.8	21.1	20.9	21.4	21.7	22.1	21.6	21.8
	25	38	99	140	277	439	277	69	142	201	71	25	1803
Ibadan:	32.9	34.2	34.4	33.6	31.6	29.8	27.9	27.8	29.3	30.4	31.9	32.4	31.4
7.26°N; 3.54°E	20.7	21.4	22.6	22.5	22.2	21.7	21.1	21.1	21.1	20.9	21.1	20.7	21.4
	10	23	89	137	150	188	160	84	178	155	46	10	1229
Bouaké	33.4	34.6	35.2	34.8	32.8	30.7	29.4	28.7	29.7	31.1	32.0	32.8	32.1
7.41°N; 5.02°W	20.4	21.7	21.8	21.7	21.7	21.1	20.6	20.3	20.6	20.8	20.7	20.6	21.6
	10	38	104	147	135	152	81	117	208	132	38	25	1186
Tamale:	35.7	37.9	38.4	36.8	34.6	32.3	30.6	30.1	30.8	33.2	35.2	34.0	34.2
9.24°N; 0.53°W	19.1	21.3	23.3	23.2	22.6	21.6	21.6	21.0	20.9	21.4	20.9	20.8	21.5
	3	8	56	81	119	140	140	206	226	94	18	5	1090
Kano:	29.8	32.2	35.4	38.2	37.4	34.7	30.7	29.5	31.1	34.2	33.6	30.6	33.1
12.02°N; 8.32°E	13.4	15.3	18.8	22.4	23.7	23.3	21.7	20.9	20.8	20.1	16.4	13.8	19.2
	0	0	3	8	69	114	203	315	130	13	0	0	853
Tambacounda:	34.9	37.2	38.9	41.0	39.6	36.1	32.2	30.7	31.6	32.2	33.4	33.9	35.4
13.46°N;	14.9	16.1	19.5	21.2	24.2	23.1	21.9	21.9	22.3	21.9	17.4	15.2	19.9
13.11°W	0	0	0	3	28	175	193	305	216	81	3	0	1006
Kayés:	35.2	38.2	40.9	44.0	43.4	40.0	34.6	32.3	33.4	35.1	37.6	34.4	37.4
14.24°N;	16.7	18.9	22.0	25.0	27.7	25.9	23.9	22.8	22.9	23.3	20.2	17.8	22.3
11.26°W	3	0	0	0	25	97	160	241	188	43	0	0	757
Agadés:	28.0	33.1	38.1	42.0	44.4	44.0	40.9	38.7	41.0	40.6	34.6	32.3	38.1
16.59°N;	10.1	11.7	16.2	20.4	24.8	24.7	23.2	22.7	22.6	19.8	15.1	11.8	18.6
7.56°E	0	0	0	0	5	8	48	94	18	0	0	0	175
Atar:	30.4	32.8	33.9	38.7	39.7	42.3	43.2	42.1	42.0	38.3	33.1	29.5	37.2
20.31°N;	12.1	12.9	16.8	19.5	21.8	26.7	25.1	25.9	26.0	23.1	16.9	13.3	20.0
13.04°W	3	0	0	0	0	3	8	31	28	3	3	0	79

(First row gives mean daily maximum temperatures in degrees Celsius; second row gives mean daily minimum temperature in degrees Celsius; third row gives total monthly precipitation in millimetres; final column gives the annual figures.)

(b) CONTRAST BETWEEN CLOSED CANOPY FOREST AND FOREST CLEARING (FROM LONGMAN AND JENIK, 1974)

Measure- ment	Position in forest	Time of day 2000	2200	2400	0200	0400	0600	0800	1000	1200	1400	1600	1800
Relative	Canopy	94	95	91	96	93	97	92	85	71	69	71	77
humidity (%)	Clearing	87	91	94	94	93	95	94	70	51	45	60	75
Saturation	Canopy	1.6	1.3	2.2	0.9	1.7	0.7	1.9	4.4	9.3	10.9	10.4	7.4
deficit (millibars†)	Clearing	3.4	2.4	1.4	1.4	1.7	1.3	1.4	10.4	22.1	24.9	16.0	8.0
Dewpoint	Canopy	20.9	20.8	19.6	19.6	19.3	19.3	19.7	20.7	19.4	20.9	20.8	20.8
(°C)	Clearing	20.5	20.4	19.5	19.3	19.1	18.9	19.7	20.4	19.6	17.9	20.7	20.2
Evaporative	Canopy	0.1	0.1	0.0	0.0	0.0	0.0	0.0	0.1	0.2	0.3	0.3	0.2
power (ml/2h)	Clearing	0.1	0.0	0.0	0.0	0.0	0.0	0.3	1.2	2.2	2.1	1.2	0.4

† 1 millibar = 1000 dynes per cm^2. Measurements taken during a dry spell on 6—7 January, 1967 in tropical lowland forest at the Ankasa Forest Reserve, Ghana.

whole year; however under more open forms of land use, such as cultivation and regenerating bush, drying out was rapid to considerable depths soon after the onset of the dry season (McCulloch, 1962; papers in Pereira, 1962).

The effects of atmospheric temperature changes are also mediated through the spatial structure of the vegetation, as are the effects of wind which may also be of importance to moisture stress and plant response, especially, but perhaps not exclusively, in the more arid areas of the tropics. There is, in fact, only minimal interchange between the atmosphere above, and that below, the canopy of a closed forest (*see for example* Hopkins, 1965). The contrast between closed forest and woody fallows, on the one hand, and more open savanna and grass–herb fallows on the other, in these respects is considerable, even in areas of similar macroclimate, where forest and savanna are contiguous (Moss and Morgan, 1977). There would seem to be a great need for much more study of the microclimate and of the soil water and soil air, both under self-propagated plant communities and under various crops and crop combinations in the tropics.

TROPICAL SOILS

Interactions such as those outlined above make the variability of tropical soils of great practical significance. No doubt there are broad general patterns of weathering which are common to many broad zones of the tropics, and other general features in which tropical soils as a group differ from those of temperate latitudes. But the fact of broad temperate/tropical differences should not lead to the erroneous assumption of tropical uniformity. Some of the main properties in which tropical soils display variability are shown in *Table 3.2*. These principally relate to parent material differences, to biological community relationships, and to soil water characteristics and have been grouped as litho–edaphic, bio–lithic,

Table 3.2 FACTORS IN THE SOIL AND REGOLITH WHICH DISPLAY
VARIABILITY IN TROPICAL AREAS

Type	*Variable factors*

I. Litho-edaphic character (properties which relate to parent material character, and thus to weathering processes and lithology, especially mineral content and grain size)	Character of mineral clay
	Proportion of mineral clay
	Presence of silt-size particles
	Size and proportion of angular gravel, especially quartz
	Size and proportion of concretionary material: predominantly iron; sometimes manganese or titanium
	Layer heterogeneity:
	    depth of rock layer below surface
	    depth of duricrust below surface (thick layer of hard crust)
	    depth of hardpan below surface (thin crust over softer layer)
	    depth of compact stone line below surface
	    depth of marked textural discontinuity below surface
	    depth and thickness of compact or indurated layer
	    depth and thickness of nutrient-rich (weathering) layer
	(These properties are of major edaphological significance only if they occur within the potential rooting volume of the plants)

II. Bio-lithic interface (properties which relate to the interaction between the organic and parent material factors in the solum, and the biological processes which occur in the soil)	Organic matter content
	Degree and stability of mineral/organic association
	Stability of organic complexes
	Rate of mineralization
	Value of Decomposition Constant (higher values mean a higher rate of restoration of organic matter levels on resumption of organic matter supply after period of cessation)
	Character of organic matter (little detailed work has been done on this for most tropical soils)
	Form of root system (this is basically conditioned by genetic factors, modified by the properties of the solum with special reference to profile heterogeneity)
	Biological activity of rhizosphere
	    presence and activity of mycorrhizae
	    presence and activity of nodulating bacteria and other symbionts
	Presence and activity of free-living bacteria
	(It is possible that both termites and earthworms may be important as active components of the soil fauna, but little work has been done on this in the tropics)

III. Hydro-lithic interface (properties relating to the holding and movement of water in the solum; thus also relating to aeration)	Water table levels and their seasonal variability Water-holding characteristics (dependent upon the properties of the particles and their arrangement in structures and fabrics) Drainage facilitation (dependent upon the properties of the particles and their arrangement in structures and fabrics) (These properties are especially important in the climatic regimes prevalent over much of the tropics, where periods of heavy rain are succeeded by periods of more or less intense aridity)

This grouping is especially designed to emphasize those edaphological properties which display variability from place to place, often over very short distances, in tropical latitudes. A more theoretical and specific categorization can be found in Moss (1978).

and hydro–lithic respectively. Some of these properties cannot be considered important from a pedological point of view, certainly at the higher levels of soil grouping, but they are of fundamental significance in terms of plant/environment relations. Pedology is essentially an historical study, and it is not surprising that its assignation of relative importance to particular properties should differ from that demanded by edaphological criteria, which is basically functional in orientation. It is the strong pedological orientation of many soil and land resources studies in the tropics which has so often limited their immediate relevance and unequivocal application to the practical problems of agricultural management and development planning, a fact not infrequently mentioned by practising agriculturalists. An edaphological model for land resource assessment has been developed elsewhere (Moss, 1980).

From *Table 3.2* it will be seen that the soil properties of most significance relate to the activity of the mineral clay, the proportions and arrangement of different sizes and kinds of particle, the degree of heterogeneity of the horizons in the solum, and that these interact with the inherent properties of the plants, genetically, physiologically and symbiotically, to facilitate or inhibit organic development and function. The practical importance of this, in relation to contiguous areas of forest and savanna agriculture, has been reported elsewhere (Moss, 1977; Moss and Morgan, 1977), and its possible relevance to the whole forest–savanna problem has also been debated (Moss, 1976).

Three main points emerge. First, the importance of organic matter to the chemical activity of the soil, and to water relationships. Second, the significance of soil organisms to nutrient behaviour and supply. And, third, the vital function of the development of the root system to the whole gamut of soil/plant relations.

Organic matter levels

Organic matter levels depend upon the rate of supply from the plants, the rate of mineralization, and the character and stability of organic compounds in the soil, both discretely, and in mineral–organic complexes. The first depends upon the bulk of the plants, and especially their nonlignous parts; the second is generally rapid, except in montane regions with lower temperatures, but depends for its continuity upon the presence of adequate amounts of soil water for microbial activity, which in its turn depends upon macroclimatic variability as it is mediated by the structure of the plant community. The third is as yet little understood, but considerably greater amounts of organic matter are held in the soil, even under cultivation, than was once thought, and levels are readily restored when supplies are returned after an interruption (Vine, 1954, 1968; Laudelout, 1962).

Soil organisms

The importance of soil organisms to nutrient behaviour and supply increasingly appears to be quite fundamental to understanding of tropical soils.

The importance of free-living nitrogen-fixers has long been understood (Nye and Greenland, 1960), but the importance of symbiotic relationships to nutrient status is only now beginning to emerge, after the pioneering work of Ellenberg (1959a,b). The importance of fungi, particularly mycorrhizal, has been a matter of debate for some time, but recent work has tended to emphasize their importance (Longman and Jenik, 1974; Janzen, 1975). The role of the mesofauna, particularly earthworms and termites, is not yet understood with any certainty. Termites of many species and many modes of life are almost ubiquitous, whereas earthworms show a much more intermittent distribution, which may be related to organic matter variability in both amount and kind.

Development of the root system

The development of the root system is a highly critical variable in the complex of soil/plant relations. It is clearly a genetic, physiological character, and varies from species to species. *Table 3.3* summarizes the grouping by Longman and Jenik (1974) of root system types in mature dicotyledonous tropical forest trees. Trees with a strong development of

Table 3.3 ROOT SYSTEM TYPES IN TROPICAL FOREST TREES (FROM LONGMAN AND JENIK, 1974)

Type	Root system	Where found
1	Thick horizontal surface roots (frequently associated with large spurs or buttresses); weak vertical sinkers; no tap root	Characteristic of many emergent species throughout rain forests in Africa, Asia and South America
2	Thick horizontal surface roots; well-developed vertical sinkers; well-developed tap root	Characteristic of many canopy forest trees, including some important secondary forest and regrowth species
3	Weakly developed surface roots; well-developed system of oblique roots associated with a well-developed tap root	Characteristic of many smaller forest trees, including regrowth species; also typical of many large woody climbers
4	Numerous significant stilt roots; weakly developed underground roots; no sizeable tap root	Characteristic of species adapted to waterlogged sites, but also found in some species from drier locations

Most seedlings of forest trees develop a significant tap root quickly on germination and early growth, but develop the features noted above as they become more mature. Many successful, rapidly growing exotic species, like *Leucaena leucocephala*, have root systems of Type 2 or Type 3.

horizontal surface-feeding roots, with or without a tap root, are typical of the emergent species, and some main canopy taxa, whereas smaller under-storey and secondary forest types, together with large woody climbers, display a system which taps the deeper layers of the solum. This is especially significant in the light of the fact that woody fallows are known to pump up nutrients from depth, and that they are most effective during the early stages of the woody fallow succession (Nye and Greenland, 1960; Laudelout, 1962; Vine, 1968). The development of the root systems of tropical plants is not, however, simply a function of plant physiology: it depends also on the rooting medium. Roots are sensitive to barriers, such as hardpans, rock layers, and even sudden changes of soil texture from light to heavy. They also respond to concentrations of nutrient sources, and to aeration and waterlogging (Lyr and Hoffman, 1967; Rogers and Head, 1969; Moss, 1977). Such differential root development in response to variable soil character may be crucial to the survival of some plantation tree crops subjected to severe biotic stress (Moss, 1963).

VARIABILITY OF TROPICAL ECOSYSTEMS

These considerations lead clearly into an examination of the importance of the variability of tropical ecosystems. The conventional zonal grouping of tropical ecosystem types, although bringing out some important features in relation to ecological character, tends to obscure the variability which in fact exists. *Table 3.4* sets out some of the variability which is to be found within

Table 3.4   TYPOLOGY OF TROPICAL ECOSYSTEM TYPES BASED ON STRUCTURE AND PHYSIOGNOMY

Zonal types	Characteristics	Subtypes	Species
I Closed forests	Dominated by trees with a closed canopy; usually with more than one tree layer; under cultivation normally produce a woody regrowth which initiates a succession of communities leading back to structured closed forest	1. Evergreen forests of equatorial latitudes and low altitudes	No dominant species; floristically very rich in woody species; poor ground flora
		2. Semideciduous forests of low latitudes and low altitudes	No dominant species; rich in woody species, some deciduous; poor ground flora
		3. Simple structure forests of low latitudes and low altitudes	One, two or three dominant species; single-layer canopy structure
		4. Montane forests of low latitudes and high altitudes	Usually dominated by a few species; often rich ground flora
		5. Freshwater swamp forests of low latitudes	With characteristic species, but including species from associated forest on higher ground; less complex structure
II Canopy woodlands	Dominated by trees and shrubs with a closed canopy, but lower than in forests; usually with one tree layer only; under cultivation does not produce a woody regrowth, but fallow usually characterized by grasses and herbs, which remain dominant for a considerable period of the succession leading back to woodland	1. Mangrove swamp forest	Often monospecific in relationship to salt tolerance and tidal inundation; poor on associated species; principally cleared for flood rice cultivation
		2. Savanna woodland	Sometimes dominated by one or two species, but often multispecific; closed single-layer canopy, but sometimes shows well-developed shrub layer which amounts to an understorey
		3. Thorn woodland	Usually dominated by species of *Acacia*, and found in semiarid regions; with a closed light canopy of touching crowns

Table 3.4 *continued*

Zonal types	Characteristics	Subtypes	Species
III Open woodlands	Variable density of trees and shrubs, with a discontinuous canopy, usually with individual trees discretely located; the ground layer is more or less continuous and displays marked seasonality, consisting of grasses and herbs, with some low shrubs; cultivation tends to reduce the number of trees, and the fallow succession consists entirely of grasses, herbs and some low shrubs; tree regeneration is slow or absent; fire is a significant factor in the ecology	1. Savanna parkland	Relatively rich in tree species though sometimes dominated by one or two; trees frequent and a significant element in the community; ground layer continuous and relatively rich in species, though often dominated by one or two
		2. Open savanna	Ground layer dominant with only a few isolated trees; ground layer very variable in response to a complex of ecological factors; tree species often dominated by taxa of use to man, or of no use, even as fuel
		3. Thorn scrub	Very open canopy or isolated trees, usually *Acacia* spp; discontinuous ground layer with much bare soil; ephemeral spp of importance in some areas; in semiarid regions
IV Herb and grasslands	Exclusively dominated by low plants, chiefly herbs and grasses; some communities are grassy, others herbaceous; within the graminaceous and the herbaceous types, different species and species combinations produce widely differing heights and patterns of vegetation; found in all climatic zones, and at a wide range of altitudes	1. Lowland or steppe grassland	A variety of types, depending upon the dominant species—elephant grass, bunch grass, short grass, and so on; usually associated with a range of other grass and herb species; herbs may be locally very important or dominant
		2. Mountain grassland	Short grassland restricted to higher altitudes; analogous to temperate types, like the veld in S. Africa
		3. Montane herb communities	Dominated by large herbs; restricted to high altitudes in the equatorial mountains

The grouping does not presuppose any particular origin for the types proposed; the interest is in function, and especially nutrient cycling, volume of soil exploited by roots, organic matter production and type, and so on; hence tree density must be a fundamental differentiating variable, and is in fact the key factor in the grouping.
Tree density in savanna types may well be the result of cultivation and fire, as well as exploitation for fuelwood; thus savanna woodland (II.2), savanna parkland (III.1), open savanna (III.2), and lowland grassland (IV.1), may well represent, at least in some instances, a sequence of types relating to the degree of human impact, directly by cultivation, and indirectly by fire.
Physiognomy and structure clearly bear some relation to floristics, but floristic differences are very difficult to characterize, and indicate more subtle environmental and anthropogenic influences than those of importance to the present grouping.
There are numerous other types of very limited extent which have not been included, such as alpine communities in high mountains, and localized bog communities; on the scale of the present survey their significance is strictly limited, but more locally they may be of considerable importance.

the main physiognomic zonal types. The strong contrast between forest and woodland ecosystems and savanna communities must be emphasized at the outset. This has been discussed fully elsewhere (Moss, 1977; Moss and Morgan, 1977), and the effects of the forest canopy on microclimate, soil temperature and soil moisture, have been pointed out, in contradistinction to savannas with a more or less open tree canopy. The characteristic of importance with reference to soil organic matter is, however, the twofold aspect of nutrient cycles, in which there is a closed cycle between plants, litter and the upper layers of the soil, and the incorporation into that cycle of nutrients brought up from depth by woody plants with a developed root system of type C (*Table 3.3*), and possibly some also of type B. This is illustrated in *Figure 3.1*. Such an accession of nutrients into the plant–litter–topsoil cycle depends, of course, upon the degree of development of the plant root system, *and* upon the presence of weathering material in the solum which is able to provide a supply of nutrients, especially cations, trace elements, and possibly phosphorus from a weathering complex containing apatite. Thus, in considering soil viability in tropical closed forest regions, it is necessary to take into account variability at depths of 3–5 m and more, as well as in those layers closer to the surface.

It is important to emphasize that this cyclic system is such that its activity, brought about by the effect of floristic diversity upon ecological function, ensures that leaching losses of nutrients from the system are very low (Nye and Greenland, 1960; Laudelout, 1962). It also explains why actual amounts of nutrients held in the soil are often low when measured by standard methods: they are actively engaged in processes of change and cyclic transfer in the soil solution, as was suggested by Baeyens as long ago as 1949. This poses a genetic problem which pedologists have yet to solve.

Savanna ecosystems are fundamentally different from this in their function. The tree cover is intermittent and any 'pumping-up' of nutrient elements is more or less localized to the small area around isolated individual trees. This has been shown to be especially significant to the calcium balance of the soil (Nye and Greenland, 1960), which appears to depend upon the woody component of the vegetation. Furthermore, not only is the grass–herb component inherently much less efficient in storing nutrients—and consequently a less efficient fallow than tropical woody vegetation—but some species of Andropogoneae appear to inhibit nitrate formation, and to present some crop plants with a nitrate deficiency at the outset of the cultivation cycle (Vine, 1968). *Figure 3.2* illustrates these relationships. Savanna woodland with a closed canopy is probably more akin to closed forest in its function with respect to nutrients, although little is known about the root systems of savanna trees.

Kowal and Kassam (1978) have recently reviewed in considerable detail the nutrient and water relations of soils in African savannas. Their full and well-documented account repays close study, but one element in their discussion is very clearly emphasized: the role of soil organic matter in maintaining nutrient supplies, in stimulating microbiological activity, and in maintaining beneficial physical properties. They also refer to the low levels of organic matter in many savanna soils, compared with those in tropical forest soils. Furthermore, the effectiveness of applications of inorganic

49

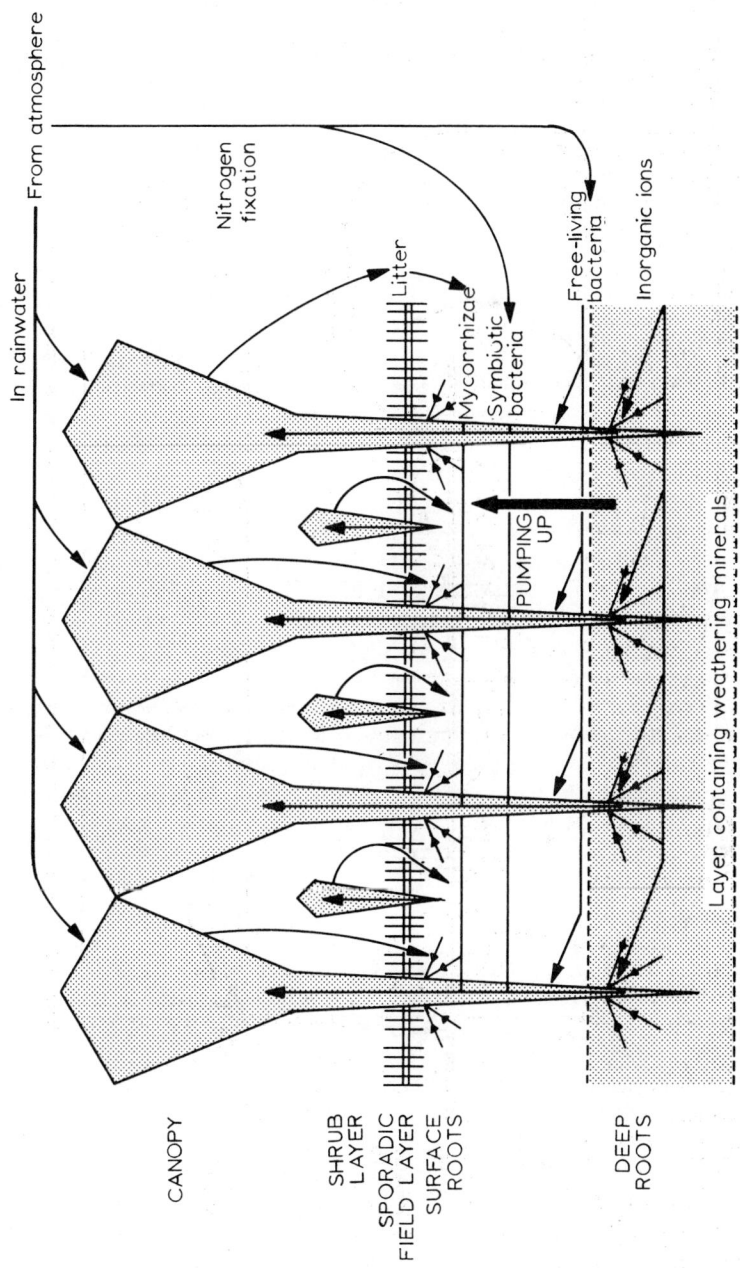

Figure 3.1 Diagrammatic representation of organic matter cycles in a closed-canopy tropical forest

50

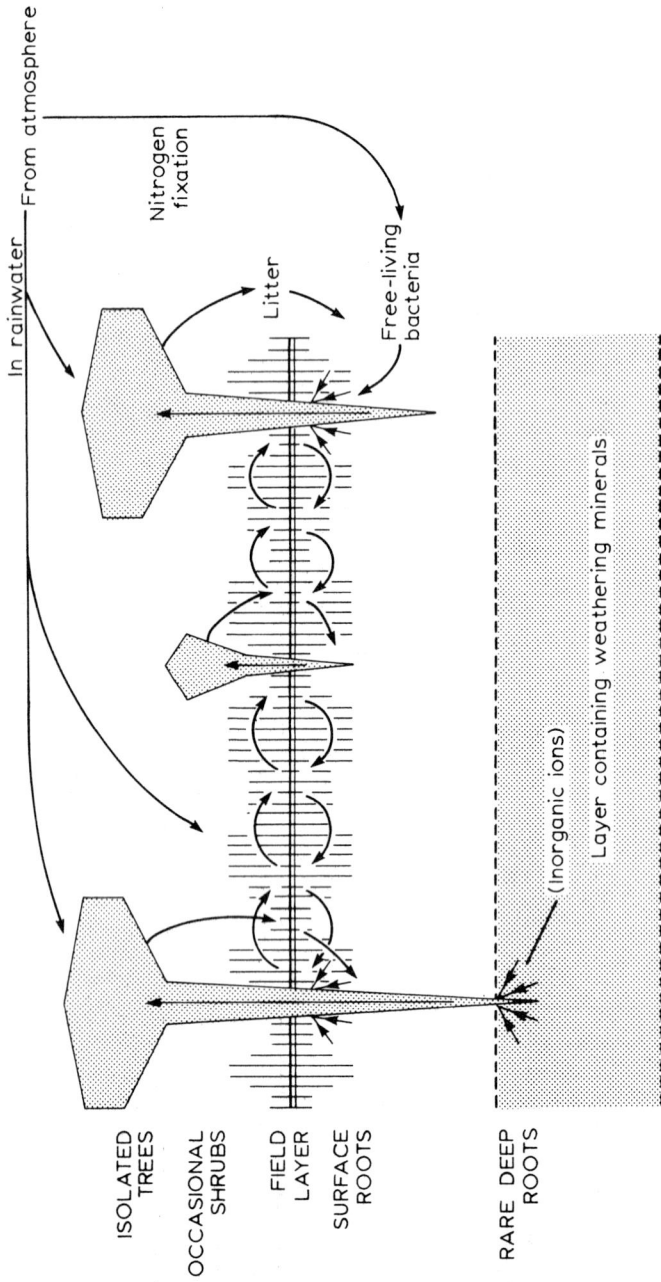

Figure 3.2 Diagrammatic representation of organic matter cycles in an open savanna

fertilizers is intimately linked to organic matter levels. While perhaps over-estimating the uniformity of savanna soils, especially in their ecosystematic relationships, their conclusions are especially important in relation to the argument of this paper.

It is important to point out the importance of the mammalian component to savanna ecosystems. This is clearly seen in game parks, but is no less important in relation to cattle, especially in the drier savannas. The arguments concerning game farming as a desirable form of land use in tropical savannas are not relevant to the present discussion, but it is necessary to point out that such a use precludes more intensive systems, which may well be feasible and much more productive and profitable.

From this consideration of tropical environments and ecosystems, apart from the great diversity to be found in low latitudes, two main points emerge. In the first place, the importance of the organic elements in the soil to chemical reactions, to microbiological activity, to structural characteristics, and to water relations, is very clear over large areas. However, within that general requirement there is considerable variability of mechanism, and a great many aspects remain to be elucidated, such as the role of the mesofauna. Second, the part played by plants in the ecosystem in maintaining nutrient status is also very clear. Not only do they supply material for decomposition, but also symbiotic relationships with bacteria and mycorrhizae are vital to the nutrient relationships of the soil layers. There is evidence that all major nutrients are, to a greater or lesser extent, involved in such relationships. Thus organic relationships in tropical soils are not simply desirable additions to soils which would function adequately without them, but essential to the functioning of the nutrient system of the plant growth medium.

The adaptation of traditional agricultural systems

The husbandry element of established agricultural systems is, in general, well adapted to these ecological conditions. A typology of these systems is set out in *Table 3.5*. It is not necessary to consider the adaptations of each in detail, but shifting agriculture, rotational bush fallowing in both closed forests and savanna, permanent tree crop plantations, and various systems involving livestock, need to be examined; some mention must also be made of some systems of permanent cultivation, which are broadly stable, usually because of considerable organic inputs.

Shifting agriculture

Shifting agriculture is a system whereby plots of land are cultivated for a short period of up to 3 years, and then they are allowed to revert to a self-propagated bush succession while the cultivators move on to new clearance, usually moving their settlements at the same time. Reversion to bush is usually rapid, and new cultivation occurs only after a prolonged period of fallow. Plot boundaries are not fixed, although the territorial limits of the land allocated to a particular seminomadic group is usually well defined.

Table 3.5 TYPOLOGY OF SOME TROPICAL AGRICULTURAL SYSTEMS ON AN ECOLOGICAL BASIS

Agricultural system	Description	Type	Cultivation characteristics
I Bush fallow systems	Systems in which the natural regeneration of self-propagated plants in successional communities is the restorative agent in respect of nutrients, organic matter, water conservation and microclimate	1. Transient agriculture with woody regrowth	Cultivation takes place on cleared patches which are ephemerally located
		2. Transient agriculture with grass-herb regrowth	Cultivation of ephemerally located cleared patches; fallow dominated by grasses and herbs
		3. Rotational fallowing with woody fallow	Cultivation of fields on a regular rotational pattern; fallow succession dominated by woody plants
		4. Rotational fallowing with grass-herb fallow	Cultivation of fields on a regular rotational pattern; fallow succession dominated by grasses and herbs
II Permanent cropping systems	Systems in which crop plants are grown continually on the same plots of land, either several crops in rotation or succession, or a single crop for several years; disturbance of microclimate and of water relations is permanent; nutrients and organic matter are replaced by inputs from outside, if at all	1. Degraded bush fallow	Permanent cultivation with a minor input of chemical fertilizer; develops by progressive shortening of the fallow, under pressure of agricultural demand, usually from an adjacent increasing urban population
		2. Irrigated agriculture: (i) padi rice	Permanent cultivation of rice under flooded conditions; dependent upon the effective control of water in the fields; fertility maintained by nutrient inputs from floodwater and microbiological activity, and of algae
		(ii) other crops	Dependent upon the maintenance of the delicate balance between water inputs and quality, and soil properties; sometimes using flood water from major rivers, but also from major or minor water control schemes; nutrient supply dependent on material in water, organic inputs, or chemical fertilizers; not necessarily flooded; sometimes involves rotations of crops

Table 3.5 *continued*

Agricultural system	Description	Type	Cultivation characteristics
		3. Garden agriculture	Crops grown on land more or less adjacent to settlements; usually vegetables etc, rather than staples; dependent on organic wastes from settlement for nutrient inputs
		4. Intensive organic agriculture	Crops grown on land around some large towns (e.g. Kano) and villages, dependent upon inputs of organic wastes (e.g. night soil) for the maintenance of soil nutrient status
		5. Permanent perennial crops	
		(i) tree crops	Plantations, small to extensive, of trees with a closed canopy; e.g. cocoa, oil palm, rubber etc; closed nutrient cycles, with gradual general loss to system; pest and disease control usually necessary for consistent yields.
		(ii) shrubs and herbs	Plantations, small to extensive, of low shrubs and herbs, usually with a more or less open canopy; e.g. sisal, bananas, tea, etc; closed but less complete nutrient cycles than tree crops, with general loss to system; pest and disease control usually necessary.
III Pastoral systems	Systems in which the keeping of grazing animals is the dominant feature, and animal products the dominant economic yield	1. Nomadic pastoralism	Grazing of cattle by movement over a very extensive range; the herds follow the availability of pasture and water, and often the location of markets; general environmental pressure low, but local pressures may be very high; grazing may be territorially controlled, or may be free.
		2. Sedentary pastoralism	Grazing of cattle in association with fixed settlements or limited territories; some

Table 3.5 *continued*

Agricultural system	Description	Type	Cultivation characteristics
			migration or transhumance may be present; environmental pressure may be high locally, especially if territories are limited in size or resources: control generally by herding rather than fencing.
		3. Ranching	Grazing of cattle extensively on a fully commercial basis; control both by herding and fencing of territories; environmental pressure generally controlled in order to preserve adequate range quality.
		4. Game farming	Grazing of a balanced mixture of ungulates extensively; control both by herding and fencing of territories; high animal biomass per hectare, but balanced resource pressure; ecologically desirable but economically and culturally problematic.
IV Mixed farming	Systems combining animal and plant production in a single organizational unit; only of local importance in most tropical areas, though locally very significant, especially in Asia; involve considerable control of local patterns of land use; animal dung potentially of considerable use in maintaining nutrient levels, but frequently used as fuel; biogas could provide both energy and fertilizer, but potentially of only limited application because of limitations imposed by the levels of organic inputs required, and the need for a substantial dung component for most		

Table 3.5 *continued*

Agricultural system	Description	Type	Cultivation characteristics
	simple plants appropriate to the tropical village situation; attempts to introduce mixed farming systems in areas where there is no such tradition have usually proved abortive, despite the *a priori* advantages from a production point of view.		
V Agro-forestry systems	Systems in which agricultural production, either plant or animal, is combined with tree planting for fuel, poles or timber, or for other tree products, such as gum arabic, cashew nuts, etc; the trees may also serve as soil improvers, or for fodder, or simply as shade or soil surface protectors.	1. Taungya systems	Systems in which trees are planted with food crops which serve as a nursery for the young tree seedlings; the aim is to produce a forestry or tree crop plantation; makes the early years of plantation establishment productive, and reduces labour cost of establishment
		2. Intercropping systems	Trees mixed with food crops in rows or other patterns; usually species used are soil improvers, and soil nutrient levels are thus maintained; spacing is critical in relation to competition between crop and tree; choice of tree species is also fundamental to success
		3. Tree–pastoral systems	Trees associated with grazing land, usually as shelterbelts or windbreaks; species used generally provide fodder (e.g. *Acacia albida*); in arid areas, suitable tree planting may provide erosion protection and assist water conservation
		4. Associative systems	Patches of trees, as woodlots, windbreaks, shelterbelts or hedgerows, mixed in with normal pattern of agricultural land use, but not incorporated into it as part of the productive system
		5. Fallowing systems	Trees planted as a woody fallow in a normal system of rotational fallowing;

Table 3.5 *continued*

Agricultural system	Description	Type	Cultivation characteristics
			replacing the self-propagated fallow; soil-improving leguminous trees (e.g. *Leucaena leucocephala*) may be planted, either sole, or in association with other planted or self-propagated species

It will be seen that the temperate distinction between agriculture and forestry does not readily apply in tropical latitudes, and the typology would rightly be completed by including forestry systems as a final separate category. This would include:

Agricultural system	Description	Type	Cultivation characteristics
VI Forestry systems	Systems in which wood, for lumber, pulp, small timber, poles or fuel, is the main product for economic use; long-term production systems usually self-sufficient in nutrients, but requiring at least some pest and disease control; plantations usually require also specialized establishment techniques.	1. Managed woodlands	Self-propagated woodlands managed for wood production; many valuable tropical timbers can only be produced in this way
		2. Plantations (i) Single-crop plantations	Producing only thinnings in the early years, and then a single timber crop at maturity; most timber or pulp can be produced only by this means
		(ii) coppices	Producing a repeated crop by coppicing or pollarding; suitable for fuelwood or poles, and occasionally small timber; species choice important

This table is based largely upon African experience, with which the author is most familiar. It probably applies satisfactorily to tropical Asia and South America also, but some expansion may well be needed.

Plots are usually small and a wide variety of crops are grown on each cleared patch. Such cultivation constitutes only a minimal disturbance of natural cycles and is usually combined with an extensive use of the resources of fallow areas, constituting an almost closed subsistence economy in which trading of any kind, although not completely absent, is of very minor significance to the human communities involved. In some areas, however, pressures on the system have been such as to bring about a breakdown of the checks and balances which are built into it.

Rotational bush fallowing

Rotational bush fallowing probably represents a development of shifting agriculture in response to increasing pressure on land resources and breakdown of the subsistence basis of the economic system. It is also connected with a greater sophistication of the social system, and a developing of more complicated political structures. Land is held in fixed plots, which are cultivated in rotation; complex cropping patterns are common, and maximum use is made of micro-environmental differences, both natural, and created by the mounds, ridges and other microhabitats, which are the product of husbandry (*see for example* Morgan, 1954; Moss, 1974; Moss and Morgan, 1977; Dupriez, 1980). Fallow periods are generally shorter than under shifting cultivation, and not infrequently what are in essence 'taungya' (*see Table 3.5*) methods are used to establish plantation plots of permanent tree crops. Nevertheless, with a cultivation period of 3 years and a fallow of not less than 7 years, the system is stable and productive under forest (Vine, 1968), and on soils with a higher clay content and a significant proportion of 2:1 clays, even shorter fallows may be sustained. *Table 3.6* shows yield levels of some crops. Most often, however, such soils are favoured for permanent plantation cropping (Smyth and Montgomery, 1962). Here again the key to the effectiveness of the fallow in maintaining soil nutrient levels and crop yields seems to depend upon the maintenance of organic matter levels and the development of a deep root system to bring up nutrients from lower layers in the solum. It is also important to notice that under a hoe cultivation many living roots and stools are left in the ground during the period of cropping, and that weeding is restricted to the removal only of completely useless adventitious plants—many self-sown plants are allowed to develop because they have medicinal, technical or nutritional uses. This constitutes an element in the maximization of the value of the labour of forest clearance, which is considerable and of the order of 100–150 man-days per hectare. Mixed cropping is almost universally practised, and is also a way of maximizing labour value, although it is clear that the sum total of the value of yields from all crops considerably exceeds the value of the yield from sole cropping in similar conditions (Kassam and Kowal, 1973; Andrews and Kassam, 1975). It may be that the maintenance of a dense plant cover also has a beneficial effect in maintaining soil organic matter levels during the cropping period.

In savanna areas the fallow is dominated by grasses and herbs; the tree component of the community shows regeneration patterns which are not related to cultivation cycles and which operate on a much longer time-scale (Adejuwon, personal communication). The herbaceous fallow is much less able to restore nutrient and organic matter levels than the woody fallow communities of forest areas (Nye and Greenland, 1960; Laudelont, 1962; Vine, 1968), although in many situations it is able to sustain satisfactory yields of food and cash crops (*Tables 3.6* and *3.7*). The reasons for this lesser effectiveness seem to be related to the lesser biomass of the plant cover, to the lesser rooting volume of the plants, and to the absence of deep-rooting species which can exploit weathering layers lower in the solum as sources of nutrients. Clearance is easier, but mixed cropping is common, for reasons similar to those applying in closed forest areas (Charreau, 1974; Baker and Norman, 1975; Abalu, 1977). The important point to be made is that there is no inherent environmental constraint on using trees and other deep-rooted plants in fallows in savanna areas. The dominance of the grass–herb fallow is largely due to the combined effects of savanna burning on tree seedling establishment, and the lower reproductive vigour of many savanna tree species.

Table 3.6   CROP YIELDS (t/ha, UNPROCESSED PRODUCT) UNDER
INDIGENOUS CULTIVATION IN THE TROPICS; WEST AFRICAN EXAMPLES
(AFTER KOWAL AND KASSAM (1978), LEAKEY AND WILLS (1977) AND
HARTLEY (1968))

Crop	General yield level	Maximum yield level	Comments
Rice			
upland cult.	2	7	Better areas
upland cult.	0.6–0.8	2.5–2.6	Lower rainfall areas
padi (W. Africa)	1–3	7–8	General level
padi (U. Volta)	—	12–14	3 crops; irrigated
Sorghum:			
peasant cult.	0.5–0.9	1–2	
improved cult.	2.5–3.5	4.0–4.5	Improved varieties
Maize:			
peasant cult.	0.5–1.1	1.5	
Govt farms	3–8	10+	Improved varieties
(Maize yields are very variable because of the considerable diversity of the varieties available, response to climatic factors, and sensitivity both to husbandry and to the effective matching of variety to environmental conditions)			
Millet:			
peasant cult.	0.3–0.7	1–1.5	
Govt farms	2.0–2.5	2.5–3.5	Improved varieties
Wheat:			
peasant cult.	0.9–1.6	2.5	Sahel and Sudan savanna
Govt farms	3–4	5+	Improved varieties
Cowpeas:			
peasant cult.	0.2–0.6	1	
Govt farms	1.5–2.5	3.5+	Improved cultivars
Groundnuts:			
peasant cult.	0.4–0.7	1	Unshelled weights
managed plots	3.0–3.5	4–5	Pest control measures
Soya beans:			
peasant cult.	0.3	0.5	Limited area grown
Govt farms	2.5–3.0	3.0–4.0	Experimental only
Bambara groundnut:			
peasant cult.	0.7–0.8	1	Shelled nuts
managed plots	2.0–2.6	3	Improved husbandry
Cassava:			
peasant cult.	5–10	15–25	Local cultivars
Govt farms	30–50	60–75	Improved cultivars
Yams:			
peasant cult.	3–10	10–15	
Govt farms	20–50	55–70	Local cultivars
(Yam yields, as well as cassava production, vary very considerably from location to location in response to climatic differences, especially water supply and the length of the effective growing season. Yield figures given are for fresh tubers, not dry material)			
Sweet potatoes:			
peasant cult.	0.5–5.0	9–10	
Govt farms	20–40	50–70	Local cultivars

Table 3.6 *continued*

Crop	General yield level	Maximum yield level	Comments
Cocoyams:			
peasant cult.	1.5–5.0	7	
commercial cult.	25	35	Nonirrigated
commercial cult.	35–55	75	Irrigated/intensive
(These high figures obtained in commercial cultivation outside West Africa have been obtained only on experimental farms in that region)			
Potatoes:			
peasant cult.	6–8	10	
exp. plots	15–25	25–30	Improved husbandry (and some fertilizers)
Bananas:			
(Yields from peasant cultivation are difficult to obtain, and commercial cultivation is usually highly efficient. Commercial yields, with proper management and fertilization, are very high, comparable to those obtained from tropical root crops under efficiently managed conditions)			
Tomatoes:			
peasant cult.	0.7–10	15–16	
Govt farms	50–80	90	Improved cultivars
Onions:			
peasant cult.	15–25	30–40	
exp. plots	50–80	90–100+	Improved cultivars
Cotton:			
peasant cult.	0.5	1.0	Seed cotton weights
exp. plots	1.5–2.0	2.5–3.0	Seed cotton weights
Sesame:			
peasant cult.	0.2–0.3	0.5	
improved cult.	0.7–1.2	2.0	Improved management
Tobacco:			
peasant cult.	0.3–1.7	1.8	
improved cult.	1.5–2.0	2.2–2.6	Improved husbandry/ fertilizers
Sugar cane:			
peasant cult.	20–25	30	
exp. plots	90–120	130–170	Improved management
Cocoa:			
peasant cult.	0.9–2.5	2.6	No fertilizer
improved cult.	3.0	3.5	With fertilizer
(Yields of cocoa vary considerably with the density of shading, and the degree of success of pest and disease control, as well as with fertilizer applications)			
Oil palm:			
peasant cult.	2.9–12.0	14.0	
improved cult.	5.8–14.5	15.6	With fertilizer
(Oil palm yields vary very considerably in response to climatic factors, especially the intensity of the dry season, and yields are thus more constantly higher in regions without a significant dry season)			

Figures are approximate, and most mask a very wide range of individual values which represent a response to husbandry, pests and diseases, and site conditions. They do, however, indicate some order of magnitude for actual yields from peasant farms, and also the degree to which yields could be increased with improvements in husbandry, without a marked development of mechanization.

Permanent tree crop cultivation

This is common practice in closed forest areas in the tropics, whether as peasant cultivation or as commercial exploitation by large units. Superficially such a use of land might seem to provide a simplified simulation of a closed forest canopy, but not only has it been shown that such systems show a gradual depletion of soil nutrients as the plantation develops (Hartley, 1968) but also organic matter levels in the upper layers of the soil are significantly lower than those under forest and fallow, over similar soil profile conditions (Smyth and Montgomery, 1962). This may be due to the fact that many peasant plantations have been established through food crop cultivation, and therefore begin their life with soil conditions similar to those at the *end* of a cultivation cycle, with lowered organic matter and nutrient levels. Nevertheless, such systems have produced successfully for many years, and even without fertilizers (although prone to pests and diseases) maintained world supplies of cocoa and palm oil for many decades. Rubber and coffee have, however, been grown on larger commercial plantations with the use of fertilizers and biocides for considerably longer. Thus these systems also have achieved a degree of ecological stability by maintaining adequate levels of organic matter in the upper layers of the soil, and, at least in some cases, by exploiting nutrient sources in the lower layers of the solum.

Permanent cropping

Successful systems of permanent cropping, apart from padi rice, are not very extensive in many parts of the tropics, especially in Africa. Whereas upland rice forms part of the normal cropping systems of rotational fallowing, padi rice involves generally a degree, sometimes a considerable degree, of water control, which is vital to its productivity. Properly managed, it is a productive and stable production system producing consistent, although relatively low, yields in the absence of nutrient inputs (Geertz, 1963). Nitrogen fixation by blue-green algae, nutrient inputs in the flood water, and the inherent nutrient status of the alluvial soils of the river flood plains, combine to maintain an adequate level of crop nutrition (Chabrolin, 1977; Kowal and Kassam, 1978). Controlled fertilizer applications, especially of nitrogen, and to a lesser extent, potash, produce highly significant increases in yield. Flood plain agriculture, as in the inland Niger Delta, and the flood plains of many rivers, is an extension of the techniques of flood plain rice cultivation to other crops. Nutrient balances seem to depend upon similar mechanisms. Coastal areas of mangrove swamp in West Africa are also used for rice cultivation, and although special husbandry problems arise because of the inundation by salt water, nutrient balances seem to be maintained by similar mechanisms.

Garden cultivation

Garden cultivation of vegetables is a common feature of the areas very close to tropical villages, dependent for their fertility upon organic wastes from the settlement. This same principle is extended to much larger areas around large cities in African savannas, where the organic wastes of the urban population are used to maintain the nutrient status of the surrounding close-settled agricultural zone (Mortimore and Wilson, 1965).

Under increasing population pressure, and demands for agricultural products both national and international, permanent cropping may replace rotational fallowing systems with severe deleterious effects on yield and soil properties. The destruction of tropical closed forests with the killing and extraction of roots and stools to prevent regeneration may also result in loss of soil. The problem is not so much the destruction of the forest, but the relatively unstable and unproductive ecosystems by which it is being replaced. However, this radical clearance is occasioned by economic demands, largely of two kinds—the demand of North America for beef for hamburgers (which makes Brazilian beef production profitable enough to warrant extensive forest clearance for extra grazing land), and the demand by Japan for wood chips for wood pulp (which makes much Asian and Australian clearance profitable). Cut these two demands and the rate of clearance of tropical forests would fall very dramatically. Such a breakdown of fallow systems leads to an inability of the soil to provide plant nutrients, to radical changes in the water balance of the soil, markedly increased leaching of inorganic ions, and a permanent reduction in levels of soil organic matter. It is perhaps not without significance that these developments are also associated with the large-scale application of the mechanical technology of temperate agriculture to tropical arable crops. Such husbandry systems can be sustained only by massive inputs of chemical fertilizers and biocides.

Pastoral systems are generally much less intensive systems of land use. Mixed farming is characteristic mainly of areas like the Kenya Highlands and parts of Zimbabwe where the influence of Europeans has been both considerable and direct. Pastoral and arable agriculture are traditionally separated by many African cultures. Purely pastoral systems are generally extensive and present few nutrient problems, although overgrazing can become very significant, particularly in more arid areas—though perhaps more locally concentrated and less extensive than has been commonly supposed (Kowal and Kassam, 1978).

Traditional systems of agriculture are thus generally ecologically stable, but giving only low or moderate yields of arable crops, and sustaining only low grazing intensities. These were adequate for past demands but can no longer meet the requirements of local pressures of population, either rural or urban, or of the demands of international markets. Much work over the past 20 years has shown that modest inputs of fertilizers containing the main nutrients, and the moderate use of biocides, especially in permanent tree-crop plantations, can produce considerably increased yields without any deleterious effects (Hartley, 1968; Richardson, 1968; Leakey and Wills, 1977; Kowal and Kassam, 1978). More drastic modifications of traditional systems have generally not been so successful.

This poses a question as to why the traditional systems are relatively stable and productive. It is suggested that this is the result of the organic matter levels which they maintain, by a diverse set of complex mechanisms, and of the nutrient cycles which they support in relation both to topsoil and subsoil, involving the organic matter and the microorganisms and meso-fauna associated with it. These features of the agricultural ecology then provide guidelines which can suggest possible lines of further development, in relation to agricultural production.

Towards an ecologically based tropical agricultural development

Two basic principles must, it is suggested, affect the rational appraisal of the development of the technology of tropical agricultural development: first, it is necessary that increased ecological knowledge and understanding should be the basis of a sound development, and not an inhibition of it; second, agricultural technology must aim at an intensification of land use, by shortening fallows and increasing their use and effectiveness, and by increasing productivity from land actually under cultivation. In pastoral systems it must aim at improving quality and increasing carrying capacity.

Development along temperate lines, involving the elimination of fallows, sole crop cultivation, and large-scale mechanization can result only in the evolution of systems which can be characterized as 'tropical field hydro-ponics', in which the soil becomes simply an anchoring medium for the crop plants, and in which nutrient supply is maintained by high-energy inputs of chemicals, and competition is eliminated by high-energy inputs of biocides. As oil-based energy becomes increasingly more expensive, such a system becomes economically more finely balanced, although there seems to be no inherent environmental reason why it should not be sustainable, in the same way that hydroponics is sustainable—at the cost of energy and money.

Alternative ways of developing are, however, already being developed, for example at the International Institute for Tropical Agriculture (IITA) at Ibadan in Nigeria (*see* Chapter 15). A number of basic principles, especially relating to woody plants, may be suggested as important guidelines in development:

1. Land in the tropics needs to produce energy as well as crops. A high proportion of both rural and urban dwellers in the tropics rely on wood, as direct fuel or as charcoal, as their basic energy source. Supply problems are acute in many areas (Morgan and Moss, 1980a,b, 1981), and increasing production problems are to be expected in the near future. Incorporation of quick-growing trees or fallows would therefore mean that the uncultivated land was itself productive.
2. Suitably selected tree species can be very effective soil improvers in a wide range of tropical environments. Associated with this there is a wide range of tropical legumes which are woody, such as *Leucaena*, *Sesbania*, and *Calliandra*. *Acacia albida* is another significant tree which will grow in dry environments and which acts as a soil improver, as well as providing fodder.
3. Many tree species have products other than fuel and fodder—gum arabic, nuts, fruits, and so on—thus increasing the useful productivity of the fallow.
4. Trees, as well as forming a significant fallow, can be grown on rela-tively poor land, for fuel at least, and on spare land, however inextensive, including hedgerows, shelterbelts and woodlots.

5.   Trees, with care in establishment, can often be used on degraded and poor land, although shallow soils present special problems.
6.   Woody plants with tap roots and a deep root system are ideal for maximizing nutrient use and organic matter cycling.
7.   Many tropical cultivators are already familiar with aboriculture, and possess the necessary skills to develop planted woody fallows.
8.   A very wide range of trees is available, suited to the widest variety of tropical environmental conditions. Careful choice of species and provenance is, in fact, the most significant step in development if agroforestry is envisaged (Combe and Gewald, 1979), and a large number of trials are at present being conducted in various parts of the tropics.
9.   Trees can be incorporated into agricultural systems in a wide range of ways (FAO, 1977); this is not restricted simply to use as fallows.

Table 3.7   YIELDS OF FORAGE CROPS IN THE TROPICS (FROM KOWAL AND KASSAM (1978) AND LEAKEY AND WILLS (1977))

Crop	General yield level	Maximum yield level	Comments
Andropogon spp	4– 7	9	
Brachiaria spp	6– 9	10	
Cenchrus ciliaris	5– 6	—	
Chloris gayana	10–14	—	
Cynodon spp	10–14	16	
Digitaria spp	5– 8	—	
Hyparrhenia spp	6– 8	—	
Melinis minutiflora	10–13	—	
Panicum maximum	10–25	—	
Pennisetum purpureum	10–25	—	
Setaria sphacelata	12–15	—	
Tripsacum laxum	10–15	—	

LEGUMES normally planted with grasses are:

Calopogonium mucunoides;   *Centrosema pubescens;*
Glycine wightii;   *Macroptilium atropurpureum;*
Pueraria phaseoloides;   *Stizolobium deeringianum;*
Stylosanthes humilis;   *Stylosanthes gracilis.*

Yield figures refer to experimental plots of pure stands of the grasses; in the natural rangeland with mixed species the following figures give an indication of the productivity:

Andropogon/Aristida	1.0–2.0
Aristida/Cenchrus	0.2–1.0
Brachiaria	0.5–1.0
Schizachyrium/Schoenfeldia	0.5–1.0
Schizachyrium/Loudetia	0.5–2.0
Schoenfeldia	0.5–1.0
Loudetia	0.1–2.0
Pennisetum/Schizachyrium	1.0–2.0
Hyparrhenia	2.0–3.5
Andropogon	1.6–4.5
Loudetia/Aristida	1.0–2.0

Productivity of the field layer of natural savanna is very variable in relation to factors other than species composition, such as the frequency, timing and intensity of fire.

10. Mixed cropping is desirable for a number of reasons, and is not necessarily incompatible with increasing mechanization of an appropriate kind, although temperate principles need considerable modification if they are to be applicable to the kinds of husbandry system which need to be developed in the tropics, if the soil resources are to be used to the full.

Then, it is necessary to make two more general observations. First, the diversity of tropical environments must be recognized, and the most appropriate use must be made of each local situation, in order that it may make its appropriate contribution to economic wellbeing. It does not follow that agroforestry is *the* answer, nor yet that arable agriculture is the universal aim of development; pastoral systems clearly have an important place in view of the widespread shortage of protein in the tropics—but trees have an important role in them also. It is argued, however, that organic matter balances are crucial, and that any stable system of land use must maintain them. Agroforestry is one very important way of achieving that aim. Composting may be another, but in many situations the labour requirements may preclude its use without very drastic modification of the work patterns of the local farmers.

Second, there seems to be no empirical reason why some chemical fertilizers and biocides should not be incorporated into the systems developed. Indeed, 'topping-up' may be highly desirable and produce considerable increases in yield for relatively small inputs without destroying the stability of the system. Such developments will be economically justified as well as ecologically tolerable.

Finally, it is important to realize that the problems of development are basically social and economic, rather than ecological. Such questions as how are innovations to be introduced effectively and how do they spread, are absolutely crucial. Then, it is necessary to build on what is there and not to sweep it away, and this requires real understanding. The fundamental economic question is how to make agriculture profitable, and that question is more economic than technical in most developing countries.

References

ABALU, G. O. I. (1977). A note on crop mixtures under indigenous conditions in Northern Nigeria *J. Develop. Studies*

ANDREWS, D. J. and KASSAM, A. H. (1975). Importance of multiple cropping in increasing world food supplies. *Proc. Agron. Soc. Amer. Symposium of Multiple Cropping, 24–29 Aug., 1975*. Knoxville, Tennessee; American Society of Agronomy

BAEYENS, J. (1949). The bases of classification of tropical soils in relation to their agricultural value. *Commonwealth Bur. Soils Tech. Comm. 46*, pp. 99–102. Farnham Royal, Bucks; CAB

BAKER, E. F. I. and NORMAN, D. W. (1975). Cropping systems in Northern Nigeria. In *Proc. Cropping Systems Workshop, 18–20 Mar. 1975*, pp. 334–361. Los Baños, Philippines; IRRI

CHABROLIN, R. (1977). Rice in West Africa. In *Food Crops of the Lowland*

Tropics, pp. 7–26. Eds C. L. A. Leakey and J. B. Wills. Oxford; Oxford University Press

CHARREAU, C. (1974). Systems of cropping in the dry tropical zone of West Africa, with special reference to Senegal. *Proc. Int. Workshop on Farming Systems, 18–21 Nov. 1974*, pp. 287–302. Hyderabad; ICRISAT

COMBE, J. and GEWALD, N. (1979). *Guia de Campo de los Ensayos Forestales del CATIE en Turrialba, Costa Rica*. Turrialba, Costa Rica; Centro Agronomica Tropical de Investigacion y Enseñanza

DUPRIEZ, H. (1980). Cultures, associées ou Monocultures? Validité du savoir paysan. *Environment Africain; Études et Recherches 50*. Dakar; ENDA: London; IAI

ELLENBERG, H. (1959a). Über den Wasserhaushalt tropischer Nebeloasen in der Küstenwüste Perus. *Berichte Geobot. Forsch Inst. Rübel*, (1958), pp. 47–74

ELLENBERG, H. (1959b). Typen tropischer Urwälder in Peru. *Schweiz. Zeitschrift für forstwesen*. **3**, pp. 1–19

GEERTZ, C. (1963). *Agricultural Involution: the Processes of Ecological Change in Indonesia*. Los Angeles; University of California Press

HARRISON CHURCH, R. J. (1980). *West Africa*, 8th Edition, ch. 3. London; Longmans

HARTLEY, C. W. S. (1968). The soil relations and fertilizer requirements of some permanent crops in West and Central Africa. In *The Soil Resources of Tropical Africa*, pp. 155–183. Ed. R. P. Moss. Cambridge; C.U.P.

HOPKINS, B. (1965). Vegetation of the Olokemeji Forest Reserve: III. The microclimates with special reference to their seasonal changes. *J. Ecol.*, **53**, 125–138

JANZEN, D. (1975). *Ecology of Plants in the Tropics*. Studies in Biology No. 58. London; Arnold

KASSAM, A. M. and KOWAL, J. M. (1973). Productivity of crops in the Savanna and Rain Forest zones in Nigeria. *Savanna*, **2**, 39

KOWAL, J. and KASSAM, A. M. (1978). *Agricultural Ecology of Savanna: a study of West Africa*. London; Longmans

FAO (1977). Forestry for ideal community development. FAO Forestry Paper 7. Rome; FAO

LAUDELOUT, M. (1962). *Dynamics of Tropical Soils in Relation to their Fallowing Techniques*. Rome; FAO

LEAKEY, C. L. A. and WILLS, J. B. (1977). *Food Crops of the Lowland Tropics*. Oxford; Oxford University Press

LONGMAN, K. A. and JENIK, J. (1974). *Tropical Forest and its Environment*. London; Longman

LYR, M. and HOFFMAN, G. (1967). Root growth in forest trees. *Int. Rev. For. Res.* **2**, 181–237

McCULLOCH, J. S. G. (1962). Effects of peasant cultivation practices in steep stream-source valleys: measurements of rainfall and evaporation. *East Afr. Agric. For. J.*, **27**, 115–117

MORGAN, W. B. (1954). The approach to regional studies in Nigeria. *Research Notes, Dept. of Geography, University College, Ibadan, Nigeria.*, **6**, 10–18

MORGAN, W. B. and MOSS, R. P. (1980a). *Rural Energy and Environment in West and East Africa*. London; Developing Areas Study Group (mimeographed)

MORGAN, W B. and MOSS, R. P. (1980b). *Rural Energy Systems in the Humid Tropics*. Tokyo, Japan; United Nations University

MORGAN, W. B. and MOSS, R. P. (1981). *Fuelwood and Rural Energy Production and Supply in the Humid Tropics: with Special Reference to East and West Africa, India and Thailand*. Tokyo, Japan; United Nations University

MORTIMORE, M. J. and WILSON, J. (1965). *Land and People in the Kano Closed Settled Zone*. Occasional Paper No. 1, Dept. of Geography, Ahmadu Bello University, Zaria, Nigeria

MOSS, R. P. (1963). Soils, slopes and land use in a part of Southwestern Nigeria: some implications for agricultural development in inter-tropical Africa. *Transactions and Papers, Institute of British Geographers*, No. 32, pp. 143–168

MOSS, R. P. (1974). Soils and soil management. In *West Africa*, ch. 5, pp. 76–92. Ed. R. J. Harrison Church. London; Longmans

MOSS, R. P. (1976). Reflections on the relations between forest and savanna in West Africa. *Savannas. A Symposium of the Biogeography Study Group*. University of Hull Dept. of Geography, Dec. 1976

MOSS, R. P. (1977). Soils, plants and farmers in West Africa: 2. Further examination of the role of edaphic factors and cropping patterns in South-West Nigeria. In *Human Ecology in the Tropics*, pp. 59–77. Eds J. P. Garlick and R. W. J. Keay. London; Taylor and Francis

MOSS, R. P. (1978). *Concept and Theory in Land Evaluation for Rural Land-use Planning*. Occasional Paper No. 5, Dept. of Geography, University of Birmingham

MOSS, R. P. (1980). Ecological constraints on fuelwood production in the humid and sub-humid tropics. *Colloque: l'Énergie dans les Communautés Rurales des Pays du Tiers Monde. (UNU/CEGET), 5–10 May*. Bordeaux; Centre d'Études de Geographie Tropicale du CNRS

MOSS, R. P. and MORGAN, W. B. (1977). Soils, plants and farmers in West Africa: 1. A consideration of some aspects of their relationships with special reference to contiguous areas of forest and savanna in South West Nigeria. In *Human Ecology in the Tropics*, pp. 27–57. Eds J. P. Garlick and R. W. J. Keay. London; Taylor and Francis

NYE, P. H. and GREENLAND, D. J. (1960). *The Soil Under Shifting Cultivation*. Commonwealth Bur. Soils. Tech. Commn. 51

PEREIRA, M. C. (1962). Hydrological effects of changes in land use in some East African catchment areas. *East Afr. Agric. For. J.*, **27** (Special Issue)

RICHARDSON, H. L. (1968). The use of fertilizers. In *The Soil Resources of Tropical Africa*, ch. 7, pp. 137–154. Ed. R. P. Moss. Cambridge; Cambridge University Press

ROGERS, W. S. and HEAD, G. C. (1969). Factors affecting the distribution and growth of roots of perennial woody species. In *Root Growth* Ed. W. J. Whittington. London; Butterworths

SMYTH, A. J. and MONTGOMERY, R. F. (1962). *Soils and Land Use in Central Western Nigeria*. Ibadan; Govt. W. Nigeria

VINE, H. (1954). Is the lack of fertility of tropical soils exaggerated? In *Proc. 2nd Inter-Afr. Soils Conf.*, **1**(26), pp. 389–412

VINE, H. (1968). Developments in the study of soils and shifting agriculture in tropical Africa. In *The Soil Resources of Tropical Africa*, ch. 5, pp. 89–119. Cambridge; Cambridge University Press

4

PLANT–MICROBIAL INTERACTIONS

J. M. LOPEZ-REAL
Wye College (University of London), Ashford, Kent, UK

The rapid depletion of fossil fuel supplies and the consequent large increases in oil-based commodities over the past decade has focused world attention on reducing escalating energy costs. One area of particular importance to agriculture, the production of nitrogenous fertilizers, has seen huge increases in costs during this period. This major problem and the future demand for high-protein plant food has led to increased investment in biological nitrogen fixation research.

However, study of the nitrogen fixation process had not been neglected previously. Biologists and agriculturalists have long been interested in the process and have studied in depth its chemistry and biochemistry and the ecology, physiology and agronomy of the various organisms and plant associations that make use of its benefits. The benefits of these plant associations, exemplified by the legume–*Rhizobium* symbiosis, had long been recognized in the pre-chemical phase of man's agricultural development. It is still recognized and highly valued by those growers and farmers who are biological (organic) in their approach. In the developing nations of the world, such associations are vitally important both for maintenance of soil fertility and for sources of high-protein food.

The purpose of this chapter is to present a brief overview of the main symbiotic associations that are involved in nitrogen fixation, followed by an examination of some recent research advances that may enhance biological nitrogen fixation, with reference principally to the legume–*Rhizobium* symbiosis. The last section will deal with the potential that exists within the tropical legumes for increased production of food, fuel and forage.

It is now fully recognized that the ability to reduce nitrogen to ammonia through the action of the nitrogenase enzyme complex is an attribute unique to the Prokaryotae-bacteria, actinomycetes and the blue-green algae (Cyanobacteria). That many of the microorganisms possessing this attribute have developed intimate associations with higher plant hosts has been recognized and utilized for many years. The list in *Table 4.1* shows the present known range of microbial–plant nitrogen-fixing associations. The latest additions to the list are the recently discovered 'associative' symbioses (Dobereiner, 1974) whose potential impact in nitrogen input is currently being reassessed (Neyra and Dobereiner, 1977). The quantity and significance of these sources of biological inputs of nitrogen into agricultural and natural ecosystems have taxed the energies of researchers for many years. A sample range of nitrogen fixation estimates from such symbiotic associations is shown in *Table 4.1*.

Table 4.1 NITROGEN-FIXING PLANT–MICROBIAL ASSOCIATIONS

Type of association	Microorganism	Host plant	Potential nitrogen supply (N kg/(ha.y))
Symbiotic	*Rhizobium* (bacterium)	Leguminosae (Papilionoideae; Mimosoideae; Caesalpinioideae)	50–500
	Frankia (actinomycete)	Nonlegumes *(Alnus; Myrica; Casvarina; Ceanothus)*.	9–362
	Cycads		
	Anabaena	Pteridophytes *(Azolla)*	18
	Nostoc (cyanobacteria)	Bryophytes *(Anthoceros; Blasia)*	—
	Nostoc	Angiosperms *(Gunnera)*	72
Associative	*Azotobacter paspali*	Bahia grass *(Paspalum notatum)*	0.2–16
	Azospirillum lipoferum *Azospirillum*	Gramineae	

Of those systems listed above, the major research effort has been directed towards the *Rhizobium*–legume symbiosis. This is hardly surprising, because forage and grain legumes constitute one of the most important types of agricultural crop in present-day use. Even so, this important and beneficial association has been only partly explored. Of approximately 13 000 known species of legumes, only about 100 have been grown commercially and less than 20 are used extensively (National Academy of Sciences, 1979). Through their ability to utilize marginal land and improve soil nitrogen levels, symbiotic nitrogen-fixing legumes must have a central role in any biological approach to agriculture.

Symbiotic associations between plants and microorganisms are, by their nature, complex interactions and therefore liable to abort or malfunction at several stages in their development. The formation of a functioning legume root nodule depends upon a very definite sequence of events between the bacterium and its host plant. The process is started by the multiplication of the appropriate *Rhizobium* strain in the soil surrounding the host legume root. Attraction, attachment and finally entry by the bacterium, via an infection thread, then occurs. This stimulates nodule formation by the root followed by infection of the cells by the rhizobia leading to bacteroid formation, nitrogenase production and nitrogen fixation. Several stages in this complex process are liable to disruption from external environmental influences acting either on the plant host or the bacterium. The most obvious and simplest failure results from the absence of the appropriate strain of *Rhizobium* in the soil. The picture may be further complicated by the presence of the correct strain which starts nodulation and then fails to fix nitrogen efficiently, providing little or no nitrogen to the host. Such ineffective nodules are a drain on the host plant's resources because they are consuming carbon from photosynthate but failing to provide nitrogen for the plant in return. These situations can be overcome by incorporating the appropriate strain into the seed before sowing. Such a procedure is a well-established practice in many parts of the world for enhancing nitrogen fixation of legume crops (Burton, 1967; Vincent, 1974). The production, quality control and application of legume inoculants is now an important branch of 'rhizobiology' and probably one of the most effective mechanisms for improving biological nitrogen fixation in the short term (Roughley, 1970).

Further enhancement of the process can occur through the selection not only of the appropriate *Rhizobium* strain but also for those strains screened for high nitrogen-fixation rates as assessed by acetylene reduction measurements or plant growth determinations. The legume can then be treated before sowing with the most efficient appropriate strain of *Rhizobium* available. Microbial geneticists have attempted to advance this process further by selecting mutant *Rhizobium* strains with a greater capacity to reduce acetylene. Maier and Brill (1978) have identified such mutants of *R. japonicum* on soybean roots. Significantly more nitrogen was fixed by the mutants than by plants inoculated with the wild-type strain under laboratory controlled conditions. No report has yet been published of the potential of these mutants in a field situation. Such mutants will inevitably face the same problems as other inoculants, of having to compete successfully with locally adapted indigenous strains of *Rhizobium* that may be ineffective in nitrogen fixation.

Extensive research into the biochemistry of nitrogenase has revealed properties of the enzyme that have important implications in the potential enhancement of biological nitrogen fixation. An apparently unavoidable reaction carried out by the enzyme, concomitant with nitrogen reduction, is the production of hydrogen gas. Such a reaction is dependent upon ATP and therefore involves a loss of energy efficiency in the system. Evolution of hydrogen from soybean nodules was first observed by Hoch, Little and Burris (1957) although the extent of such energy losses in legumes was not truly appreciated until the survey carried out by Schubert and Evans (1976). With most symbionts, including soybeans, only 40–60 per cent of the electron flow to nitrogenase was transferred to nitrogen; the remainder was lost through hydrogen evolution. The cowpea legume (*Vigna sinensis*) was the only one tested that did not show this characteristic. It is now known that some rhizobial bacteroids possess a second enzyme-catalyzed reaction involved in hydrogen metabolism. This is a unidirectional hydrogenase which catalyzes the oxidation of hydrogen in the presence of an appropriate acceptor, and ATP is generated as a result (Evans *et al.*, 1977).

The presence of hydrogenase therefore increases the overall efficiency of the nitrogen-fixing system. It is therefore possible to screen for *Rhizobium* strains that possess the hydrogen uptake enzyme. Albrecht *et al.* (1979) investigated the effects of such *Rhizobium japonicum* strains on soybeans and found that they fixed significantly more nitrogen (31 per cent more) and produced greater yields, 24 per cent more total dry weight, than plants inoculated with strains lacking the hydrogen-uptake capacity. The authors advocated that the hydrogenase system be incorporated as a desirable characteristic for *Rhizobium* strains used as field inoculants. Similar beneficial effects of the hydrogenase system have been reported in a strain of cowpea *Rhizobium* (Schubert, Jennings and Evans, 1978).

The amount of nitrogen fixed in the *Rhizobium*–legume symbiosis varies not only with the competence of the bacterial strain but also with the host legume and a range of environmental factors. Temperature, water relations, salinity and pH are all major environmental factors influencing the efficiency of nitrogen fixation. Internal constraints and limitations are imposed from nodule respiration, an efficient vascular system, nitrogenase and ultimately the supply of photosynthate. Fixation is also an expensive process in terms of energy use in the form of ATP and therefore phosphate supply to the legume may become a limiting factor in fixation efficiency.

There is still considerable controversy concerning the extent to which a nodulated legume can be supplied with its nitrogen requirements solely from bacteroid fixation, and still provide a high yield or even meet the nitrogen needs of the plant. The whole question of energy efficiency of symbiotic nitrogen fixation has recently been excellently reviewed by Phillips (1980). There are several stages in the development of the symbiosis that can become critical in the absence of any external nitrogen supply. One crucial period arises at the very beginning of legume growth and development—the early seedling stage. At this time, before and during infection by *Rhizobium* but before production of a fully operational nitrogenase, the plant is dependent on seed nitrogen and soil nitrogen reserves. There are numerous reports in the literature advocating the use of small amounts of combined nitrogen for the stimulation of nodulation during this period of potential nitrogen starvation, and when nodule development places a heavy drain on plant resources (Sprent, 1979). The severity of such a drain will, of course, vary with growth rates of the different plant parts (Sprent, 1979), the legume variety and the status of soil nitrogen levels. A biological approach to agriculture would need the careful evaluation of such soil nitrogen levels to avoid both the problem of nitrogen stress in early legume growth and the presence of elevated nitrogen levels that would inhibit the start and development of the symbiotic association.

Problems also arise in later growth stages when there may be competitive effects for photosynthate supply between nodules and other developing regions of the plant. That photosynthesis is probably the major limiting factor acting on symbiotic nitrogen fixation has been amply shown by experiments involving defoliation (Halliday and Pate, 1976), depodding or deflowering (Lawn and Brun, 1974; Lawrie and Wheeler, 1974) in which photosynthate supply to the nodule is enhanced by removal of alternative energy sinks. Direct effects on photosynthetic rates by shading (Lawrie and Wheeler, 1973; Halliday and Pate, 1976; Sprent, 1976), planting density (Sprent, 1976), supplemental lighting (Lawn and Brun, 1974) and increased carbon dioxide levels (Hardy and Havalka, 1976), have all been shown to have a marked effect on nitrogen-fixation rates.

Graham and Halliday (1976) have concluded that, in general, those treat-ments that enhance photosynthesis or eliminate energy sinks also enhance fixation. In a study on soybeans, Harper (1974) concluded that nitrate may be needed for the later stages of vegetative growth but that nitrogen fixation may be essential during the period of pod fill as the plants are then unable to take up and reduce nitrate. This work suggested that soybean yields are greatest when both nitrate and atmospheric nitrogen are available to the plant. That potentially high seed yields are possible from symbiotic fixation has been shown by the investigation of cowpea (*Vigna unguiculata*) by Eaglesham *et al.* (1977). In pot-grown nodulated plants, symbiotic fixation supplied 83–89 per cent of the plants' total nitrogen content at all stages. Investigations such as these, however, are certain to show considerable variation between legume species, the environmental conditions and the effect of residual levels of nitrogen in the soil. In *Phaseolus vulgaris*, for example, the total growth period is considered too short for the plant to fix enough nitrogen for its needs. Under the conditions tested by Graham and Halliday (1976) in Colombia, fixation in *Phaseolus vulgaris* contributed less than 50 per cent of the total plant nitrogen. A crucial period in the development of this legume with respect to nitrogen demand occurred in the post-flowering period—a time of leaf-nitrogen decline and nitrogen-fixation decline as a result of increased nitrogen demand for bean development and pod fill. Graham and Halliday (1976) obtained maximum yields when such nodulated plants were supplemented with nitrogen fertilizer 50 days after germination.

To what extent can a biological approach overcome the difficulties presented by such complicated interacting factors? In a real sense this presents us with a major paradox that arises from the attempt to integrate a biological system into the context of chemical agriculture. The need for nitrogen supplementation has to contend with the fact that certain levels of combined nitrogen will reduce nodulation, nodule activity and lead to accelerated nodule senescence (Sprent, 1979). It may, of course, be possible to select strains of *Rhizobium* for inoculation that will provide for nitrogen-fixation inputs in the presence of lower levels of nitrogen fertilizer. In many ways, however, this would not necessarily be the main point of attack for the biological agriculturalist. As Sprent (1979) has pointed out, the question of whether nitrogen-fixation potential of a legume is sufficient for maximum yield or whether crops should have fertilizer nitrogen added is unfair to the nodule. Selection in crops for high-yield characteristics has invariably taken place in plots given nitrogen fertilizer. Selection may therefore have been against nitrogen-fixing ability and the *Rhizobium* in consequence has little chance of maximizing inputs. The emphasis therefore should switch from solely *Rhizobium* selection to the selection of host-plant phenotypes that may enhance nitrogen fixation. Such research into potential improvement of the plant genotype has been very limited. That host-plant genetic factors are important in the symbiotic relationship was shown long ago by Wilson (1940) and Aughtry (1948). Nutman (1957) had reported that the inheritance of ineffectiveness in *Trifolium pratense* (red clover) was under the control of two recessive genes. The same author found that variability for nitrogen fixation in subterranean clover was too low for an effective selection programme (Nutman, 1961). Working with lucerne cultivars and an Australian strain of *R. meliloti*, Gibson (1962) concluded that selection for increased nodulation was feasible. Seeton and Barnes (1977) found that crosses among lucerne clones with high acetylene-reduction rates produced progenies with acetylene-reduction rates more than twice those of progenies from crosses of low acetylene-reduction-rate clones. Selection for nitrogen fixation in lucerne was also investigated by Duhigg, Melton and Balten Sperger (1978). Plants with high and low acetylene-reduction rates were selected. These plants were intercrossed, with the result that the progeny of high selections showed an 82 per cent increase in acetylene reduction, 57 per cent increase in dry weight of top growth and a 60 per cent increase in total nitrogen in top growth compared with the original cultivar.

A study of intraspecific variability for nitrogen fixation in cowpea (*Vigna unguiculata*) was reported by Zary and Miller (1977). Tenfold differences in nitrogen fixation efficiency were found among host-plant genotypes following application of a standard mixed strain of *Rhizobium* inoculant. These differences were obtained whether the criterion used was nodule number, mass or nitrogenase activity. These studies serve to show the potential that exists for selection of high nitrogen-fixing genotypes.

Further interesting evidence for such genetic potential is provided by an investigation into the growth and development of bush and climbing, indeterminate cultivars of *Phaseolus vulgaris* (Graham and Rosas, 1977). Ten cultivars of each type were compared for a range of growth parameters and acetylene-reduction rates. Climbing beans were superior to bush bean cultivars in acetylene-reduction activity, and botanically these are the more primitive cultivars. The authors also examined varietal differences in carbohydrate supply and availability in the nodule. The high nitrogen-fixing climbing varieties tended to hold more of their carbohydrate in the soluble form. The primitive climbing beans are those predominantly grown by small farmers unable to afford nitrogen fertilizer inputs, and these findings are therefore of considerable practical importance. Nitrogen accumulation in leaves of bush beans was found to exceed that of climbing cultivars during the early growth period. As bush beans are normally monocropped with more fertilizer than climbing beans, the authors considered that this could have reflected unconscious selection by agriculturalists for varieties responsive to nitrogen fertilization. Breeding for a combination of yield and nitrogen fixation will probably assume a greater importance in future years.

Other aspects of plant development may also be genetically improved to enhance biological nitrogen fixation. The conflict between energy sinks and photosynthate supply has already been mentioned. A decline in nitrogenase activity usually occurs during early pod filling (Harper, 1974; Sinclair and De Wit, 1975). Such a phenomenon may result from degradation of photosynthetic enzymes (e.g. ribulose 1,5-bisphosphate carboxylase) in the leaves to supply amino acids for the developing high-protein seeds (Sinclair and De Wit, 1975). Such loss in photosynthetic capacity leads to reduced photosynthate supply and hence a drop in nitrogen fixation in the root nodules. Abu-Shakra, Phillips and Huffaker (1978) however, have identified soybean cultivars with extended ribulose 1,5-bisphosphate carboxylase activity in the leaves and hence acetylene-reduction activity in the root nodules throughout seed maturation. Such potentially useful cultivars await field testing before full evaluation of these host physiological differences. Lower-leaf loss of *Phaseolus vulgaris* appears to be critical for nodule activity in later stages of growth in the tropics. Graham and Halliday (1976) have advocated the selection of varieties with a canopy structure permitting longer life in the lower leaves to promote nodule nitrogen fixation.

The high demand of legumes for ATP and therefore phosphorus in order to effect nitrogen fixation has already been briefly mentioned. In biological agriculture the management of soil phosphate levels is as critical as that of nitrogen. Biological inputs of phosphate can take place through mycorrhizal associations, a phycomycetous fungus and a plant host. Infection of the roots by the fungal hyphae takes place and phosphate uptake is thought to be enhanced by the increased area of soil available to the plant for phosphate supply by the hyphae ramifying through the soil.

Extensive research into such vesicular–arbuscular (VA) mycorrhiza has been carried out over the last three decades (Sanders, Mosse and Tinker, 1975). Phosphorus uptake by higher plants can therefore be increased in phosphate-deficient soils through the presence of roots infected with VA mycorrhizal fungi. The association is astonishingly widespread among higher plants. It is the norm rather than the exception for such a root association to exist. As in the case of nitrogen and *Rhizobium,* the presence of certain levels of soluble phosphate will prevent the symbiosis taking place. Agricultural applications of VA mycorrhizas have been hampered by the inability to grow the fungi concerned axenically in the laboratory and hence to produce the substantial quantities required for extensive field testing. However, the testing that has been carried out shows promise for future application of mycorrhiza in agriculture (Mosse, 1976).

As has already been stated, plants will, in the normal course of events, provided that phosphate levels are low enough, form a mycorrhizal association. Such is the case with legumes and some interesting data have been reported for studies of these tripartite systems—legume host plant, *Rhizobium* and VA mycorrhizal fungus (Daft and El-Giahmi, 1974; Mosse, 1976). Nodulation of legumes with effective rhizobia occurred only in very phosphate-deficient soils when plants were mycorrhizal. Schenk and Hinson (1976) examined nodulated and non-nodulated isolines of soybean infected with *Endogone* mycorrhiza. Nodulated plants showed significantly increased seed yield.

The problems and difficulties of utilizing biological systems for nitrogen inputs in conventional chemical agriculture are apparent when one considers the dependence of modern methods on chemical control of pests and disease. This presents us with another paradox, because the use of such compounds is likely to influence the carefully established and effective symbiotic associations. Clearly, legume–*Rhizobium* symbioses are non-target organisms but their current use as seed inoculants exposes them quite drastically to the effects of pesticides, herbicides and fungicides, which are often applied as seed dressings.

The effects of such chemicals on the legume symbiosis and other non-target microorganisms has been thoroughly and excellently reviewed by Anderson (1979). The published results vary considerably with the soil type, the time of application and strain sensitivity. Anderson (1979) evaluated herbicidal effects on rhizobia and nodulation. He concluded that, in general, they were harmful to rhizobia. Overall, there appeared to be a net negative effect, so that herbicide applications to legumes or to soil intended for legume growth should be carefully evaluated. The same degree of variability is found with fungicides, with little correlation between laboratory and field evaluation. Captan, thiram, ceresan and dichlone appeared to be toxic to most rhizobia. Staphorst and Strijdom (1976) evaluated the effects of 13 fungicides on *Vigna unguiculata.* Although thiram was the most toxic *in vitro* fungicide to the bacterium, it had no effect on nodulation. The data compiled by Anderson (1979) confirm the variable effect on rhizobia, nodulation and nitrogen-fixing efficiency. The situation is perhaps best summed up by those agencies who supply legume inoculants, who invariably recommend that no pesticide at all should be used.

The mechanisms so far described, that may potentially increase symbiotic biological nitrogen fixation in the future, represent the culmination of many years of research involving sophisticated technology and considerable capital investment. For a large part of the world, however, access to fertilizers or the purchasing power to procure them, has hardly ever existed. The majority of smallholding farmers in the developing countries have always, of necessity, farmed biologically and generally in soils poor in nitrogen or phosphate, or both. Legumes in such agricultural systems already assume a position of major importance, not only for their ability to grow on poor marginal soils, but also for their important contribution to alleviating protein malnutrition.

Inoculation of legume seed with appropriate *Rhizobium* strains has been successfully carried out in many developing countries, enabling legumes to be introduced into new areas. Problems can occur, because such a seemingly simple task requires an elaborate technological infrastructure, expertise and investment. That such difficulties do exist has been recognized by the international scientific institutes.

Invariably, inoculants that are produced are more representative of legumes (e.g. soybeans) that are of a high international commodity value. Thus, though 65–70 per cent of soybean seed planted in Brazil is inoculated, less than 0.5 per cent of *Phaseolus vulgaris* bean seed is treated (Araujo, 1974). Far fewer reports are published of inoculation studies with indigenous legumes. It is here, perhaps, that a major impact from biological nitrogen fixation can be made in developing countries.

The recent excellent publications by the United States National Academy of Sciences (NAS) (1975, 1977, 1979) have served to highlight the enormous potential that exists in the tropics for exploiting neglected plant species. The family Leguminosae shows the most promise for producing the greatly increased supplies of vegetable protein that will be required in the future. Almost every developing country suffers from chronic protein deficiency and the 'Green Revolution', with its emphasis on cereals, has often led to decreased legume production (NAS, 1979). In addition, Leguminosae encompass a wide range of genera, not only herbaceous annual crops but also vines, shrubs and forest trees. In such a wide and diverse family it is perhaps not surprising to find examples which show enormous potential for food, forage, fuel and reclamation uses. Add to this their ability to nodulate and fix nitrogen and to grow over a wide range of soil types and climatic conditions, then their neglect is astonishing. A few of the most interesting tropical legumes selected by the NAS (1979) are listed in *Table 4.2*. It is apparent that many tropical legumes can be truly classified as multipurpose plants. One of the most impressive in this respect is *Leucaena leucocephala*, now extensively grown in Malawi and the Philippines (NAS, 1977). The winged bean (*Psophocarpus tetragonalobus*) is gaining in prestige and recognition throughout the tropics where test cultivations have been carried out (NAS, 1975). This plant produces edible leaves, shoots, flowers and pods as well as seeds whose composition virtually duplicates that of soybeans. In addition, some varieties produce edible root tubers with

Table 4.2 UNDEREXPLOITED LEGUMES OF POTENTIAL FUTURE
AGRICULTURAL IMPORTANCE (FROM NAS, 1975, 1977, 1979)

Genus	*Common name*	*Characteristics*
Psophocarpus	Winged bean	High-protein seed, tuber, edible leaves
Leucaena	Leucaena	Multipurpose—forage, firewood, timber, erosion control, fertilizer
Lupinus mutablis	Tarwi	High-protein seed, oil, green manure
Voandzeia subterranea	Bambara groundnut	Adapted to arid regions; high seed protein
Prosopis spp	Mesquite	Drought-resistant trees and shrubs. Foliage, pods and beans all sources of fodder
Sesbania grandiflora	—	Very fast-growing (4.3–8 m/y). Forage, firewood, food, green manure
Calliandra callothyrus	—	Very fast-growing (7 m/y). Fuelwood source— 'fuel crop'. Erosion control

protein contents averaging 20 per cent compared with cassava, potatoes, sweet potatoes (1 per cent) and yams (3–7 per cent). *Psophocarpus* nodulates extensively and freely in nearly all regions tested so far.

The Bambara groundnut (*Voandzeia subterranea*) is a further example of a highly valued indigenous crop that has received scant attention from breeders and agriculturalists, despite producing a nutritious food (24 per cent seed protein; 60 per cent carbohydrate) and being cultivated throughout Africa. The plant derives its name from a district near Timbuktu on the fringe of the Sahara Desert (NAS, 1979). It is therefore admirably suited for hot, dry regions unsuitable for peanuts, maize or sorghum. Another group of legumes well adapted to arid regions are members of the genus *Prosopis*. Some are aggressive weeds but six species are recognized as being the source of useful forage and wood on the poorest soils in extremely low rainfall areas. They usually need an annual rainfall of 250 mm but some specimens have been well established where annual rainfall is as low as 75 mm. One of their most obvious attributes in relation to this is the production of spreading lateral roots and very deep tap roots, reaching depths of over 20 metres. Pods produced by the plant are high in carbohydrate (largely sucrose) while the seeds are rich in protein (34–39 per cent protein). Recently, considerable interest has been shown in their potential use as desertification control plants and they are being evaluated in the Sudan for this purpose. Felker and Bandurski (1979), in an interesting study, examined a range of food plants for examples that would fit a predetermined set of criteria: plants whose physiological, ecological, and morphological characteristics would obviate the need for tillage, irrigation and fertilization and yet provide high yields of protein-rich food. An idiotype was developed on the above basis and food plants assessed. Three legumes were found to have the full range of characteristics and, of these, *Prosopis* was the model example. The authors recommended such leguminous tree orchards as an agricultural system requiring minimal inputs of fossil fuel, machinery and capital.

Sesbania grandiflora and *Calliandra callothyrsus* are included as examples of the potential use of tropical tree legumes as sources of fuelwood. These legumes, which nodulate prolifically, have very high growth rates and are frequently being evaluated as potential fuel crops. Both legumes will coppice readily and *Calliandra* has been grown in Indonesia as an annual fuel crop. This brief review cannot do justice to the potential that exists in these and other tropical legumes and the interested reader is directed to the NAS publications cited in the bibliography.

Such legumes as those described and listed by the NAS are truly under-exploited. They all require investment in agronomic and allied research for the full benefits to be attained. As such, their exploitation represents an enormous challenge to the world and to the developing countries in particular.

An examination of these publications not only reveals the tremendous future potential of many of these tropical legumes but also the extreme diversity of the group. There are legumes adapted to almost every conceivable range of soil type and climatic region, from the arid zones to the hot and humid tropics. Perhaps the next move should be towards a general concept of 'appropriate biology' within the context of alternative agricultural systems. The challenge there would be to make and maximize productive use of arid or other regions by the effective development and breeding of indigenous or similarly adapted legumes from other parts of the world.

References

ABU-SHAKRA, S. S., PHILLIPS, D. A. and HUFFAKER, R.C. (1978). Nitrogen fixation and delayed leaf senescence in soybeans. *Science*, **199**, 973–975

ALBRECHT, S. L., MAIER, R. J., HANKS, F. J., RUSSELL, S. A., EMERICH, D. W. and EVANS, H. J., (1979). Hydrogenase in *Rhizobium japonicum* increases nitrogen fixation by nodulated soybeans. *Science*, **203**, 1255–1257

ANDERSON, J. R. (1979). Pesticide effects on non-target soil micro-organisms. In *Pesticide Microbiology*, pp. 313–501. Eds I. R. Hill and I. J. L. Wright. New York; Academic Press

ARAUJO, S. C. (1974). Reuniao necronal sobre nodulacao e fixacao de nitrageno en *Phaseolus vulgaris* L. *Vicosa*, **44**, 48

AUGHTRY, J. D. (1948). *Effect of Genetic Factors in* Medicago *on Symbiosis with* Rhizobium. Cornell University Agriculture Exp. Stn. Mem. 280

BURTON, J. C. (1967). *Rhizobium* culture and use. In *Microbial Technology*. Ed H. J. Peppler. New York; Reinhold Publishing Co.

DAFT, N. J. and EL GIAHMI, A. A. (1974). Effect of *Endogone* mycorrhiza on plant growth. VII. Influence of infection on the growth and nodulation in French bean (*Phaseolus vulgaris*). *New Phytologist*, **73**, 1139–1147

DOBEREINER, J. (1974). In *The Biology of Nitrogen Fixation*, pp. 86–120. Ed. A. Quispel. Amsterdam; North-Holland Publ. Co.

DUHIGG, P., MELTON, B. and BALTEN SPERGER, A. (1978). Selection for acetylene reduction rates in "Mesilla" alfalfa. *Crop Science*, **18**, 813–816

EVANS, H. J., RUIZ-ARGUESCO, N. T. JENNINGS, N. T. and HANUS, F. J. (1977). In *Genetic Engineering for Nitrogen Fixation*, p. 333. Ed. A. Hollaender. New York; Plenum

EAGLESHAM, A. R. J., MINCHIN, F. R., SUMMERFIELD, R. J., DART, P. J.,

HUXLEY, P. A. and DAY, J. M. (1977). Nitrogen nutrition of cowpea (*Vigna unguiculata*). III. Distribution of nitrogen within effectively nodulated plants. *Expl. Agric.*, **13**, 369–380

FELKER, P. and BANDURSKI, R. S. (1979). Uses and potential uses of leguminous trees for minimal energy input agriculture. *Econ. Bot.*, **33**, 172–184

GIBSON, A. H. (1962). Genetic variation in the effectiveness of nodulation of lucerne varieties. *Aust. J. Agric. Res.*, **13**, 388–399

GRAHAM, P. H. and ROSAS, J. C. (1977). Growth and development of indeterminate bush and climbing cultivars of *Phaseolus vulgaris* L. inoculated with *Rhizobium*. *J. Agric. Sci. Camb.*, **88**, 503–508

GRAHAM, P. H. and HALLIDAY, J. (1976). Inoculation and nitrogen fixation in the genus *Phaseolus*. In *Exploiting the Legume–* Rhizobium *Symbiosis in Tropical Agriculture*. Eds J. Vincent, A. Whitney and J. Bose. USAID College of Tropical Agriculture Miscellaneous Publication 145. University of Hawaii, Niftal Project

HALLIDAY, J. and PATE, J. S. (1976). The acetylene reduction assay as a means of studying nitrogen fixation in white clover under sward and laboratory conditions. *J. Br. Grassld. Soc.*, **31**, 29–35

HARDY, R. W. F. and HAVALKA, U. D. (1976). In *Symbiotic Nitrogen Fixation in Plants*, pp. 421–439. Ed. P. S. Nutman. Cambridge; Cambridge University Press

HARPER, J. E. (1974). Soil and symbiotic nitrogen requirements for optimum soybean production. *Crop Sci.*, **14**, 255

HOCH, G. E., LITTLE, H. N. and BURRIS, R. H. (1957). Hydrogen evolution from soybean root nodules. *Nature, Lond.*, **179**, 430–431

LAWN, R. J. and BRUN, W. A. (1974). Symbiotic nitrogen fixation in soybeans. I. Effect of photosynthetic source–sink manipulations. *Crop Sci.*, **14**, 11–16

LAWRIE, A. C. and WHEELER, C. T. (1973). The supply of photosynthetic assimilates to nodules of *Pisum sativum* L. to the fixation of nitrogen. *New Phytologist*, **72**, 1341–1348

LAWRIE, A. C. and WHEELER, C. T. (1974). The effects of flowering and fruit formation on the supply of photosynthetic assimilates to the fixation of nitrogen. *New Phytologist*, **73**, 1119–1127

MAIER, R. J. and BRILL, W. J. (1978). Mutant strains of *Rhizobium japonicum* with increased ability to fix nitrogen for soybean. *Science*, **201**, 448–456

MOSSE, B. (1976). Role of mycorrhiza in legume nutrition. In *Exploiting the Legume–Rhizobium Symbiosis in Tropical Agriculture*. Eds J. Vincent, A. Whitney and J. Bose. USAID, College of Tropical Agriculture Miscellaneous Publication 145. University of Hawaii, Niftal Project

NATIONAL ACADEMY OF SCIENCES (1975). *Underexploited Tropical Plants with Promising Economic Value*. Washington;

NATIONAL ACADEMY OF SCIENCES (1977). *Leucaena: Promising Forage and Tree Crop for the Tropics*. Washington, D.C.; NAS

NATIONAL ACADEMY OF SCIENCES (1979). *Tropical Legumes: Resources for the Future*. Washington D.C.; NAS

NEYRA, C. and DOBEREINER, J. (1977). Nitrogen fixation in grasses. *Advances in Agronomy*, **29**, 1–38

78   *Plant–microbial interactions*

NUTMAN, P. S. (1957). Symbiotic effectiveness in nodulated red clover. III. Further studies on inheritance of ineffectiveness in the host. *Heredity*, **11**, 157–173

NUTMAN, P. S. (1961). Variation in symbiotic effectiveness in subterranean clover, *Trifolium subterraneum* L. *Aust. J. agric. Res.*, **18**, 381–425

PHILLIPS, D. A. (1980). Efficiency of symbiotic nitrogen fixation in legumes. *A. Rev. Pl. Physiol.*, **31**, 29–49

ROUGHLEY, R. J.(1970). The preparation and use of legume seed inoculants. *Pl. Soil*, **32**, 675

SANDERS, F. F., MOSSE, B. and TINKER, P. B. (1975). *Endomycorrhizas*. London; Academic Press

SCHENK, N. C. and HINSON, K. (1976). Response of nodulating and non-nodulating soybeans to a species of *Endogone* mycorrhiza. *Agron. J.*, **65**, 849–850

SCHUBERT, K. R. and EVANS, H. J. (1976). Hydrogen evolution: A major factor affecting the efficiency of nitrogen fixation in nodulated symbionts. *Proc. Natn. Acad. Sci. USA*, **73**, 1207–1211

SCHUBERT, K. R., JENNINGS, N. T. and EVANS, W. F. (1978). Hydrogen reactions of nodulated leguminous plants. II Effects on dry matter accumulation and nitrogen fixation. *Pl. Physiol.*, *Lancaster*, **61**, 398–401

SEETON, M. W. and BARNES, D. K. (1977). Variation among alfalfa genotypes for rate of acetylene reduction. *Crop Sci.*, **17**, 783–787

SINCLAIR, T. R. and DE WIT, C. T. (1975). Photosynthate and nitrogen requirements for seed production by various crops. *Science*, **189**, 565

SPRENT, J. L. (1976). In *Symbiotic Nitrogen Fixation in Plants*. Ed. P. S. Nutman. Cambridge; Cambridge University Press

SPRENT, J. L. (1979). *The Biology of Nitrogen Fixing Organisms*. London; McGraw-Hill (UK) Ltd

STAPHORST, J. L. and STRIJDOM, B. W. (1976). Effects on rhizobia of fungicides applied to legume seed. *Phytophylactica*, **8**, 47–54

VINCENT, J. M. (1974). In *Biology of Nitrogen Fixation*, pp. 265–342. Ed. A. Quispel. Amsterdam; North Holland Publishing Co.

WILSON, P. W. (1940). *The Biochemistry of Symbiotic Nitrogen Fixation*. Madison, Wisconsin; The University of Wisconsin Press

ZARY, K. W. and MILLER, J. C. (1977). Intraspecific variability for dinitrogen fixation in cowpea (*Vigna unguiculata*). *HortScience*, **12**, 402

5

INFLUENCE OF AGRICULTURAL PRACTICE ON SOIL INVERTEBRATE ANIMALS

D. S. MADGE
Department of Biological Sciences, Wye College (University of London), Ashford, Kent, UK

Introduction

Invertebrate animals live in considerable numbers in most undisturbed soils and their diversity is remarkable (Wallwork, 1976). The soil structure, its fertility and even its formation greatly depend on them. Many of the soil fauna live by feeding on decaying plant materials, breaking them down into their organic and inorganic constituents and incorporating the end-products into the soil (Madge, 1966; Wallwork, 1970; Edwards and Lofty, 1977). Where they are few, the soil structure is often poor, with an undecomposed layer of organic matter lying on its surface.

Soil animals vary greatly in size and form. The most important groups of them include earthworms (lumbricids), potworms (enchytraeids), round-worms (nematodes), slugs and snails (molluscs), and representatives of several arthropod classes: free-living (cryptostigmatic) and parasitic (meso-stigmatic) mites, millipedes (diplopods), woodlice (isopods) and several orders of insects, in particular springtails (collembolans), the larvae of beetles and flies, ants and, in the tropics, termites. Less important (and more obscure) groups include the symphylids, chilopods, pauropods (myriapods), and the protozoans.

While most soil invertebrate animals are generally beneficial to the soil, a few of them are serious crop pests. To eradicate them or control the numbers of these and other pests living in the crops above, farmers and contractors apply a vast and varied array of potent pesticides which sooner or later reach and become mixed with the cultivated soil. How do these chemicals affect the soil animals? How does the influence of various other agricultural practice, mechanical and chemical, affect them? The subject is very diverse and to answer these questions requires selection, generalization and simplification.

Pesticides

Pesticides comprise a large number of various chemical substances used to control arthropod pests and infectious diseases, improve the fertility of soils, and prevent the growth of weeds. They include insecticides, herbicides, nematocides and fungicides. In 1975, the sales of pesticides in the UK

79

amounted to £140 million for herbicides, £35 million for insecticides, and £15 million for fungicides (Kornberg, 1979). Ideally, a pesticide should eradicate only the target pest at which it is aimed but often this has not been possible. Unfortunately, most pesticides are biologically active chemicals and hence frequently will also eliminate or reduce a large proportion of the beneficial nontarget soil fauna. When the pesticide is toxic the soil is usually rendered sterile for days or a few weeks, after which it will gradually become recolonized as its toxicity disappears. Other pesticides are less toxic but will persist in the soil for months or many years, and thus the animals in the soil continually come into contact with the chemicals as they move about. This second group of pesticides often have long residual deleterious effects and, as a result, the animals recolonize the soil only slowly.

INSECTICIDES

On the basis of their chemical composition and origin, insecticides can be divided into four major categories: inorganic insecticides, botanical insecticides, fumigants, and synthetic organic insecticides. The last group are now the leading insecticidal products and are manufactured on a large scale at relatively low cost. They are divided according to the chemical families to which they belong: the organochlorines, the organophosphates and the carbamates.

The organochlorines are persistent pesticides and concern about their possible long-term effects, combined with the resistance to them by soil arthropod pests, has led to their restriction or phasing out. Alternative, less persistent insecticides, the organophosphates and carbamates, were developed which superseded the organochlorines. Insecticides act by contact (external surface) or systemic (swallowed) action, or a combination of both, and are applied as dusts or liquid formulations.

Generalizations on the effects of insecticides at normal rates of application to agricultural land on populations of invertebrate soil animals are misleading, because their effects often vary within the main groups of insecticides and the main groups of soil fauna. Moreover, different investigations do not always agree on the influence of some of these chemicals on soil animals. The following examples illustrate the effects of insecticides on some groups of invertebrate soil animals (*mainly from* Thompson and Edwards, 1974).

Mites (Acarina)

Organochlorines: aldrin either had no effect on mites or reduced all groups except the predators. DDT killed some species while others increased in numbers because of mortality among predatory mites (DDT was toxic to the predators and not to others). HCH (BHC) was more toxic than DDT. Heptachlor and chlordane were moderately toxic, while endrin and isobenzan were highly toxic. Organophosphates: chlorfenvinphos, diazinon, disulfoton, fenitrothion, parathion, phorate and thionazin were all effective. Because these chemicals are particularly toxic to predaceous mites, increases in numbers in other species have sometimes been reported.

Springtails (Collembola)

Organochlorines: aldrin had no effect on populations or sometimes decreased their numbers. DDT applications usually increased populations of springtails because it killed the predaceous mites that attacked them. Chlordane, endrin, heptachlor and isobenzan were all toxic. Most workers have reported that HCH decreased the numbers of springtails. Organophosphates: chlorfenvinphos, diazinon and trichlorphon generally increased populations of springtails because of the death of predatory mites. Disulfoton, fenthion, fonofos, menazon and phorate decreased their numbers. Parathion and thionazin either increased or decreased populations of springtails.

Earthworms (Lumbricidae)

Organochlorines: aldrin, dieldrin, DDT, HCH and isobenzan were all relatively ineffective. Chlordane, cynazine, endrin and heptachlor usually killed earthworms. Organophosphates: diazinon, malathion and menazon were not toxic. Chlorfenvinphos, disulfoton, dyforate, fenitrothion and parathion were only mildly toxic to them, while fensulfothion, fonofos and phorate were highly toxic. Carbamates: aldicarb, benomyl, carbaryl, carbofuran, dazomet, mildothane and propoxur were all very toxic.

Nematodes (Nematoda)

Organochlorines: aldrin, DDT, HCH, and heptachlor were either ineffective or slightly increased their numbers because predators were killed. Organophosphates: chlorfenvinphos, chlorpyrifos, diazinon, dimefox, phosphamidon, thionazin and phorate were all moderately toxic.

Effects of insecticides

Insecticides can have diverse effects on populations of soil invertebrate animals. Groups of them, or even individual species within a group, may become selectively affected, or the interaction between predator and prey may become altered. Some soil animals may accumulate the insecticide in their body tissues, or the insecticide may alter its chemical characteristics on reaching the soil and then produce unexpected effects on the animals. These effects will be discussed at greater length below.

Insecticides are sometimes *selective* in their action. Species or groups of soil animals susceptible to them may be eliminated either when the pesticide is applied annually or when it persists for several years while other animals remain unaffected. In an experiment, cultivated soils were treated with DDT and aldrin, the soil invertebrates recovered, and the number of species from different groups identified and compared with those in untreated soil (Edwards and Lofty, 1969). In arable soil, a total of 66 species of various groups of soil animals were recovered from untreated plots, while in DDT- and aldrin-treated soils the number of species decreased to 48 and 40 respectively. In pasture soil, 148 species were collected from the control plots, while 109 and 99 species were found in DDT and aldrin-treated plots respectively. Although DDT and aldrin diminished the total number of soil arthropod species by about one-third and one-quarter respectively, it was not clear which groups of soil fauna were most affected by the insecticides.

The influence of insecticides on populations of soil invertebrates has been clearly illustrated with regard to their *predator–prey relationship* (Edwards, 1969). Following applications of soil insecticides, springtails multiplied to levels consistently far higher than those in previously untreated soil. These striking increases were accompanied by considerable decreases in the number of predatory mites. As the insecticide lost its activity in the soil, the low number of predatory mites again rose to its previous level, while the high population level of springtails returned to its normal level more slowly.

The effect produced by the pesticide is a reflection of a relaxation of predator pressure exerted by the parasitic mites. The imbalance in predator–prey relationship became obvious only after the soil was treated with DDT, and not with any other persistent organochlorine (*Figure 5.1*).

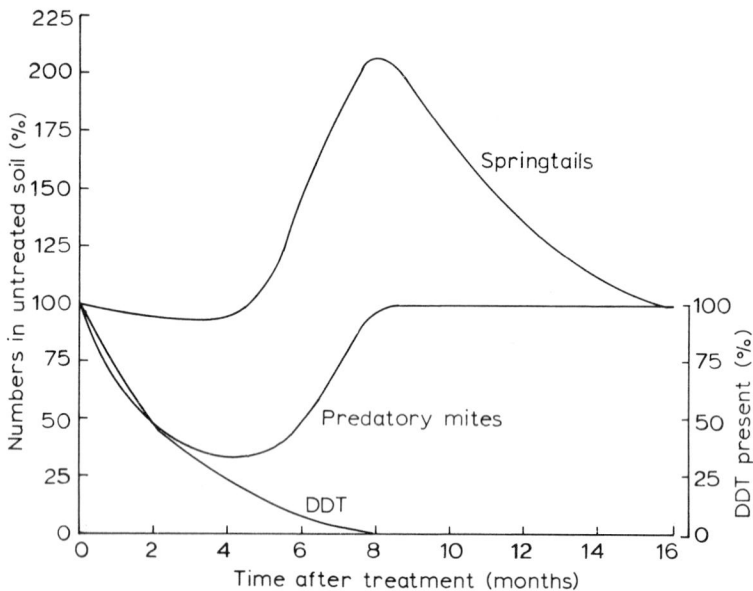

Figure 5.1   Effect of DDT on predator–prey interaction (after Edwards, 1969)

Soil parasitic mites were thus highly susceptible to DDT and were selectively eliminated, whereas springtails were unaffected by the pesticide. Springtails have a tissue enzyme, DDT-dehydrochlorinase, that metabolizes the relatively toxic DDT into nontoxic DDE. Other observations have shown, however, that there was an increase in the population of springtails before any significant decrease in the number of predatory mites, suggesting that the presence of DDT stimulated the springtails to multiply, by an unknown mechanism.

When less-persistent organophosphorous insecticides were applied to soils, the populations of springtails and of free-living mites both increased while the population of predatory mites decreased (Edwards, 1969). Thus, the larger numbers of free-living mites and springtails were usually limited more by predator pressure than by lack of available food: when their predators were removed, they both multiplied rapidly.

Earthworms are not very susceptible to persistent insecticides and can live in soil containing significant amounts of the pesticide. The insecticides become absorbed from the soil as they pass through the worms' intestines and the residues accumulate in their tissues, frequently at concentrations higher than those in the soil in which they live (*Figure 5.2*). There is a linear

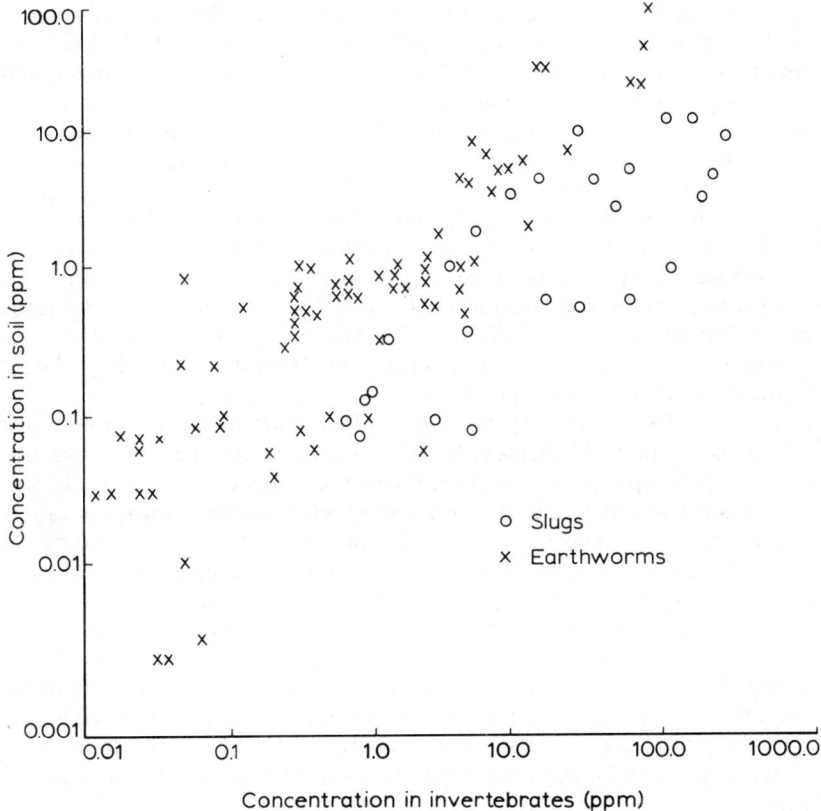

Figure 5.2 Concentration of insecticide residues from soils into earthworms and slugs (after Thompson and Edwards, 1974; based on data from several workers)

relationship between the amounts of pesticides in the soil and the amounts in the worms (Edwards, 1976). The average amount of all organochlorine residues in different soils studied was 1.5 ppm and that in the worms in these soils was 13.8 ppm, a concentration factor of about ninefold for all organochlorine insecticides and doses.

Earthworms do not concentrate all organochlorines from the soil to the same extent. The concentration was highest using DDT and dieldrin, and fell progressively with chlordane, endrin, aldrin and heptachlor. Generally, the uptake of insecticides in worms living near the soil surface, particularly in cultivated soils, was greater than in those living deeper in the soil. Organochlorine insecticides were taken up only slowly into the tissues of the worms and some insecticides were broken down to other materials in their bodies. Thus, chemical analyses showed that their tissues contained a greater proportion of DDT metabolites, especially DDE.

Less work has been done on the uptake of organophosphate and carbamate compounds in the tissues of soil invertebrates. The carbamate aldicarb was rapidly absorbed in large amounts in the body of earthworms. Uptake depended on the moisture content of the soil. In a range of 26–37 per cent soil moisture in different experiments, 85 per cent of the worms survived in uncontaminated soil with 31 per cent optimal moisture, whereas only 40 per cent of them survived in soil treated with aldicarb (Edwards *et al.*, 1977). In contrast, the organophosphate phorate accumulated in earthworms much more slowly and in only small amounts.

Slugs, too, are able to concentrate organochlorines in their tissues, and at a much higher concentration than in the surrounding soil than the corresponding concentration in earthworms in the same soil (Thompson and Edwards, 1974) (*Figure 5.2*). Dieldrin, endrin, and DDT and its metabolite have all been recovered in substantial amounts in the tissues of slugs. Whereas earthworms concentrated more DDE than TDE, slugs did the opposite, and hence the metabolism of DDT in these invertebrates differed. Sublethal concentrations of insecticide residues have also been reported in the tissues of several other invertebrates, including mites, springtails, beetles and caterpillars.

The *newer, less persistent organophosphate and carbamate insecticides* have recently been substituted for the older, more persistent organochlorines. Development of resistance with these newer insecticides in soil arthropods is less likely and only in those species that have one generation or less each year. However, little is yet known of the potential biological dangers of these less persistent compounds and the assumption that they are less hazardous than the more persistent insecticides may be untrue, as illustrated by the following example (Harris, Thompson and Tu, 1972).

Dasanit (fensulfothion) is an organophosphorous insecticide–nematocide used mainly for soil arthropod pests and nematode control. Experiments showed that dasanit persisted in the soil for about 4 weeks, after which its activity dropped rapidly. Dasanit was degraded to dasanit sulfone which persisted for much longer than dasanit, its activity lasting for at least 32 weeks.

The effect of dasanit on earthworm populations was determined in a field trial and its activity compared with that of DDT (*Figure 5.3*). Three weeks after treatment, the number of earthworms in the DDT- and dasanit-treated plots decreased to about one-third and four-fifths respectively, compared with earthworm numbers in untreated plots. Bodies of earthworms surviving the dasanit treatment contained residues of 0.25 ppm of dasanit and 22.5 ppm of dasanit sulfone; DDT residues were very high, and averaged 128 ppm. After 52 weeks, earthworm populations had increased to normal in both the treated plots; residues of DDT had decreased to 8 ppm while residues of dasanit or dasanit sulfone were not detected.

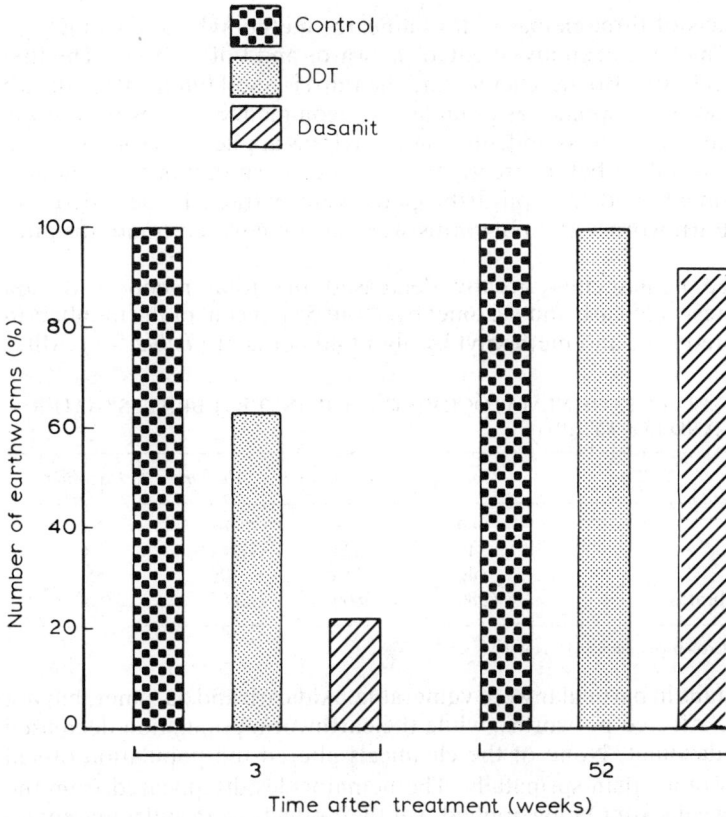

Figure 5.3 Effects of dasanit and DDT on earthworms in pasture (after Harris, Thompson and Tu, 1972)

Other field experiments showed the effects of DDT and dasanit on arthropods. Twenty weeks after initial treatment with DDT the total number of springtails rose markedly, because of the death of predatory mites. Treatment with dasanit, however, caused a marked decrease in springtails, a concentration of 0.04 ppm of dasanit sulfone appearing to be sufficient to suppress the springtail population. These results clearly showed that the dasanit sulfone, the degraded product of dasanit, was mainly responsible for the biological effects on the nontarget soil organisms. Thus, it cannot be assumed that simply because an insecticide is less persistent it will have less deleterious biological side-effects, since unsuspected degradation products may be insecticidally active and more persistent in the soil.

NEMATOCIDES

Paradoxically, in the soil there are more populations of nematodes than of any other multicellular group of soil animals, yet less work has been done on the effects of nematocides on soil fauna than for any other pesticide.

The effects of three nematocides, aldicarb, methomyl and dazomet, on the soil fauna have been investigated (Edwards and Lofty, 1971). The first two chemicals are also insecticides and the third is a soil fumigant. Aldicarb and dazamet were applied as granules at recommended doses to a newly ploughed cultivated grassland, and methomyl was applied as a solution. Soil samples were taken before treatment and once each month for 5 months after treatment, and the soil arthropods were extracted, identified and counted. Earthworms and potworms were also recovered 5 months after treatment.

All three nematocides greatly decreased the total numbers of soil invertebrates—aldicarb and dazomet by about 55 per cent of the numbers in the untreated soil, and methomyl by about 40 per cent (*Table 5.1*). Mites

Table 5.1   EFFECTS OF NEMATOCIDES ON SOIL INVERTEBRATES (AFTER EDWARDS AND LOFTY, 1971)

Soil fauna	Control	Methomyl	Aldicarb	Dazomet
Earthworms	69	78	65	24*
Potworms	345	328	141*	78*
Insects	146	216	126	155
Total invertebrates	6980	3860*	2667*	2885*

*Significantly different from control at 5% probability level

and springtails in particular were vulnerable. Aldicarb and dazomet, but not methomyl, affected potworms, while the earthworm population decreased only with dazomet. None of the chemicals altered the population of soil arthropods other than springtails. The nematocides disappeared from the soil, 8–10 weeks after being applied, yet their effects were still apparent on populations of mites and springtails after 5 months, although most of the other invertebrates recovered sooner.

Fumigants, for example D-D, methyl bromide and methomyl are nematocide–insecticides used to control nematodes and soil pathogens. When injected into the soil, they act as toxic gases and have been shown to be lethal to virtually all organisms in the soil (Edwards, 1969). However, deep-living soil animals, for example symphylids and diplurans, were unaffected. The action of fumigants persisted for only a few weeks, after which the animals living deeper in the soil repopulated the soil from below. Unaffected eggs in the treated soil hatched, while small soil arthropods in air currents recolonized the soil. However, the process was slow and fumigated soil may take up to 2 years to recover its normal soil animal community.

FUNGICIDES

Fungicides are used to control crop pathogens damaged or killed by infection and contagious disease. They are applied as seed dressings, soil fumigants, or foliar sprays.

There are very few reports on the effects of fungicides on soil invertebrate animals. Captan, dicloran, formaldehyde, quintozene and thiram were all ineffective on populations of soil animals; carbendazim was very toxic to earthworms. Earthworms were virtually missing from soils in apple orchards treated with heavy applications of copper sulphate, resulting in a poor soil crumb structure and a thick superficial layer of undecomposed organic matter (Raw, 1962). From experimental apple orchard soil sprayed with benomyl and thiphanatemethyl for several years, an average of 22.7 and 5.5 earthworms/m^2 respectively were recovered, while 94.8/m^2 were collected from unsprayed controls. In a commercial orchard, benomyl sprays reduced the average number of earthworms in a single season from 272.4 to 29.4/m^2 (Wright, 1979).

HERBICIDES

Herbicides are applied regularly each year on large areas of cropped land, and a substantial proportion of these compounds reach the soil. Their main purpose is to control weeds and hence prevent weed competition with the crop for nutrients, light and water, which might reduce crop yields. There are about 250 different kinds of herbicides in current use, and nearly £150 million was spent on them in the UK in 1975, three times the expenditure on insecticides and fungicides combined (Fryer, 1977).

Herbicides are divided into two broad groups: the total, or non-selective herbicides, which destroy all the vegetation, and the selective ones, now used extensively, which will kill some plants but not others. Alternatively, herbicides can be classified according to their contact (scorching), translocated (absorbed by leaves), or residual (absorbed by roots) mode of action. Some act either as growth regulators or inhibitors; others as photosynthesis inhibitors.

Many herbicides break down in the soil in a few weeks; a few (for example atrazine, dimuron, monuron and simazine) can persist in the soil for a year or longer. Herbicides were originally applied as dusts but are now generally used as liquid formulations, either to the foliage of weeds or to the soil. Herbicides are apparently not concentrated in the body tissues of soil animals, unlike some insecticides.

Relatively few investigations have been made on the influence of herbicides on soil animals. Applications of 2,4-D, dalapon, linuron, MCPA, paraquat, tri-allate and 2,4,5-T at normal doses had little influence on the soil fauna, while monuron and TCA affected soil animals at large doses only; only DNOC and simazine were lethal to some groups (Edwards, 1970).

Herbicides can affect invertebrate soil animals in two ways. First, directly, by killing them on contact; secondly, indirectly, by destroying weeds and so eliminating the decaying plant debris on which many soil animals feed. A combination of both direct and indirect effects is also possible. However, the direct or indirect effects of herbicides on the soil fauna have often not been distinguished. Edwards and Stafford (1979) found that simazine had a direct effect on populations of soil arthropods, while chlorthiamid affected earthworm populations indirectly; however, cyanazine acted both directly and indirectly on populations of springtails.

Generally, the only herbicides known to affect soil invertebrate animals directly at normal doses are the triazines—atrazine cyanazine and simazine—and also DNOC. The indirect effects of other herbicides are more likely to exert a greater influence on the soil fauna than the herbicides themselves. Herbicides probably do not constitute a serious danger to the soil animal community, because their effects are often slight and they are therefore unlikely significantly to alter soil fertility.

Fertilizers

There are two categories of fertilizers: the inorganic mineral ('NPK') or artificial fertilizers, and the organic manures. Other substances may also be added to the soil to counteract its acidity or alkalinity, to assist in plant growth. The chemical elements nitrogen, phosphorus and potassium are essential for plant growth. During the past three decades the use of artificial fertilizers in the UK, especially of nitrogenous fertilizers, has increased sharply. In the UK in 1975, almost 100 000 tonnes of nitrogen were applied to farmland as artificial fertilizer, over twice the combined weights of phosphorus and potassium (Kornberg, 1979).

Organic manures directly affect some members of the soil fauna by increasing their food supply and so allowing them to multiply, frequently to very high levels. Thus, some 15 million arthropods have been recovered in manured soil compared with five million in untreated soil (Morris, 1922). Adding manures to soil may also increase the diversity of the fauna by attracting them to the food source, for example earthworms, potworms, nematodes, springtails, and dung and rove beetles. As the manure gradually disappears, the species composition of the soil fauna will alter. Plots treated with a liquid-manure irrigation resulted in many more earthworm populations compared with nonirrigated plots.

The effects of *artificial fertilizers* are less striking but generally also tend to increase the populations of soil fauna, particularly of mites. The animal life in the soil is stimulated directly through increased plant root growth and plant remains, and is not a direct result of the fertilizers. Nitrogenous fertilizers favour the increase of large numbers of earthworms but high levels decrease the population. A clear correlation has been found between the amounts of nitrogenous fertilizers used in leys and the weight of earthworms recovered from these soils. Most other inorganic fertilizers have little influence on earthworm populations, although overall increases in numbers have sometimes been reported. Lime is beneficial to earthworms because some species avoid acidic soils, while ammonium sulphate is unfavourable to others because the soil is made more acid thereby.

The long-term effects of artificial fertilizers on the invertebrates in grassland soil have been studied by recovering the fauna from Park Grass, one of the 'classic' experiments at the Rothamsted Experimental Station, which has been continually fertilized since 1856 (Edwards and Lofty, 1975a). The larger the dose of nitrogen applied, the more adversely affected were populations of many soil animals, particularly earthworms, myriapods and wireworms. In contrast, large applications of fertilizers increased numbers of some other animals, especially potworms, springtails and proturans.

The effects of artificial and organic fertilizers on the soil fauna have been compared (Edwards and Lofty, 1969). Much higher populations of mites and springtails were recovered from plots treated with dung, compared with plots treated with inorganic fertilizer, while the numbers of soil arthropods from artificial fertilizer plots were slightly higher than those from unfertilized plots (*Figure 5.4*).

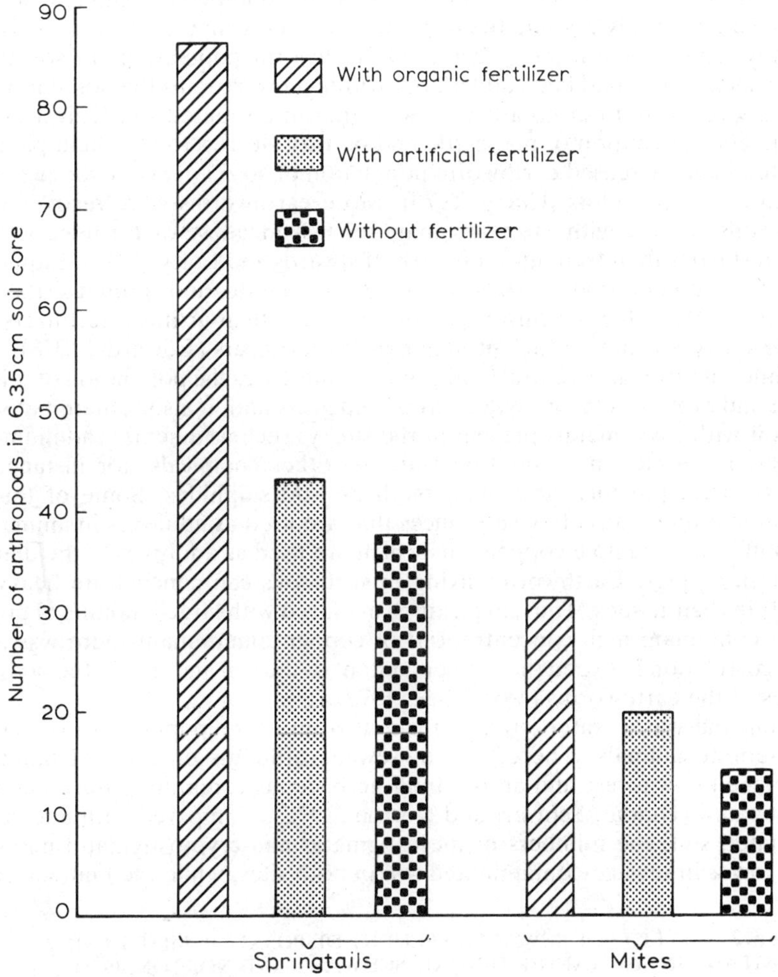

Figure 5.4 Effects of fertilizers on soil microarthropod populations (after Edwards and Lofty, 1969)

Fertilizers in the wrong place and in wrong concentrations act as *pollutants*. Agriculture can pollute land (and water) in several ways. Minerals from untreated farm waste materials, such as manures and slurries from livestock and silage effluents, and sewage and seepage from rubbish dumps and spoil heaps, can all accumulate in farm and waste land, pollute the soil and so influence the animal population living there.

There are few reports on the effects of animal wastes on soil invertebrate animals. The only practical method of disposing of animal slurry in some farms is to spread it over farmland. About 170×10^6 t of undiluted excreta are produced by farm animals in the UK each year, almost one-third of which is voided indoors and the remainder returned to farmland by grazing animals. The excreta contains about one-fifth of the nitrogen, one-quarter of the phosphorus, and three-quarters of the potassium potentially available in organic form to farmers (Kornberg, 1979). The addition of animal wastes to the soil generally favours the population of many invertebrate animals. Initially, slurry significantly decreased earthworm populations in soil to which such wastes had been added, but within a few months the population not only recovered but greatly exceeded that in untreated soil (Edwards, 1979). The application of pig, cattle and poultry slurries to grassland plots resulted in an increased earthworm population of 40–53 per cent above the untreated control plots (Curry, 1976). More earthworms were recovered from soils treated with farmyard manure, fish meal, sewage sludge and animal slurries than from untreated soil (Edwards and Lofty, 1979). Liquid manure treatment produced larger populations of mites and springtails than sewage sludge, which in turn was more effective than artificial fertilizers. Cattle slurry was more efficient than mineral fertilizers (Edwards, 1978).

While slurry on agricultural land generally increases the population of soil fauna and promotes the growth of crops and grassland, it also contaminates the soil with toxic metals present in the slurry, such as arsenic, cadmium, copper, lead, mercury, and zinc and also other chemicals, for instance benzoic acid, phenols, ammonia, methane and sulphides. Some of this contamination is caused by substances that are used as additives in animal feedstuffs, for example copper salts which are used as additives in the diet of fattening pigs. Earthworms living in such soils can concentrate heavy metals in their tissues. Thus, in pastures manured with large amounts of pig slurry containing high concentrations of copper contaminant there was a close correlation between the copper content of the soil and that in the body tissues of the earthworms (Von Rhee, 1977).

Municipal waste water irrigation can also influence populations of soil invertebrate animals. The effects of sewage effluents on the soil fauna in a mixed oak forest and an old field herbaceous community have been investigated (Dindal, Schwert and Norton, 1975). Compared with uncontaminated soil, the numbers of mesostigmatid and cryptostigmatid mites were lower in sewage-contaminated soil in both sites, while the number of

Table 5.2 EFFECTS OF SEWAGE EFFLUENT DISPOSAL ON MITES AND SPRINGTAILS IN SOIL (AFTER DINDAL, SCHWART AND NORTON, 1975)

Taxonomic Group	Mixed oak forest community				Old field herbaceous community			
	Mean number of individuals		Mean number of species		Mean number of individuals		Mean number of species	
	Treated	Control	Treated	Control	Treated	Control	Treated	Control
Prostigmata	102	108	7	21	443	483	12	36
Mesostigmata	11	50	6	9	22	38	4	7
Cryptostigmata	8	199	6	17	24	196	8	18
Collembola	74	236	17	10	232	105	11	17

prostigmatid mites remained unaffected (*Table 5.2*). The species diversity of all three groups of mites decreased in polluted soil from both sites. The population of springtails decreased in polluted forest soil but increased in polluted grassland soil; species diversity, in contrast, was greater in contaminated forest soil than in contaminated grassland soil. Earthworm populations generally increased in both sites affected by sewage irrigation, implying a greater incorporation of organic matter and a greater aeration of the soils in the polluted sites.

Cultivation

Soil cultivation may be accomplished either by turning the soil over by ploughing, or by loosening and mixing the soil by harrowing and rotavating it. The use of implements for working the land markedly alters the soil structure, making it easier for some soil animals to survive and more difficult for others.

The effect of human disturbance of the natural layers of the soil on groups of soil fauna is strikingly demonstrated when the soil animals of fields and meadows are compared (*Table 5.3*). The total number of individuals found in tilled fields was reduced by one-fifth to one-tenth of that in uncultivated meadows.

Table 5.3   REDUCTION IN SOIL FAUNA IN CULTIVATED LAND (AFTER SCHALLER, 1968)

Animal group	Field	Meadow	Ratio
Nematodes	2 000 000	10 000 000	1 : 5
Mites	30 000	180 000	1 : 6
Springtails	15 000	90 000	1 : 6
Potworms	4 000	40 000	1 : 10

In undisturbed woodland soil the invertebrate animals are concentrated in the litter layer and fewer are found deeper; in pasture (uncultivated) soil they are mainly found in the superficial organic matter; in arable (cultivated) soil they are found more evenly distributed in the topsoil and subsoil (Edwards and Lofty, 1969). Groups of soil animals differ in their vertical distribution in the soil and this will affect their susceptibility to cultivation. Thus, many insect larvae, some centipedes and millipedes, some springtails and many mites were recovered in superficial soil, while springtails and other insects, millipedes, and earthworms were found living deeper, and others like symphylids, diplurans and pauropods, deeper still.

The effects of *ploughing* a pasture will affect the populations of soil animals. Soil samples were taken from a 300-year-old pasture, the pasture was then ploughed and disked, and samples taken again 2 weeks later and the soil animals recovered (Edwards and Lofty, 1969). Free-living (cryptostigmatid) mites, springtails, chilopods and pauropods were all generally affected by the cultivation, whereas parasitic (mesostigmatid) mites, symphylids, diplopods, proturids and most insects were less affected by it (*Figure 5.5*).

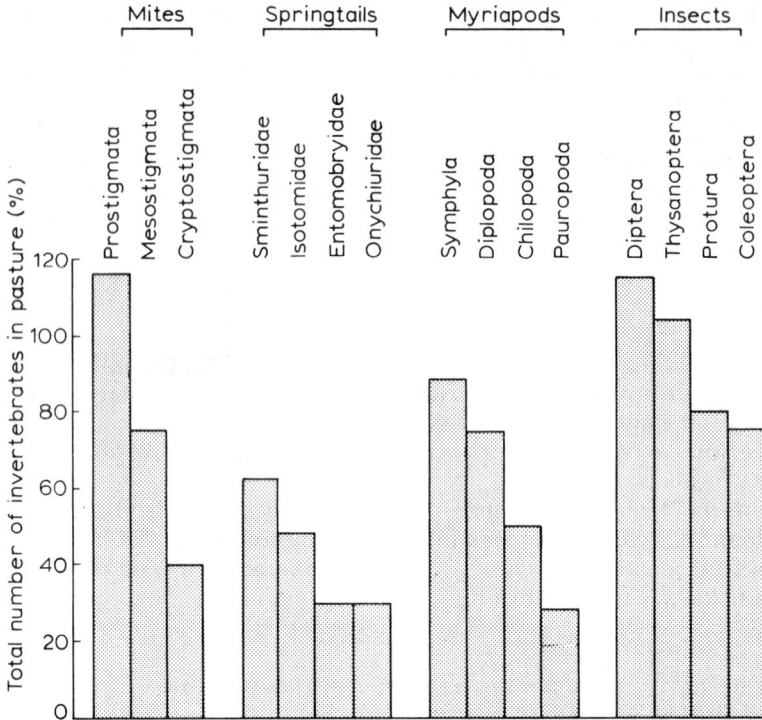

Figure 5.5 Effect of ploughing pasture on numbers of soil arthropods (after Edwards and Lofty, 1969)

The effects on the soil invertebrate populations of ploughing up an old pasture and reseeding it, have also been investigated in long-term cultivation experiments (Edwards and Lofty, 1975b). Some plots were ploughed, rolled, disked five times, harrowed and reseeded (most cultivations); others were ploughed, rolled, disked twice, rolled and reseeded (fewest cultivations); others were left undisturbed (*Figure 5.6*). Over several years, the population of all major groups of soil arthropods decreased significantly, particularly parasitic mites and surface-dwelling springtails, the extent of the decrease depending on the amount of cultivation. In contrast, cultivation increased the earthworm populations.

Ploughing is no longer regarded as essential. Increasingly today, cultivation of arable land in the UK is being kept to a minimum by first spraying the land with a herbicide to kill the existing vegetation and then sowing the uncultivated land by *direct drilling* (or *slit seeding*), a technique allowing seeds to be sown with a special drill that cuts slits into which the seeds are dropped without otherwise disturbing the soil.

The effects of direct drilling have been assessed on populations of soil invertebrate animals. Unploughed soil was first treated with the herbicide paraquat, which is inactivated on reaching the soil, and then direct-drilled plots were seeded with wheat. After 2 and 6 months the soil animals were compared with those in unploughed plots (Edwards and Lofty, 1969). Parasitic mites and surface-living springtails were little affected by the treatment after 2 months, but the numbers of parasitic mites and deep-living springtails were reduced. However, after 6 months the number of soil animals both in unploughed and ploughed land was virtually identical, suggesting that the effect of ploughing was transitory and would not persist into a second season.

Long-term studies have also been made (Edwards and Lofty, 1975b). Although overall changes in soil invertebrate animals were relatively small, consistently greater populations of small arthropods and earthworms were recovered from direct-drilled soil than from that which had been conventionally cultivated (*Figure 5.6*). A recent survey of the earthworm

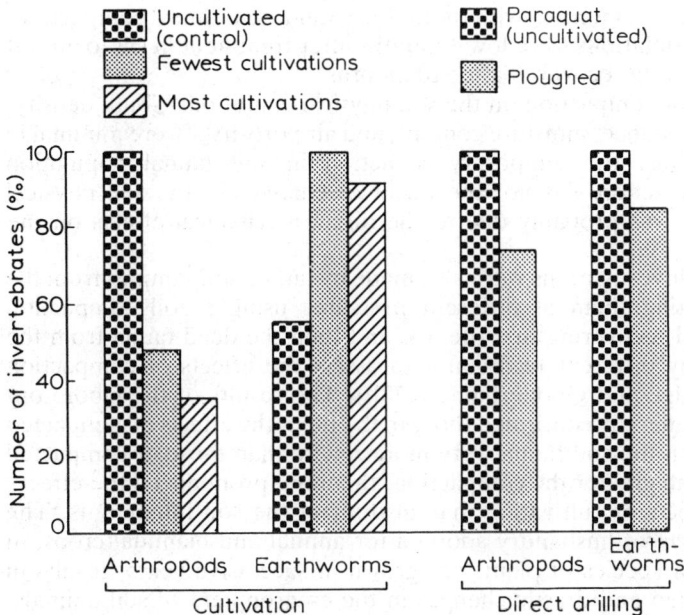

Figure 5.6 Effects of cultivation and reseeding an old pasture (*left*) and direct drilling and ploughing (*right*) on soil arthropods and earthworms (after Edwards and Lofty, 1975b)

population in seven direct-drilled fields gave an average of 72.8 worms/m², and that in seven ploughed fields, 32.1 worms/m², mainly because of the large number of deep-burrowing species in the direct-drilled soil (Edwards and Lofty, 1979). When cultivation was stopped, the population of soil fauna soon returned to the previous levels, the small arthropods being carried mainly by wind currents. Hence, the overall effects of cultivation on the soil fauna are probably not very great, although direct drilling generally favoured populations of animals.

Some soil structure problems which have occurred recently, particularly that of compaction, have been associated with the use of large and heavy implements when the soil is excessively wet. The effects of *compaction* of agricultural soils on the soil fauna have been studied (Aritajat, Madge and Gooderham, 1977a). Silt loam and clay grassland soils at Wye College Estate were compacted either once or ten times by using a heavy tractor in early winter and the invertebrate soil animals were recovered at intervals.

Two weeks after the silt loam soil had been compacted once, the population of soil fauna declined slightly, particularly the springtails, compared with the initial uncompacted soil. With ten times compaction, the number of soil animals including mites, springtails and other insects, greatly decreased. One month after treatment, the soil faunal population compacted once had generally recovered, but in the soil compacted ten times the population of soil mites had not. With clay soil, the effects of compaction on all faunal groups were less marked after 2 weeks, but more so after 1 month. The soil invertebrate animals thereafter gradually increased in both sites. In silt loam soil, the fauna recovered after 3 months from the once compacted soil and after 6 months from the ten times compaction. In clay soil, the fauna recovered after 3 months with both once and ten times compaction. Earthworm populations were low 4 months after treatment in silt loam soil only, but they fully recovered after 10 months.

The effects of compaction on the soil physical conditions—bulk density, mechanical resistance, moisture content, and air porosity—were minimal in both sites. Thus, the temporary reduction in soil faunal population following compaction was not the result of changes in the soil physical conditions but was probably due to the direct mechanical effects on the animals.

Laboratory investigations were also made by taking soil samples from the field, compacting them at different pressures using a soil compaction apparatus, and recovering first the live and then the dead fauna from the soil samples by different techniques to assess the effects of compaction (Aritajat, Madge and Gooderham, 1977b). The results of the laboratory experiments generally confirmed those in the field: the effects of compacted silt loam soil on the soil fauna were more severe than those of compacted clay soil and the greater the compaction, the more pronounced the effect.

Indirect effects of cultivation may also affect the soil inhabitants. The *rotational system* of husbandry adopted for annual and biannual crops, in which different species of plants are grown in successive years, results in frequent, sudden and drastic changes in the environment of soil animals. Large populations of a few plant-feeding species associated with the crop that build up are frequently and quickly eliminated. Such a cultural operation will eradicate certain species but create favourable microhabitats for others.

Traditional rotational systems are nowadays tending to be replaced by *monoculture,* or growing only a single crop, especially cereals. Monoculture tends to eliminate those soil animals associated with other plants. Generally, monoculture has less drastic effects on the soil fauna than the rotation system.

The practice of *burning* straw after harvesting is now widespread in England and Wales, and about half the total cereal straw is now burnt annually. The effects of removing straw by bailing, spreading straw, burning spread straw, and burning straw in rows, on invertebrate animal populations in the soil underneath have been studied (Edwards, 1977; Edwards and Lofty, 1979). The surface-living soil fauna, notably springtails, mites and millipedes, were all drastically reduced by burning the straw, but deeper-living small arthropods and earthworms were unaffected. However, repeated straw burning over 4 years decreased the earthworm population because it diminished the organic matter in the soil on which they fed. Burning did not appear to have any direct effects on the soil invertebrate animals; eliminating the superficial organic matter by burning may have indirect effects.

Irrigation, drainage, and *soil aeration* all presumably have some influence on the soil fauna, although these effects may be temporary, allowing the survival of the most resistant species. Very little information is available on the effects of such agricultural operations on soil animals. Presumably, they will introduce new species and encourage the multiplication of permanent soil fauna by creating favourable microhabits for them.

Retrospect and prospect

Figure 5.7 summarizes the overall influence of agricultural practice on the population of soil invertebrate animals. On balance, the impact of human

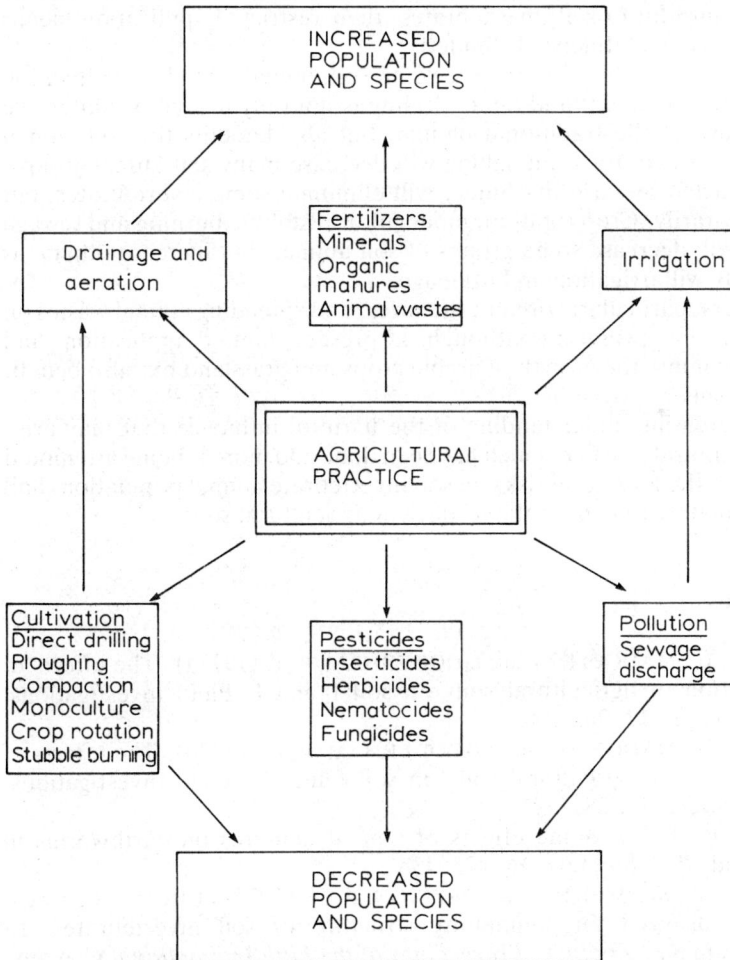

Figure 5.7 Influence of agricultural practice on soil invertebrate animals (after Edwards and Lofty, 1969)

activities on them appears to be more deleterious than beneficial. However, many of these adverse effects are minimal and sometimes transitory, so that soil fertility is unlikely to become seriously altered by the impact of these effects on the animals.

Insecticides generally do not usually have drastic overall effects on soil invertebrate communities. The older, persistent insecticides have been adversely criticized recently, yet some of them are still of considerable benefit to mankind. The newer insecticides are less persistent, more specific and efficient, have minimal side-effects, and their risks as environmental pollutants are small compared with the benefits that they offer. Nevertheless there is little room for complacency. Because some insecticides are particularly dangerous chemicals they must be used prudently, with stricter regulatory controls, and long-term assessment.

Herbicides are widely used but there is little evidence that they markedly influence the soil animal population. Although nematocides and fungicides are often harmful to soil invertebrates, their restricted application means that their adverse influence is limited.

Crop cultivation has diverse effects on soil animals. Modern cultivation by herbicide treatment and direct drilling is not only less labour-intensive than the use of the traditional plough, but also benefits the soil animal community. In contrast, ploughing will decrease many soil faunal groups. Soil compacted by farm machinery will eliminate some invertebrates, but only temporarily. Crop rotation, monoculture, stubble burning and sewage effluents will decrease some groups of soil animals but increase others, as presumably will irrigation and drainage.

Fertilizers, particularly organic manures, and spreading animal wastes on farm land, offer promising although, at present, limited applications and not only promote the growth of arable crops and grassland but also benefit soil organisms.

With increasing understanding of the harmful influence that may arise from agricultural practice, much has been done and more is being attempted to minimize the long-term risks on soil invertebrate animal populations and hence ultimately to benefit the fertility of agricultural soils.

References

ARITAJAT, U., MADGE, D. S. and GOODERHAM, P. T. (1977a). The effects of compaction of agricultural soils on soil fauna I. Field investigations. *Pedobiologia*, **17**, 262–282

ARITAJAT, U., MADGE, D. S. and GOODERHAM, P. T. (1977b). The effects of compaction of agricultural soils on soil fauna II. Field investigations. *Pedobiologia*, **17**, 283–291

CURRY, J. P. (1976). Some effects of animal manures on earthworms in grassland. *Pedobiologia*, **16**, 425–438

DINDAL, D. F., SCHWERT, D. and NORTON, R. A. (1975). Effects of sewage effluent disposal on community structure of soil invertebrates. In *Progress in Soil Zoology. Proceedings of the Fifth International Congress of Soil Zoology, Prague, 1973*, pp. 419–427. Ed. J. Vanek. Amsterdam; Junk

EDWARDS, C. A. (1969). Soil pollutants and soil animals. *Scient. Am.*, **220**, 88–99

EDWARDS, C. A. (1970). Effects of herbicides on the soil fauna. In *Proceedings of the Tenth Weed Control Conference, Brighton*, pp. 1052–1062

EDWARDS, C. A. (1976). *Persistent Pesticides in the Environment* (2nd edition), p. 170. Ohio, USA; CRC Press

EDWARDS, C. A. (1977). Investigations into the influence of agricultural practice on soil invertebrates. *Ann. appl. Biol.*, **87**, 515–524

EDWARDS, C. A. (1978). OILB/WPRS Report on integrated control of soil pest Working Group. *Pedobiologia*, **18**, 228–230

EDWARDS, C. A. (1979). Soil biology and land use. *Nature, Lond.*, **281**, 339–340

EDWARDS, C. A. and LOFTY, J. R. (1969). The influence of agricultural practice on soil microarthropod populations. In *The Soil Ecosystem*, pp. 237–247. Ed. J. G. Sheals. Systematics Association Publication No. 8

EDWARDS, C. A. and LOFTY, J. R. (1971). Nematicides and the soil fauna. In *Proceedings of the Sixth British Insecticides and Fungicides Conference, Brighton*, pp. 158–166

EDWARDS, C. A. and LOFTY, J. R. (1975a). The invertebrate fauna of the Park Grass plots I. Soil fauna. In *Report of the Rothamsted Experimental Station for 1974, part 2*, pp. 133–154

EDWARDS, C. A. and LOFTY, J. R. (1975b). The influence of cultivation on soil animal populations. In *Progress in Soil Zoology. Proceedings of the Fifth International Congress of Soil Zoology, Prague, 1973*, pp. 399–407. Ed. J. Vanek. Amsterdam; Junk

EDWARDS, C. A. and LOFTY, J. R. (1977). *Biology of Earthworms* (2nd edition), p. 333. London; Chapman & Hall

EDWARDS, C. A. and LOFTY, J. R. (1979). Direct drilling and earthworms. In *Report of the Rothamsted Experimental Station for 1979, part 1*, pp. 89–90

EDWARDS, C. A. and STAFFORD, C. J. (1979). Interactions between herbicides and the soil fauna. *Ann. appl. Biol.*, **91**, 125–146

EDWARDS, C. A., LOFTY, J. R., FRENCH, M., BRIGGS, G. G. and BROWN, R. (1977). Pesticides and the soil fauna. In *Report of the Rothamsted Experimental Station for 1977, part 1*, p. 99

FRYER, J. S. (1977). Recent developments in the agricultural use of herbicides in relation to ecological effects. In *Ecological Effects of Pesticides, Linnean Society Symposium Series No. 5*, pp. 27–45. Eds G. F. H. Perring and H. Mellanby. London; Academic Press

HARRIS, C R., THOMPSON, A. R. and TU, C. M. (1972). Insecticides and the soil environment. *Proc. Ent. Soc. Ontario*, **102**, 156–168

KORNBERG, SIR HANS (1979). *Royal Commission on Environmental Pollution. Seventh Report: Agriculture and Pollution.* Cmnd 7644. London; HMSO

MADGE, D. S. (1966). How leaf litter disappears. *New Scientist*, **32**, 113–115

MORRIS, H. M. (1922). The insect and other invertebrate fauna of arable land at Rothamsted. *Ann. appl. Biol.*, **9**, 282–305

RAW, F. (1962). Studies on earthworm populations in orchards. *Ann. appl. Biol.*, **50**, 389–403

SCHALLER, F. (1968). *Soil Animals.* USA; University of Michigan Press

THOMPSON, A. R. and EDWARDS, E. A. (1974). Effects of pesticides on non-target invertebrates in freshwater and soil. In *Pesticides in Soil and Water*, pp. 341–386. Ed. W. D. Guenzi. Madison, USA; Soil Science Society of America

VON RHEE, J. A. (1977). Effects of soil pollution on earthworms. *Pedobiologia*, **17**, 201–208

WALLWORK, J. A. (1970). *Ecology of Soil Animals*. London; McGraw-Hill

WALLWORK, J. A. (1976). *The Distribution and Diversity of Soil Fauna*. London; Academic Press

WRIGHT, M. A. (1979). Effects of benomyl and some other systemic fungicides on earthworms. *Ann. appl. Biol.*, **87**, 520–524

6

THE COMPOSTING OF AGRICULTURAL WASTES

K. R. GRAY and A. J. BIDDLESTONE
Compost Studies Group, Department of Chemical Engineering, University of Birmingham

Introduction

Composting is the decomposition of organic wastes by a mixed microbial population in a warm moist aerobic environment. The wastes are gathered into a heap in order to conserve heat, thereby raising their temperature and accelerating the basic degradation process which normally occurs slowly on the surface of the ground.

The microbiological and biochemical principles of composting, the process engineering factors involved, and the application of the process to the treatment of municipal wastes are reviewed extensively by the authors elsewhere (Gray and Biddlestone, 1971a, b; 1973; 1974). In this chapter the current practices of composting in agriculture and horticulture are examined.

The need for composting

Composting has advantages over the natural decomposition process which will often justify its use on the farm.

1. The final weight of compost is less than half that of the original wastes; the volume reduction is even greater. This can greatly reduce costs of transport and spreading in the field.
2. At composting temperatures of over 60 °C there is an effective kill of pathogenic organisms, weeds and seeds.
3. Reasonably mature compost quickly comes into equilibrium with the soil, whereas raw organic wastes can cause a period of major disruption to soil processes.
4. Compost heaps are free-standing, whereas manure slurries require tanks or pits for storage.
5. During composting, a number of wastes from several sources, both animal and plant, can be blended together.

Principles of composting

A simplified version of the overall process flowsheet for composting is shown in *Figure 6.1*. This illustrates the interactions between the organic

99

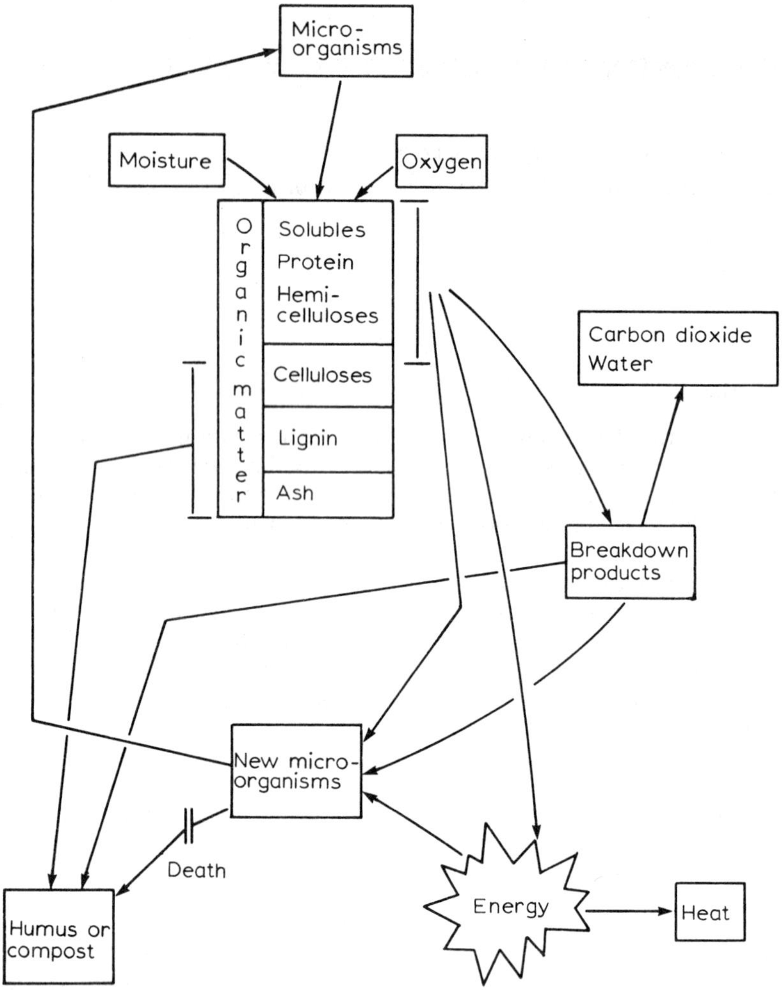

Figure 6.1 The composting process

waste, microorganisms, moisture and oxygen. The organic waste, comprising several fractions of components, normally has an indigenous population of microorganisms derived from the atmosphere, water or soil. Once the moisture content of the waste is brought to an appropriate level and the mass aerated, microbial metabolism speeds up. Energy is obtained by biological oxidation of part of the carbon; some of this energy is used in metabolism, the rest is given off as heat. The end-product, compost, is made up of the more resistant residues of the organic matter, breakdown products, dead and some living microorganisms, together with products from further chemical reaction between these materials.

Some idea of the types and numbers of organisms involved in composting is given in *Table 6.1*. Microorganisms are present throughout the process, whereas macroorganisms invade the composting mass towards the end, when the temperature approaches the ambient.

Table 6.1 ORGANISMS INVOLVED IN COMPOSTING

Organism type		Numbers per g of moist compost
Microflora	Bacteria	10^8–10^9
	Actinomycetes	10^5–10^8
	Fungi	10^4–10^6
	Algae	10^4
	Viruses	
Microfauna	Protozoa	10^4–10^5
Macroflora	Fungi	
Macrofauna	Mites, ants, termites, millipedes, centipedes, spiders, beetles, worms	

The temperature and pH changes in most composting masses follow a pattern similar to that shown in *Figure 6.2*; this may be divided into four stages—mesophilic, thermophilic, cooling-down and maturing. Initially, the mesophilic strains of microorganisms, which are present on the organic waste or in the atmosphere, start to decompose the materials; heat is given off and the temperature rises. The pH falls as organic acids are produced. Above 40 °C the thermophilic strains take over, and the temperature rises to 60 °C where the fungi become deactivated. Above this temperature the

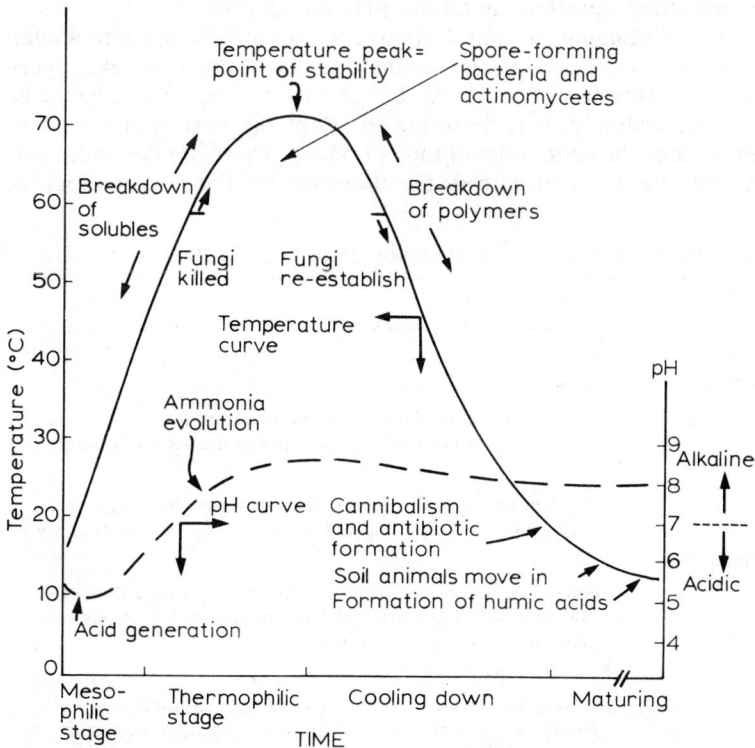

Figure 6.2 Temperature and pH variations in a compost heap

reaction is kept going by the actinomycetes and spore-forming bacteria. In this high-temperature phase the more readily degradable substances such as sugars, starches, fats and proteins, are rapidly consumed; the pH becomes alkaline as ammonia is liberated from the proteins. The reaction rate decreases as the more resistant materials are attacked; the heap then enters the cooling-down phase.

As the temperature falls, the thermophilic fungi reinvade the heap from the cooler extremities and start to attack the cellulose. Later, the mesophilic strains of microorganisms reinvade. These first three stages of composting—mesophilic, thermophilic and cooling-down—occur fairly rapidly, being completed within a few weeks. The final stage, maturing, requires several months; reactions occur in the residual organic matter to produce the stable product of humus or humic acids. Little heat is generated and the final pH is normally slightly alkaline. During this period there is intense competition for food between the microorganisms: antagonism and antibiotic formation occurs and the heap is invaded by macrofauna which contribute to breakdown by physical maceration of particles.

The composting of organic wastes is a dynamic and extremely complicated ecological process in which temperature, pH and food availability are constantly changing. In consequence, the numbers and species of organisms responsible also change markedly. The rate of progress towards the end product, humus, is dependent on a number of interrelated environmental factors. These include homogeneity of the organic wastes, particle size, nutrients, moisture, aeration, agitation, pH and heap size.

Nature is very obliging, so that if organic wastes are thrown into a heap they will be turned eventually into humus. However, this may take a very long time and offensive odours may be given off because of anaerobic conditions. Accordingly, it is desirable to adopt the best operating conditions allowed by the economics of the operation. There are now sufficient experimentally determined data in the literature on the microbiological,

Table 6.2   OPTIMUM VALUES OF MAJOR COMPOSTING CHARACTERISTICS

Characteristic	Value
C : N ratio of feed	30–35 : 1
C : P ratio of feed	75–150 : 1
Particle size	15–35 mm for agitated plants and forced aeration 35–75 mm for windrows, unagitated plants and natural aeration
Moisture content	50–60%
Air flow	0.6–1.8 m³air/day solids volatile during thermophilic stage being progressively decreased during cooling down and maturing
Maximum temperature	55 °C
Agitation	Short periods of vigorous agitation, alternating with periods of no agitation which vary in length from minutes in the thermophilic stage to hours during maturing
pH control	Normally none desirable
Heap size	Any length but not over 1.5 m high or 2.5 m wide for heaps and windrows using natural aeration. With forced aeration, heap size depends on need to avoid overheating

physical and chemical characteristics of composting for reasonably accurate process design of composting plants. Optimum values of the important characteristics are summarized in *Table 6.2*.

The practice of farm composting

Composting of organic waste materials has been practised for many centuries by farmers and gardeners in many parts of the world. Probably the outstanding example has been that of the Chinese in the river deltas who, by returning to the soil their crop residues, animal manures, human wastes and alluvial mud swept down the rivers and canals, have been able to support very high population densities for many hundreds of years. Composting, as practised by the Chinese, has probably changed very little over the centuries, being essentially a small-scale batch operation in heaps.

In agriculture/horticulture, composting is practised in a number of varying situations. These allow quite different inputs of capital investment, running costs and labour and hence require different processing arrangements. Because of the fairly low value of composts, sophisticated plants cannot be afforded and process conditions frequently fall far short of the optimum levels listed in *Table 6.2*.

The major composting situations are:

1. The garden, smallholding or nursery
2. The mixed farm
3. The completely arable farm
4. Preparation of substrates for mushroom growing
5. Treatment of manure slurries in the liquid or solid phase
6. Composting in rural development in the tropics.

SUITABLE MATERIALS FOR COMPOSTING

Most waste organic materials produced in agriculture are suitable for compost production. These are mainly animal manures and crop residues such as cereal straws. The carbon/nitrogen ratio (C/N) of manures is below the optimum: this leads to excessive ammonia loss during composting. Hence manures are best mixed with materials having a high C/N ratio, such as straws or wood wastes. *Table 6.3* gives the approximate C/N ratios of some compostable materials.

There are several other useful organic wastes but, because of the high cost of transport, their value to the farmer depends on the proximity of their sources to the composting site. Examples are allotment/smallholding/nursery wastes, apple residues from cider making, bracken, feathers and hair, food-processing wastes, grass and lawn mowings, greengrocer's and vegetable market wastes, hedge clippings, hop wastes, leaves, nettles and weeds, paper, sawdust and wood shavings, seaweed, sewage sludge from country areas, waterweed from river clearance, and wood ash. Virtually any material which has once lived will compost, but crockery, glass, metal, plastics and man-made fibres should not be used.

Table 6.3   APPROXIMATE COMPOSITION OF MATERIALS SUITABLE FOR COMPOSTING

Material	Nitrogen % dry-weight basis	C/N ratio
Urine	15–18	0.8
Dried blood	10–14	3
Hoof and horn meal	12	
Night soil, dung, sewage sludge	5.5–6.5	8
Grass	4	20
Bone meal	4	8
Brewers wastes	3–5	15
Farmyard manure	2.2	14
Water hyacinths	2.2	20
Millet, pigeon pea stalks	0.7	70
Wheat, barley, rice straws	0.4–0.6	80–100
Coconut fibre waste	0.5	300
Fallen leaves	0.4	45
Sugar-cane trash	0.3	150
Rotted sawdust	0.2	200
Fresh sawdust	0.1	500
Paper	nil	infinity

Where it can be obtained cheaply, a little soil is a very useful component of a compost heap; clay is far preferable to sand. Clay has a high base exchange capacity and will help to hold any liberated ammnonia within the heap until the microorganisms can immobilize it; nitrogen losses are thereby much reduced. The soil will also be extensively 'weathered' in the heap; the availability of its nitrogen, phosphorus and potassium and the minor/trace elements will thereby be improved.

Composting procedures in various situations

THE GARDEN, SMALLHOLDING OR NURSERY

In this situation, plant materials—crop residues and weeds—are available for composting from spring to autumn. Some manure may be available occasionally. Composting is generally done in a series of batch heaps through the year, mainly with little mechanical help and a high input of labour per tonne of product.

Where the quantity of material being handled in a batch is small, less than 1 tonne, the major problem is that of high heat loss because of the large surface : volume ratio. As recommended by Gray and Biddlestone (1976, 1978) the essentials are to compost within a walled enclosure with some degree of insulating value, to cover the wastes with a porous but insulating top blanket which sinks with the wastes, to enable air to get underneath the heap and percolate upwards by the 'chimney effect', and to prevent rain falling on the wastes, thereby causing cooling and leaching of nutrients. Another recommendation is to save up wastes and build up heaps to a full height of 1–1.25 m in one operation, rather than in thin layers at perhaps weekly intervals. By construction in one operation, much higher temperatures are achieved and a greater degree of weed-seed kill and pasteurization results. As far as possible, the wastes should be premixed thoroughly before assembling the heap.

Where temperatures of over 60 °C are achieved and the heat permeates to the edges of the heap because of good side-wall insulation, then turning the heap is not necessary. In 2–3 weeks the wastes should have broken down to a good state of maturity.

Compost bins up to 0.5 tonne capacity, in plastic, wood or metal, are available on the market. Although often rather ineffective in the past, their design is being improved with greater scientific understanding of their operation.

THE MIXED FARM

This case probably covers most of the farms currently practising biological husbandry. The organic waste handled is mainly farmyard manure (FYM), cleared out during the summer from strawed yards used for overwintering cattle. Bedding from other livestock—pigs, poultry and sheep—may occasionally be handled. In this situation, labour for composting is likely to be short and capital investment may cover only a front-end loader with manure fork and a flat-bed muckspreader.

The simplest procedure is to use a front-end loader with muck fork to lay out the wastes in a rough windrow heap about 2.5 m wide and 1.5 m high. This is not an ideal method: the muck fork does not easily open out heavily compacted manure, a neat finish to the pile is not obtained, and adequate air cannot easily diffuse to the centre bottom of such a windrow.

A better procedure is used on one of the largest organic farms in the UK (Mayall, personal communication). The FYM is loaded into a flat-bed muckspreader using a front-end loader and manure fork. The stationary muckspreader is then put to work, shredding the wastes, aerating them and throwing them out into a pile about 1.25 m high which gradually backs up against the machine; the latter is then drawn forward about 1.5 m. A series of vertical holes to ground level and 1 m apart are next made through the mass with a 75 mm diameter stake. Another section of heap is then made and the aeration holes put in. Where possible, green material is combined with the FYM, being added to the load in the muckspreader which acts as a blender. If well made, the heap warms to about 70 °C within a week, then cools to 30 °C after a month, when it is turned with a front-end loader. After 2–3 further months the compost is ready. The main problem with this approach is that it requires two tractors to lay out the windrow and uses quite a lot of manpower. As the heap rapidly sinks to less than 1 m in height, a considerable land area is involved per tonne of material handled.

In the USA the current approach appears to involve laying out a windrow of FYM about 2 m wide by 1.25 m high and dragging through it a compost turner several times a month (Demmel, 1980). One such turner is mounted on a tractor (74 600 W) and processes half the heap moving down on one flank and the remainder while returning on the other flank. It is a method which gives a good product but consumes a lot of energy and space; it is completely reliant on firm ground conditions. It has a high work-rate, handling 400–500 t/h.

A more appropriate technique is that of the Moving Windrow, first advocated and practised in Germany for municipal organic wastes by Spohn (1980) and now being adopted by farmers in Europe (Vogtmann, personal communication). This employs a flat-bed muckspreader with side-belt elevator which can lay out a windrow up to 2 m high alongside the moving machine. In practice, a small triangular-shaped windrow is laid out first; because of its dimensions this is readily permeated by adequate air for good composting. After 3–4 days, when the bulk of the high-rate oxygen demand has been met, a layer of new material is added to one side. Every few days, further layers are added to the side, with the original triangle now widening to a trapezoid shape. The heap does not need subsequent turning. The original machine had the hood, cross-auger and elevator mounted at the rear; it was loaded over the side by a muck fork. A recent modification has the elevator etc. mounted at the front, so that it can be loaded from the rear. In the UK the authors are using a muckspreader with a rotary flinging wheel to achieve the desired elevation of the wastes. In damp climates such as the UK it is highly desirable to use a concreted base for the composting area; to minimize the area and cost of concreting, attempts are being made to use windrow heaps much higher than the 1.3 m advocated by Spohn.

THE COMPLETELY ARABLE FARM

There are many UK farms in this category, particularly in the drier, warmer areas of the country. Considerable organic plant wastes are often generated—straws from cereals, rape and peas, plus other vegetable crop residues. However, these are seriously deficient in nitrogen: hence manures, manure slurries or sewage sludge need to be brought on to the farm from outside and blended with the wastes before composting. There is little sign at present of this being done.

THE PREPARATION OF SUBSTRATES FOR MUSHROOM GROWING

A number of firms manufacture substrates for mushroom growing, both for themselves and for sale to smaller enterprises. This is done by blending straws, manures and other additives and composting for about 10 days, including 2–3 turns. It is important to turn out a consistent product, so the process is carefully controlled. Sophisticated and expensive mobile machines are used for blending, stacking the windrows and turning, while the composting yard usually has a concreted base: capital expenditure is therefore high.

THE TREATMENT OF MANURE SLURRIES

Modern methods of pig and poultry husbandry, and the overwintering indoors of dairy cattle, lead to the production of manure in the form of a slurry. When spread in certain situations this material can give rise to serious pollution, by smell, by getting into water-courses and by disrupting the flora and fauna in the soil. No really effective treatment method, of wide acceptability, has been discovered yet.

Composting of such slurry can be done in either the liquid or the solid phase. With its low carbon : nitrogen ratio of about 6, considerable losses of nitrogen as ammonia take place on composting; these losses can be greatly reduced by adding extra carbonaceous material such as straw.

Liquid composting is starting to be practised in Europe. The slurry is run into an insulated tank and aerated by either a surface aerator which stirs in air, or a hollow shaft impeller which sucks in air and distributes it as fine bubbles through the slurry. The temperature rises to about 25–55 °C. The original Licom method aimed to produce a treated sludge, plus clean water of virtually drinkable quality (Grant, 1975; Terwelleger and Crauer, 1975). However, to reach this objective required at least two stages of aeration and two of sedimentation; these involved high capital expenditure, and therefore only very large livestock units could afford to instal the process. Recent emphasis has centred on using a single stage of aeration, mainly with the object of smell removal; the slurry often has chopped straw added to achieve a C/N ratio of 12, in order to reduce ammonia loss. A high degree of pathogen control and weed seed destruction is realized, even at temperatures as low as 25 °C. Foam is generated but is kept under control using a foam cutter on the agitator shaft (Vogtmann, personal communication).

In the ARCUB process of solid-phase treatment, which is being developed by the authors, the slurry is mixed with straw fluffed out from bales. The mixture is then composted under forced aeration, and drainings are recycled over the mass. The process has been tried out on a small scale in a unit handling the effluent from 60 pigs. Scale-up to a 1000-pig (100-cow) sized unit is progressing. The work has shown the need for intimate mixing of the slurry and straw; much settlement of the slurry–straw mixture takes place within 24 hours of laying out a windrow (Gray and Biddlestone, 1975).

COMPOSTING IN RURAL DEVELOPMENT IN THE TROPICS

In many developing countries with hot climates, the implementation of sound compost-making techniques is likely to prove a major factor in the improvement of food production. In these countries the average conditions differ greatly from those in the UK.

1. The hot climate rapidly oxidizes humus from the soil, and drought conditions are often present.
2. Chemical fertilizers are very expensive and often unobtainable.
3. Labour is very cheap but there is insufficient finance to set up and operate complicated machines.
4. Manures tend to be used as fuel and fresh green material as animal fodder, leaving mainly cellulose and lignin materials for composting.

In these circumstances the improved Indore process of Sir Albert Howard (1943, 1945) is still remarkably pertinent. In this technique the wastes are composted in pits 9.2 m by 4.3 m by 1 m deep. Vegetable wastes are put on in 150 mm layers, followed by manure in 50 mm layers and a sprinkling of earth, wood ashes and water. Layering is continued to a height of 1.2 m and vertical aeration vents are made. The heap is turned by hand two or three times and further water is added as necessary. The method has recently been used with marked success by Dalzell at the Medak Agricultural Centre, India, (Dalzell, Gray and Biddlestone, 1979). In a recent study tour of organic waste recycling in China, several techniques of composting were encountered (FAO, 1977). Human and animal wastes were layered into a heap with chopped plant stalks, similar to the Indore process. Horizontal and vertical bamboo poles, 100 mm in diameter, were inserted to act as aeration vents, then the whole heap was plastered with mud, about 30 mm thick. The bamboos were removed after 24 h leaving the holes for aeration. After 4–5 d the temperature reached 60–70 °C and the aeration holes were sealed with mud. The heap was turned after 2 weeks and the compost was ready in 2 months. In another method, straw, manures and green crop wastes were composted with a large excess of river silt under anaerobic conditions. Three turnings were made and the compost was ready in 3 months. Composting under such anaerobic conditions was claimed to reduce nitrogen losses to a minimum.

Reduction of nitrogen losses

A drawback to composting as a waste-treatment method is the loss of nitrogen, in the form of ammonia. This occurs in the early stages of composting as urea and the various proteins are broken down, and generally before the cellulose is degraded to sugars. In its most extreme, in the liquid composting of manure slurries of low C/N ratio at temperatures of over 60 °C, some 70 per cent of the original nitrogen can be lost. Now that chemical fertilizer nitrogen is expensive, such losses are serious.

On the other hand, high temperatures of over 55 °C for a few days are very desirable to destroy weeds and pathogenic organisms, especially where sewage sludge or night soil are being composted. When composting under completely anaerobic conditions, such temperatures are not achieved and a satisfactory kill cannot be assured. However, under anaerobic conditions little nitrogen is lost, the ammonia being held as ammonium carbonate in the CO_2-rich atmosphere.

Ammonia loss in composting is reduced by lower temperatures, adequate moisture, high C/N ratio, reducing air flow to the minimum, addition of clay soil and the addition of phosphatic material such as rock phosphate or superphosphate. It also appears that operating the heap close to anaerobic conditions will help. Some practitioners of garden composting consolidate their heaps by treading them down firmly to restrict air flow. The Chinese technique mentioned earlier is a good compromise, covering the heap with a skin of mud, allowing it to warm aerobically to 60–70 °C for several days to achieve pathogen kill, and then blocking the aeration vents to bring about anaerobic conditions for the rest of the fermentation. The large quantities of river silt and much hand labour available to the Chinese enable them to use such a technique, which would not be at all appropriate on a larger scale in

Western agriculture. Suitable alternative techniques will be necessary in the next few years if composting is to be adopted on a large scale in the UK.

Heat recovery

An actively composting mass generates considerable heat which is normally dissipated by radiation/convection from the heap surface, by warming up the air stream and by evaporating moisture. It would be particularly helpful to the large-scale adoption of composting if a significant fraction of the generated heat could be recovered. The major problems are that the maximum temperature achievable is only 60–70 °C, that the whole mass sinks considerably and probably unevenly, and that heat-transfer surfaces could affect the unloading of the compost heap by mechanical means. A few examples of heat recovery are now starting to appear.

Types of compost and their use

Composts are made from plant remains and animal manures, as described in this chapter. The Chinese also incorporate ashes from wood and coal fires, plus quantities of silt from rivers and canals (FAO, 1977). Composts from municipal wastes, refuse or refuse plus sewage sludge, are being made in mechanized plants, especially in the cities of the Middle and Far East (Gray and Biddlestone, 1973; 1974; 1980). The Beltsville process of composting dewatered sewage sludge using wood chips as a recyclable matrix is slowly being adopted by communities in the USA (Epstein *et al.*, 1976). With the increasing cost of chemical fertilizers and the need for better public health, the future must undoubtedly see a major increase in the composting of sewage sludge and night-soil before recycling these materials to agriculture.

Table 6.4 CHEMICAL COMPOSITION RANGES OF MATURED COMPOSTS

Substance	Weight % on dry basis		
Organic matter	25	–	80
Carbon	8	–	50
Nitrogen (as N)	0.4	–	3.5
Phosphorus (as P_2O_5)	0.3	–	3.5
Potassium (as K_2O)	0.5	–	1.8
Calcium (as CaO)	7.0	–	1.5
Ash	65	–	20
	← Municipal composts		
	Farm/garden composts →		

Compost is primarily a soil conditioner and, to some extent, a fertilizer. *Table 6.4* gives an indication of the composition range of composts: farm/garden composts are much higher in organic matter and the major plant nutrients nitrogen, phosphorus and potassium, while those made from municipal wastes contain more ash, calcium and trace metals.

When compost is added to the land it breaks down, releasing the major plant nutrients nitrogen, phosphorus and potassium plus minor and trace elements. It is attractive to soil fauna. The gummy constituents and fungal/actinomycete mycelia help to bind the soil particles into crumbs, while its organic components increase the water-holding capacity of the soil. These factors greatly increase the stability of the soil to wind and water erosion. One of the major uses for municipal compost in Europe is in the stabilization of steep vineyard slopes in France and Germany (Satriana, 1974).

The degree of maturity of compost can, with advantage, be chosen to suit soil conditions and the crop to be grown. During composting, some gummy materials are exuded which disappear as the material reaches maturity. Hence, on sandy soils where the formation of crumb structure is required, compost should be applied in a fairly new state, leaving the earthworms to complete the humification process and form the soil crumbs. However, this approach is possible only for well-established plants, shrubs and trees. Immature compost gives off ammonia which can badly affect tiny seedlings. Hence, where small seeds are to be planted, well-matured compost at least 3 months old should be used. According to Vogtmann (personal communication) maize, potatoes and rape require compost which is just coming to maturity at 2 months, while grassland prefers old compost.

The Chinese claim to match compost to the soil type: composts prepared from ashes are used on sandy soils, ones from sheep or cattle manures are applied to black soils, and ones based on pig manures are used on brown soils (FAO, 1977).

Ideally, the rate of compost application should be matched to plant nutrient uptake and soil structure requirement. The Chinese make large quantities of compost and apply it at about 75 t/ha with variations according to the crop. In the UK, organic farmers appear to apply up to 50 t/ha to the leys in a ley–arable rotation. When calculating compost application rates to meet plant nutrient needs it should be noted that composts vary widely in composition, depending upon the raw materials used in their preparation. Additionally, nutrients are released more slowly from composts than from very soluble inorganic fertilizers; consequently the effects of compost can last for more than one season. Dalzell, Gray and Biddlestone (1979) suggest that the percentage of the major nutrients which become available in the year of application are: nitrogen 25 per cent; phosphorus 100 per cent; potassium 80 per cent.

Reasonably mature compost is a much more friable and easily handled material than the raw organic wastes. Nevertheless, it is still a bulky substance which, because of its moisture content and texture, is not readily amenable to pneumatic conveyance. Hence, the use of composts at dressing rates of about 50 t/ha does present quite an organizational problem in transport and spreading. There are several manure spreaders of substantial capacity which will reduce the number of moves between compost stack and field to a reasonable level.

Compost should be incorporated into the top 100 mm of the soil by shallow ploughing or harrowing. If ploughed in deep, much of the nutrient and soil-conditioning value is wasted. If left on the surface, nitrogen losses will occur. In tropical climates with high soil temperatures, organic matter is very rapidly oxidized on the surface, so it is vital to incorporate the compost into the topsoil close to the start of the growing season and to make annual applications of at least 15 t/ha if possible.

References

DALZELL, H. W., GRAY, K. R. and BIDDLESTONE, A. J. (1979). *Composting in Tropical Agriculture.* Stowmarket, Suffolk, England; The International Institute of Biological Husbandry

DEMMEL, D. (1980). Farm-scale composting. *Compost Science/Land Utilization,* **21,** (1) 42–44

EPSTEIN, E., WILLSON, G. B., BURGE, W. D., MULLEN, D. C. and ENKIRI, N. K. (1976). A forced aeration system for composting wastewater sludge. *J. Water Polln. Control Fed.,* **48,** (4) 688–694

FAO (1977). China: recycling of organic wastes in agriculture. *FAO Soils Bulletin No. 40.* Rome; Food and Agriculture Organization of the United Nations

GRANT, F. A. (1975). Liquid composting of dairy manure. In *Proceedings of 3rd International Symposium on Livestock Wastes,* pp. 497–500, 505. St. Joseph, Michigan; American Society of Agricultural Engineers

GRAY, K. R. and BIDDLESTONE, A. J. (1971a). Review of composting: Part 1. *Process Biochem.,* **6,** (6) 32–36

GRAY, K. R. and BIDDLESTONE, A. J. (1971b). Review of composting: Part 2. *Process Biochem.,* **6,** (10) 22–28

GRAY, K. R. and BIDDLESTONE, A. J. (1973). Review of composting: Part 3. *Process Biochem.,* **8,** (10) 11–15, 30

GRAY, K. R. and BIDDLESTONE, A. J. (1974). Decomposition of urban waste. In *Biology of Plant Litter Decomposition,* Vol. 2, ch. 24. Eds C. H. Dickinson and G. J. F. Pugh. London; Academic Press

GRAY, K. R. and BIDDLESTONE, A. J. (1975). New slurry composting process. *Farm Buildings Digest,* **10,** (3) 5–6

GRAY, K. R. and BIDDLESTONE, A. J. (1976). The garden compost heap. *Jl. R. Hort. Soc.,* Part I, **101,** 54–544; Part II, **101,** 594–598

GRAY, K. R. and BIDDLESTONE, A. J. (1978). *Garden Compost,* 3rd edn. Stowmarket; The Soil Association

GRAY, K. R. and BIDDLESTONE, A. J. (1980). Agricultural use of composted town refuse. Paper 21 in *Inorganic Pollution and Agriculture,* pp. 279–305. Ministry of Agriculture, Fisheries and Food, Reference Book 326. London; HMSO

HOWARD, SIR ALBERT (1943). *An Agricultural Testament.* London; Oxford University Press

HOWARD, SIR ALBERT (1945). *Farming and Gardening for Health and Disease.* London; Faber and Faber

SATRIANA, M. J. (1974). *Large Scale Composting,* p. 33. London; Noyes Data Corporation

SPOHN, E. (1980). Experiences in converting sludge into humus. *Compost Science/Land Utilization,* **21,** (1) 50–51

TERWELLEGER, A. R. and CRAUER, L. S. (1975). Liquid composting applied to agricultural wastes. In *Proceedings of 3rd International Symposium on Livestock Wastes,* pp. 501–505. St. Joseph, Michigan; American Society of Agricultural Engineers

II

AGRICULTURAL METHODS

Introduction

There are many books and booklets on organic farming and horticulture, and there are many theories and techniques that biological farmers agree among themselves and practise on their farms. However, there is as yet no well-argued, authoritative textbook on methods of biological husbandry, and the subject is not taught as such in agricultural schools and colleges.

Twenty years ago it would have been impossible to consider such a textbook. There was not enough critical research in progress to support many of the claims of the organic husbandry movement, and little to recommend organic theory (although much to recommend organic practice) for a place in overcrowded curricula of agricultural teaching.

Today the situation is changing rapidly. The scientific background of organic farming is being examined on experimental farms and laboratories all over the world; theory and practice alike are being tested, and biological husbandry is developing from the results. The seven Chapters in this Section all examine some aspect of biological farming methods, investigating and developing their scientific background and examining them critically.

In Chapter 7 Dr S. G. Lisansky takes up the question of biological methods of pest control within modern monoculture systems of crop production: he draws attention to many effective biological agents already known and in use on man's behalf, and the possibility that others await discovery and exploitation as alternatives to chemical control agents. Mr J. F. Newman examines the question of how far the use of agricultural chemicals may be linked with biological husbandry in integrated systems of farming. Mr N. W. Martin and Mr J. Keable are concerned with energy saving under various systems of biological farming, examining methane production from slurry, savings involved in direct drilling, recycling of urban wastes and biological nitrogen fixation. Dr L. G. Plaskett examines quantitatively some of the ways in which existing or future developments in biotechnology may be adapted to serve the needs of biological husbandry. Dr J. Coombs explores the limiting factors that affect production of biologically orientated farming, and indicates some possible remedies. Dr J. V. Lovett and Miss J. Levitt describe their investigation of those plant-produced chemicals that inhibit the growth of neighbouring plants or give protection against parasites and pathogens. Their paper explores possibilities of exploiting this phenomenon—allelopathy—on an agricultural scale. Finally, Mr J. L. H. Chase discusses uses of seaweeds in both agriculture and stock-raising.

115

7

BIOLOGICAL PEST CONTROL

S. G. LISANSKY
Tate & Lyle Ltd., Group Research and Development, Reading, Berks, UK

Introduction

Natural agriculture is a contradiction in terms. Plants were not intended to grow in closely packed groups of identical species, neatly laid out in rows over fields extending for acres or square miles. When man constructs these highly artificial conditions by practising agriculture to produce food, shelter and clothing, he creates environments where pests and diseases can flourish. Once they are adapted to a particular crop, weeds, insects, diseases and small animals are able to multiply and spread rapidly in all directions, unchecked by the natural barriers formed by other plant species in a mixed growth. In these unnatural conditions, pests will not establish a balance with a crop but will totally destroy it. Pest control becomes absolutely essential once the basic premise of agriculture is established.

In the past, the most obvious pests could be partly controlled. Ploughing and subsequent tilling could keep weeds from competing too successfully with the crop. Scarecrows, nets, guns, traps and many other devices were used to keep birds and animals from competing too successfully with man for the end-product. Obvious pests were partly controllable because their mode of attack was clear and their life cycles understood to a certain extent. The degree of control that one could achieve was directly correlated with the amount of labour used. As long as the need for the crop, i.e. its value, was greater than any alternative value obtainable from labour, this system could work effectively to control pests. Current reports from China of very high labour inputs for pest control in agriculture are a modern example of this practice. In medieval Europe after the Black Death, labour was in short supply; its value rose and with it the price of food. Certainly this system, using a high input of manual labour, can be considered as the first and most enduring method of biological control. However, pests whose origin, life cycles, and modes of action were less obvious, such as insects and plant diseases, caused greater problems but on an irregular basis. There was virtually no defence against plagues of insects or disasters like the Irish potato blight.

It is suggested by Green (1976) that man's interest in the 'chemistry' of natural substances to cure his own ailments led to an awareness of the potential of chemicals to control agricultural pests. He reports a papyrus of 1500 BC, recording formulae for insecticides against lice, fleas and wasps. Chemicals were largely ignored for the next 3400 years until the late 1800s when the arsenical compound, Paris Green, was used against the Colorado potato beetle (still a serious pest today), Bordeaux mixture was used as a fungicide on vines, and copper sulphate and ferrous sulphate came into use as herbicides. Evolution to modern agriculture, with its heavy reliance on chemicals, was swift and accompanied rapid developments in many technical fields. The ultimate consequence of this change is not yet clear but some of its unanticipated side-effects were discussed by Rachel Carson (1963) in *Silent Spring*. In the ensuing two decades, other examples have arisen, most recently the controversy over the dioxin contaminant in 2,4,5-trichlorophenoxyacetic acid.

Biological control, by its name, is the clear counterpoint to chemical control, but after that its definition is highly varied, embracing all pest control measures which do not involve chemicals. After reviewing the meaning of 'biological control' and discussing a few examples for at least one definition, the stages to effective commercial implementation of biological control in modern agriculture will be considered in this Chapter.

Biological control

CULTURAL PRACTICE

Cultural practice is the oldest form of deliberate biological pest control. It is the baseline, the prime technique, which every good farmer uses to produce his product. However, 'fine tuning' by cultural practice is extremely complex. How can one determine the precise actions to maximize yield for each crop, with every seed quality, in soil types of varying texture and quality, with every alternative method and timing of ploughing and tillage, with possible additions of natural or artificial fertilizers at various times, with every possible planting and harvesting date, under varied conditions of residue removal or destruction, crop rotation, intercropping, with different types and positions of barriers to wind, with pretreatments like flooding and, lastly, under weather conditions that lack extremes, that are unpredictable and never follow the same pattern twice? Cultural practice in agriculture, like science itself, relies on an ability to predict the consequences of an action. General predictions may be made on the strength of many years of accumulated experience, but precise predictions are impossible, not only because of the complexity and interactive nature of the variables listed, but also because controlled experiments are not done and virtually cannot be. Therefore, although one may hear someone else's experience of pest control: 'intercrop this with that, after rotating with something else and shallow ploughing early with low nitrogen, etc.', there is a much better than 50 per cent chance that this technique will fail, and with it, the crop and the farmer's income. Chemical pesticides, whatever their undesirable short- and long-term side-effects, will 'fine-tune' a crop, and rarely fail in their prime task of protecting both crop and income. Alternatives to chemicals must have this ability, because it cannot be obtained consistently by cultural practice. Furthermore, because research into cultural practice cannot lead to profit, except for the farmer, it will be undertaken only with public money or farmer's money. A profit-making company will not invest in research from which it can anticipate no return, which is the situation with respect to any cultural practice advice without sales attached to it.

THE INTRODUCTION METHOD

'Classic' biological control, well described in 1976 by Huffacker and Messenger, is the introduction of a wholly novel organism into an isolated or new geographical area, to control a pest. The pest may be an introduction itself, but this usually has been by accident: it may thrive in an environment free from the parasites or predators with which it has co-evolved and may be checked only by the deliberate introduction of a natural enemy into the new environment. The International Centre for Integrated and Biological Control, established in conjunction with the Entomology Department at the University of California at Berkeley, has worked extensively on this principle and has a variety of successful insect control programmes to its credit. These have involved the release and establishment of small numbers of predatory parasitic insects, which then multiply unchecked and control pest species.

The introduction technique has been successfully used to control weeds: an impressive example has been the control of skeleton weed, *Chondrilla juncea* L., by a rapidly spreading rust fungus, *Puccinia chondrilla*, in Australia in 1974 (Hasan, 1974). Rabbit control by myxomatosis is another example of pest control by introduction of a hitherto unknown antagonist (Fenner and Ratcliffe, 1965).

Perhaps the main problem with such introductions is their unpredictability because of the difficulty of constructing an accurate model. In Huffacker and Messenger (1976), Davis, Myers and Hoy observe that, '. . . cats, liberated on islands for rat control . . ., have frequently exterminated birds. Thompson (1963) quotes Stonehouse as saying that cats liberated on Ascension Island in 1815 resulted in extirpation of 10 of 11 species of sea birds'.

In addition, a significant factor is that there is no profit in a 'once-and-for-all' introduction. Any pest problem which was not recurrent would not attract the research interest of a profit-making company. As with cultural practice, research and development has been, and will necessarily continue to be, funded either publicly or by user groups.

RESISTANCE

Plant resistance is also included in the definitions of biological control. Plant breeders have had many successes in producing varieties resistant to disease and occasionally to insects, although not so far to weeds. Conventional methods include finding naturally resistant cultivars, crossing these with desired cultivars and selecting the disease-resistant progeny. This may result in losing some of the more desirable properties, such as taste. Introduction of higher-yielding disease-resistant varieties can also force growers into adopting these varieties in order to remain competitive in terms of productivity versus cost; in the process, older, more traditional products can disappear from the market. Resistance to soil pathogens may also be provided by the use of resistant rootstocks and grafting.

Recent research has shown that plants can locally accumulate antimicrobial compounds (phytoalexins) in response to invasion by disease. If a way were found to induce these compounds permanently, disease resistance might be conferred on otherwise susceptible varieties. Other new research by, among others, Kuć and colleagues (Caruso and Kuć, 1979) has shown that, in the Cucurbitaceae at least, a systemic immunity to disease may be induced by challenging the first leaves of the plant with a variety of pathogens. This appears to work by a mechanism independent from that of the conventional resistance (often a hypersensitive reaction to invasion) and from that of phytoalexin production. The suggestion has been made that disease resistance may operate by a series of independent mechanisms which are at present poorly understood but which offer exciting new future possibilities. The ability to confer season-long systemic disease immunity to crop plants while they are seedlings would obviate the need for many of the fungicides and other antimicrobial chemicals in use today.

BIOLOGICAL CHEMICALS

Another definition of biological control is the use of 'natural' chemicals. Investigation of natural resistance has shown that plants produce many substances (such as the phytoalexins) which provide resistance to insects and disease. The identification of natural chemicals and subsequent formation of their derivatives can open new areas of pest control. One early example was insect control with nicotine: a more recent and more spectacular example has been the chemical alteration of the natural substance pyrethrum to produce the very effective pyrethroid insecticides. The medical use of fungal antibiotics (e.g. penicillin) and their derivatives is the major pharmaceutical success in the history of medicine and it is hoped that achievements on a similar scale can be achieved with naturally derived agrochemicals.

Natural and synthetic chemicals which do not kill insects, but either attract, repel or modify their sexual behaviour, may also be regarded as biological control agents. Attractants can be used in traps to reduce insect populations or to measure their density, indicating when the toxic sprays will be most effective, thereby rationalizing and reducing chemical use. Behaviour-modifying chemicals which reduce mating frequency, or the release of large numbers of sterile insects, control pests biologically by reducing their numbers.

The research and development of biological chemicals is being actively pursued by the agrochemical industry because such substances can be discovered, investigated, modified, developed, patented, formulated, applied and sold for profit exactly like wholly synthetic chemicals. Whether their use is or is not 'biological husbandry' may be difficult to determine, but will need to be resolved in the future.

ALTERING THE NUMBERS

In nature there is always a balance between victor and victim, predator and prey. Every change, man-made or not, will ramify and change other balances, occasionally pushing species into extinction. Agriculture in particular is a major alteration from the undisturbed or natural state. Substantial environmental niches are created, in which many pests thrive. However, even in the total absence of chemicals, pests rarely dominate totally in the long term. Weeds, insects, disease organisms and animals all have natural enemies which will check a population explosion. A too-successful weed or insect will become a vulnerable target in the same way that a crop is vulnerable, because of its quantity and uniformity. However, the ultimate control of a pest by its natural enemies is little comfort to a farmer who has seen a large percentage of his effort and livelihood destroyed. Pest control is, to him, not just tempting but essential.

Altering the numbers of the natural enemies of pests means the purposeful addition of quantities of natural enemies before the pest has caused significant damage. The pest's numbers will be lower and the peak population will occur sooner. The critical distinction between this method and the introduction method is that only those organisms already present in a particular environment are used, although the environment may be altered to favour the natural enemy by, for example, adding hedgerows or intercropping.

Although any organism could, in theory, be used in this technique, the ideal organisms are those that can be produced, or will multiply naturally, in great profusion and can be held in reserve ready for use at appropriate times. They should be capable of localized use and compatible with existing agricultural practice and machinery to facilitate their acceptance by users. They need to be fairly specific in their targets and in their period of efficiency to avoid undesirable side-effects. Many microorganisms and insects fit these conditions.

MICROBIAL PESTICIDES

Extensive research was conducted, before the Second World War, on the use of microorganisms as biological control agents. However, the spectacular successes achieved by pest control chemicals, during and after the war, gave them a dominant position in agriculture and led to the conclusion that biocontrol was technologically irrelevant. By the early 1970s, despite the relatively short historical period of chemical agriculture, people had become aware of some of the chemicals' unanticipated side-effects and interest in microbial pesticides revived. In the last decade, bacteria, fungi, viruses and other microorganisms have been developed as bioherbicides, bioinsecticides and biological disease-control agents.

BIOHERBICIDES

In the United States, weeds are estimated to cost $11 billion per year in crop losses and control measures. Crop losses account for $6 billion; chemical control for $2.3 billion, and cultural, biological and ecological methods for $2.7 billion. World-wide, nonchemical weed control includes the use of insects to control klamath weed, tansy ragwort, prickly pear cactus, lantana, and others; the use of insects, fish and mammals to control aquatic weeds and, increasingly, the use of plant-pathogenic fungi to control specific weeds.

Plant-pathogenic fungi are usually highly specific in their targets and are therefore most useful when one or two weeds are the dominant problems, or when the weed has a close phylogenetic relationship to the crop plant. The United States Department of Agriculture sponsors the Southern Regional Research Project S-136 on Biological Control of Weeds with Fungal Plant Pathogens, a 5-year programme running until September 1983, involving 23 laboratories in 13 southern states of America.

The University of Florida Plant Pathology Department has worked for some years on the control of water hyacinth and hydrilla in waterways, a world-wide problem, with *Cercospora* and *Fusarium* species (Charudattan and McKinney, 1977; Freeman, Charudattan and Conway, 1978). Another research group in Gainsville, Florida, the Bureau of Plant Pathology of the Division of Plant Industry, has developed the use of *Phytophthora palmivora* for the control of milkweed vine in citrus orchards to the point of commercialization by 1981 (Ridings *et al.*, 1978).

A group at the University of Arkansas, led by Templeton and Te Beest, have developed several pathogens for weed control. *Colletotrichum gloeosporiodes* (Penz) Sacc. f.sp. *aeschynomene* is completely effective at controlling northern jointvetch, *Aeschynomene virginica*, a leguminous weed in rice, and should be marketed in 1981 after 11 years of research. Another rice weed, *Jussiaea decurrens*, the winged water primrose, can be controlled with *C. gloeosporiodes* f.sp. *jussiaea*. An important discovery has been the potential of *C. malvarum* for control of *Sida spinosa*, a prickly weed in soya and cotton which, if not controlled, can reduce cotton yields by 17 per cent (Templeton and Smith, 1977).

BIOLOGICAL DISEASE CONTROL

In 1974, Baker and Cook devoted 350 pages and 600 references to the subject of biological control of plant pathogens. With very few exceptions, such as the control of *Fomes annosus* in tree stumps with *Peniophora gigantea* (Rishbeth, 1963) the research had not led to much practical disease control. Since then, research has advanced considerably and several bio-logical disease-control agents are now marketed. Recent excellent reviews have been published by Rodriguez-Kabana, Backman and Curl (1977) and by Gindrat (1979).

Trichoderma spp were reported to control effectively *Verticillium malt-housei*, dry bubble disease, in mushrooms (De Trogoff and Ricard, 1976) and a product has been approved for sale in France. A similar material was found by Grosclaude (1970) to protect plum trees against attack by silver-leaf disease, when applied directly to wounds via special pruning shears or when deliberately introduced to healthy plants as a prophylactic measure. It could also cause disease remission if drilled directly into silver-leaf fructifications. These uses of *Trichoderma* were developed partly by Corke and others at Long Ashton Research Station, Bristol (Corke, 1974), and were included in his review, *Microbial Antagonisms Affecting Tree Diseases* (Corke, 1978). This material is commercially available in the UK.

Experiments by Wells, Bell and Jaworski (1972) over the last ten years, and more recently by Elad, Chet and Katan (1980) have shown that

Trichoderma has potential as a preventative agent against *Sclerotium rolfsii* (Southern stem rot) and *Rhizoctonia*, both major destroyers of crops worldwide. At least three companies are experimenting with *Trichoderma* for commercial development.

A major success in plant disease control has been the discovery (New and Kerr, 1972), development (Schroth and Moller, 1976; Moore, 1977) and marketing of *Agrobacterium radiobacter* for prevention of crown gall caused by *A. tumefaciens*. When root-pruned seedlings are dipped in liquid suspensions of *A. radiobacter*, crown galls (which make seedlings unsaleable) do not develop on the newly planted seedlings, even in fields infested with *A. tumefaciens*.

Although many laboratories work with many organisms as potential disease-control agents, the complexity of the task should not be underestimated. Disease occurs above, below, and on the ground, three very different conditions of air, moisture and temperature. Varied environments and the multiplicity of other organisms present impede accurate understanding of field situations. It is likely that the next few years will see the development of additional biological disease-control agents, but this will remain the area of pest control most difficult to describe and predict.

BIOINSECTICIDES

Biological control of insects by microorganisms has been the subject of most research and development efforts in the past and is consequently the most successful aspect of biological control at present. The ability of insect diseases to control insect populations has been known for over 80 years, based on nineteenth-century observations of the growth of green muscardine fungus on sugar-cane froghoppers. This fungus, *Metarhizium anisopliae*, is currently used commercially in Brazil for the control of spittlebug in pasture and has been successfully used, together with viruses, to control rhinoceros beetles in Western Samoa. *Bacillus popillae*, the milky spore disease, has been commercially available for many years to control Japanese beetles. *Bacillus thuringiensis* has been commercially available for at least 30 years: improvements in the strain and formulation have led to a substantial increase in market size to around 1500 tonnes p.a., for the control of over 100 insect species. A new strain, *B. thuringiensis* var. *israeliensis* will soon be available for control of mosquitos, including the *Anopheles* mosquito which is important because it carries malaria. The protozoon, *Nosema locusta*, is being introduced for the suppression of grasshoppers in rangeland. Several viruses are already being marketed for control of various insects and others are near commercialization, with many more undergoing investigation. Several fungi have now been introduced: *Hirsutella thompsonii* for control of citrus mites, and *Verticillium lecanii* for aphid control. *Beauveria bassiana* is in commercial use in Eastern Europe and China, as are no less than five strains of *B. thuringiensis*, and four viruses. The fungi, *Nomurea rileyii* and *Entomophthora* spp, have been developed to the point of safety testing.

The United States Department of Agriculture supports Regional Project

S-135, *Development of Microbial Agents for use in Integrated Pest Management,* now in the middle of its second 5-year programme. Cooperating scientists are based around the country.

The US maintains a special laboratory on the biological control of insects in Colombia, Missouri, as well as a large research group in the Plant Protection Institute at Beltsville, Maryland. Similarly, Germany has a laboratory investigating biological control of insects at Darmstadt, near Frankfurt.

The level of interest can be judged by the fact that, in May 1980 at a 3-day workshop on Insect Pest Management with Microbial Agents, 50 speakers from 15 countries presented a review of the achievements of the past decade and an optimistic view of the future. This was followed by the 5th Beltsville Symposium in Agricultural Research, on Biological Control in Crop Protection, the proceedings of which are to be published soon. In October 1979, the Vienna meeting of the International Organization of Biological Control attracted 500 participants from 33 countries and at least 10 companies.

PRODUCING A MICROBIAL PESTICIDE

Microbial pesticides exist in nature: they are discovered, not invented or synthesized. Such agents have been selected by evolution and may be discovered by looking for the results of their action. A diseased insect or weed may yield a new bioinsecticide or bioherbicide: a disease-free crop may yield a new disease-control agent. Once discovered, isolated and identified, the organism is assessed in the laboratory, usually under highly favourable conditions. This enables the investigator to determine its potential as a biocontrol agent.

The field efficacy of microbial pesticides will always require a programme of improvement. If any microbe were naturally as effective as required for crop protection, the pest would be long extinct. Field conditions include wide variations in such factors as temperature, humidity, aeration and incident radiation, all of which may reduce efficacy. The objective of an improvement programme must be first, to understand as much as possible about the mechanism of action and to determine the factors likely to be efficacy-limiting, and then to select the strain of the organism most likely to have broad field efficacy.

Regardless of the nature of the organism, a process must be developed to produce it in sufficient quantity. Even if it multiplies readily once used, the inoculum must be produced in the first place. This may involve extensive insect-rearing, as in the production of insect predators, parasites and viruses, or it may involve the development of artificial media for production in fermenters. Once produced, the biological control agent must be stabilized, whether in a liquid or dry state, so that it can be stored and delivered as required. Successful completion of this step may involve very precise process conditions and specialized handling to avoid lethal damage to the organism, or the addition of protective compounds to retard loss of viability. The microbial pesticide must also be in a form that is reasonably compatible with common agricultural machinery and practice.

When a product has been discovered, proved, improved, produced and stabilized, it is necessary to obtain legal approval for its use in each country and it is advisable to protect the investment already made, by patents. Patents cannot be obtained on organisms (a recent exception was an organism that had been created by genetic engineering and was therefore considered to be an invention) but they may be obtained for specific uses for an organism, for the process by which one makes it, or for the method and ingredients used to formulate it. Protection of proprietary rights to microbial pesticides may become critically important in future as more companies consider the commercial prospects of this field. Without some form of protection, most companies will be unwilling to risk investing in biological control. Equally important is the cost of obtaining legal approval. Recently, the regulatory authorities of both the UK and the USA have rethought the safety requirements for what the EPA calls 'biorationals' and have differentiated procedures for biorationals from those for chemicals. New safety procedures do not mean lower safety standards. Biological control agents have existed in nature longer than man, unlike chemicals which are wholly new additions to the environment. There are already substantial amounts of data or experience with particular biocontrol agents, and it is therefore now considered unnecessary to treat them as newly synthesized chemicals, for purposes of safety testing. In addition, regulatory authorities have recognized that the specificity of biological control agents limits their potential markets: very expensive approval procedures would permanently deter commercial production by private companies, leaving the development of the field to public funding.

Finally, the biocontrol agent must be sold into a complex market of $10 billion/year, dominated, in terms of new products, by a small number of very large companies. It must be sold to farmers who have become accustomed to convenient products, and rapid and total pest control. It may be that the ultimate success of biological control will depend on how well and how rapidly users can be re-educated towards nonchemical pest control.

Conclusion

Biological control of pests is practised in a variety of forms. Recent advances in research, development and law and, most important, in public attitudes towards chemicals, have caused private companies and publicly funded bodies to reconsider their attitudes to biological control. There is more interest and activity concerning biological control at present than ever before. It is likely that new biocontrol agents will be progressed through the stages outlined in the previous section, but it is impossible to forecast whether they will ever obtain a substantial share of the agrochemicals market.

It is important to recall that there are no panaceas. The present approach to agricultural pests is similar to the symptomatic treatment of humans, dispensed by doctors before the disease concept was understood. The welfare of human beings was always man's prime concern, yet little reasonable understanding of how to provide that well-being was available until relatively recently in human history. Human beings are one temperature and thoroughly wet. Agricultural interactions do not occur at one temperature, nor at one moisture level, nor are they between one organism and every other, but are made up of a multitude of interactive systems. Despite this overwhelming complexity and the artificiality of agriculture, some assert that simple solutions to pest problems are possible. Whether it is the simple solution of the organic gardener or the simple solution of a chemical treatment, it is equally unlikely to produce a healthy 'patient'. The ultimate answer is likely to come from a much higher level of understanding than we have at present.

Acknowledgements

The author would like to acknowledge the assistance of Dr N. E. A. Scopes, Glasshouse Crops Research Institute, and Dr J. Coombs, in the preparation of this chapter.

References

BAKER, K. F. and COOK, R. J. (1974). *Biological Control of Plant Pathogens.* San Francisco; W. H. Freeman and Co.

CARSON, R. (1963). *Silent Spring.* London; Hamish Hamilton

CARUSO, F. L. and KUĆ, J. (1979). Induced resistance of cucumber to anthracnose and angular leaf spot by *Pseudomonas lachrymans* and *Colletotrichum lagenarium. Physiol. Pl. Pathol.,* **14,** 191–201

CHARUDATTAN, R. and McKINNEY, D. E. (1977). A Fusarium disease of the submerged aquatic weed *Hydrilla verticillita. Proc. Am. Phytopath. Soc.,* **4,** 222

CORKE, A. T. K. (1974). The prospect for biotherapy in trees affected by silver leaf. *Hort. Sci.,* **49,** 391–394

CORKE, A. T. K. (1978). Microbial antagonisms affecting tree diseases. *Ann. appl. Biol.,* **89,** 89–93

DE TROGOFF, H. and RICARD, J. (1976). Biological control of *Verticillium malthousei* by *Trichoderma viride* sprayed on casing soil in commercial mushroom production. *Pl. Dis. Reptr.,* **60,** 677–680

ELAD, Y., CHET, I. and KATAN, J. (1980). *Trichoderma harzianum.* A biocontrol agent effective against *Sclerotium rolfsii* and *Rhizoctonia solani. Phytopathology,* **70,** 119–121

FENNER, F. and RATCLIFFE, F. N. (1965). *Myxomatosis.* Cambridge; Cambridge University Press

FREEMAN, T. E., CHARUDATTAN, R. and CONWAY, K. (1978). *Biological Control of Water Weeds with Plant Pathogens.* University of Florida Water Resources Research Center Publication (45)

GINDRAT, D. (1979). In *Soil Disinfestation,* pp. 252–287. Ed. D. Mulder. The Netherlands; Elsevier Scientific Publishing Co.

GREEN, M. B. (1976). *Pesticides: Boon or Bane?* London; Elek Books Ltd.

GROSCLAUDE, C. (1970). Premiers essais de protection biologigue des blessures de taille vis-à-vis du *Stereum purpureum* Pers. *Annls. Phytopathol.*, **2**, 507–516

HASAN, S. (1974). First introduction of a rust fungus in Australia for biological control of skeleton weed. *Phytopathology*, **64**, 253–257

HUFFAKER, C. B. and MESSENGER, P. S. (1976). *Theory and Practice of Biological Control*. New York; Academic Press

MOORE, L. W. (1977). Prevention of crown gall on prunus roots by bacterial antagonists. *Phytopathology*, **67**, 139–147

NEW, P. B. and KERR, A. (1972). Biological control of crown gall: field measurements and glasshouses experiments. *J. appl. Bact.*, **35**, 279–287

RIDINGS, W. H., MITCHELL, D. J., SHOULTIES, C. L. and EL-GHOLL, N. E. (1978). Biological control of milkweed vine in Florida citrus groves with a pathotype of *Phytophthora citrophthora*. *In Proceedings of the Fourth International Symposium on Biological Control of Weeds, Gainsville, 1976*, p. 224–240

RISHBETH, J. (1963). Stump protection against *Fomes annosus*. III. Inoculation with *Peniophora gigantea*. *Ann. appl. Biol.*, **52**, 63–77

RODRIGUEZ-KABANA, R., BACKMAN, P. and CURL, E. A. (1977). Control of seed and soilborne diseases. In *Anti-fungal Compounds*, vol. 1, pp. 117–161. Eds M. R. Siegel and H. D. Sisler. New York; Marcel Dekker

SCHROTH, M. N. and MOLLER, W. J. (1976). Crown gall controlled in the field with a non-pathogenic bacterium. *Pl. Dis. Reptr.*, **60**, 275–278

TEMPLETON, G. E. and SMITH, R. J. (1977). Managing weeds with pathogens. In *Plant Disease: An Advanced Treatise*, vol. 1, pp. 167–176. Eds J. G. Horsefall and E. B. Cowling. New York; Academic Press

WELLS, H. D., BELL, D. K. and JAWORSKI, C. A. (1972). Efficacy of *Trichoderma harzianum* as a biocontrol for *Sclerotium rolfsii*. *Phytopathology*, **62**, 442–447

Workshop on Insect Pest Management with Microbial Agents, Insect Pathology Resource Center, Boyce Thompson Institute, Cornell University, Ithaca, New York 12–15 May 1980

8

THE PLACE OF AGRICULTURAL CHEMICALS IN BIOLOGICAL HUSBANDRY

J. F. NEWMAN
Biological Consultant, Jealotts Hill, Bracknell, Berkshire

Agricultural chemicals may be divided into those concerned with crop production—fertilizers—and those concerned with crop protection—herbicides, fungicides and insecticides. Although they are not strictly concerned with crop protection, the plant-growth regulators, which exert control over the rate of growth and the maturity of plants, are commonly included in the latter group.

The major fertilizer elements are nitrogen, phosphorus and potassium, and a number of other elements are necessary for plant growth in smaller or in trace quantities. Where a soil is grossly deficient in a particular nutrient element it is generally recognized that no plant growth can take place unless this deficiency is made good. It is, however, in relation to supplies of nitrogen that the greatest divergence of opinion exists, between those who favour a biologically orientated agriculture and those who regard the provision of inorganic fertilizer elements as essential. Nitrogen differs from other fertilizer elements in that the main reserves exist, not in the soil, but in the form of nitrogen gas in the atmosphere. This atmospheric nitrogen is not, of course, directly available to plants. It must be fixed, that is converted into a soluble nitrogen compound before it can be absorbed by the roots of plants.

Nitrogen supplies for plant nutrition are derived in three ways:

1. Natural fixation of atmospheric nitrogen through the action of free-living bacteria in the soil, or of symbiotic organisms associated with the roots of certain plants and particularly with the Papilionaceae;
2. Synthetic fertilizer produced by a process in which atmospheric nitrogen is converted into soluble fertilizer using energy usually derived from oil or coal;
3. Recycled organic matter. Soluble nitrogen compounds become available from the decomposition of organic matter in the soil. This organic matter may be derived from plant residues produced earlier on the same site or from material such as farmyard manure or compost imported from elsewhere. Recycling does not, of course, involve any new fixation of nitrogen but ensures more complete use of nitrogen fixed previously.

It is perfectly feasible, and in some circumstances economic, to grow plants in the complete absence of any organic matter whatsoever. This procedure is known as hydroponics. The roots of the plants are immersed in a solution of the appropriate nutrient elements, with or without an inert supporting material such as sand or vermiculite. Tomatoes and various other high-value crops under glass are produced commercially in the UK by such a process. The fertilizer solutions are pumped around the roots of the plants and such factors as pH, dissolved oxygen and concentration of fertilizer salts are monitored and adjusted automatically. Hydroponic culture is highly economic in the use of water and has the advantage that considerable savings can be effected through avoiding costly movement and sterilization of soil. It would, of course, be absurd to suggest that hydroponic culture should replace field cropping on any large scale. In seeking to provide optimum concentrations of nutrients for the growth of a particular crop, it is clearly wise to make use of as much naturally produced material as possible, always provided that it does not involve excessive transport or handling costs or undesirable distortion of the required cropping pattern. In field crops, the relatively low value of the crop limits the amount of detailed attention which can be given to it. The frequency and amount of rainfall are variable and the soil structure is important in retaining water and nutrients. Under natural vegetation, where no cropping takes place and little is removed from the system, a decomposer community of great complexity develops in the surface layers of the soil. The respiration of this community returns carbon to the atmosphere as carbon dioxide, while nitrogen is to a much greater extent retained in the body proteins of the decomposer community. In some circumstances, large reserves of organic nitrogen can be built up. In the North American prairies before their cropping, the size of this reserve has been estimated at 32 000 kg/ha. Agricultural systems based on the natural fixation of nitrogen with recycling of organic matter probably reached their peak in the well-known Norfolk four-course rotation, which included a clover/grass nitrogen fixation stage and a winter fodder root crop for the support of animals. Cooke (1967) quotes a nutrient balance for such a system and suggests that an annual deficit of 16 kg/ha of nitrogen was required to be made up through natural fixation. This system eventually failed because of economic pressures: the repeal of the Corn Laws in 1869 exposed it to competition from the virgin lands overseas, and particularly from North America, and clearly the system would be unable to compete.

Before 1973 it was generally agreed that to support a person in a reasonable degree of health required an average daily intake of 70 g protein and that the contributions from animal and vegetable sources should be roughly equal. More recently, the World Health Organization has suggested a somewhat lower daily protein intake, amounting to 37 g/d for men and 28 g/d for women. These figures, however, assume that the protein is of high quality, such as that in eggs or milk. For protein of lower quality, such as most plant protein, the requirements are 46–62 g for men and 36–48 g for women. Tanner (1968) has calculated the mean amount of fertilizer nitrogen per person necessary to produce 70 g of protein/d. The annual protein requirement per person is about 55 lb (25 kg) containing about 4 kg of nitrogen. The efficiency of production of animal protein from soil nitrogen is much lower than that for vegetable protein, approximate values being 15 per cent and 50 per cent respectively. The amount of fertilizer nitrogen needed to grow the necessary animal and vegetable food can be calculated and comes to about 17 kg annually per person. In temperate countries, such an amount is derived from natural sources by about 2 acres (0.81 ha). A leguminous crop could, of course, fix nitrogen in much greater quantities, but allowance must be made for other crops and for incomplete utilization. It is interesting to consider the amount of agricultural land available per head of human population. On a world basis it amounts to about 0.5 ha. In relatively sparsely populated areas, such as the USA, the figure rises to 1.4 ha per person, and it is understandable that large surpluses of food can be produced. In Britain the figure is 0.3 ha per person. It thus seems reasonable to suggest that, while all reasonable and practicable means should be used to make good use of naturally available plant nutrients, food supplies for a high standard of living, at least in the more densely populated and developed countries, are highly dependent upon the supplementation of natural nitrogen fixation with synthetic fertilizer.

The environmental problems associated with a high use of fertilizer, and particularly of nitrogen, are well known. They are the leaching and run-off of soluble nutrients from agricultural land and the accumulation of excess nitrate in plants at certain stages of growth. Direct surface run-off of recently applied fertilizer is not common, and is usually associated with extreme climatic conditions, as when heavy rain falls on frozen ground to which fertilizer has been applied in early spring in North America. Leaching through the soil is of greater significance. A study by Owens *et al.* (1972) on the origins of nitrates carried in the rivers Trent and Ouse in the Midlands of England considered the contributions of point and diffuse sources. Industrial and sewage effluents are point sources, while diffuse sources represent drainage from agricultural or other land. In the river Ouse, which drains predominantly agricultural land in Bedfordshire, point sources provided only 35 per cent of the total load of 7100 kg/d. In the Trent, which drains the densely populated Midland industrial area, point sources accounted for 65 per cent of the total load of 35 000 kg/d. This ratio is related to population density, and the crossover point, where effluents become more important than land drainage, can be calculated to be 3 persons/ha. The occurrence of nutrient loss from the land is not, however, very much related to the type of agriculture, biological or otherwise. Loss through leaching is known to be related to the vegetation cover. The removal of vegetation stops the uptake of available nitrogen from the soil and so leaves it available to dissolve and leach out if rainfall is sufficient. Borman *et al.* (1968) studied nutrient run-off when the forest of a small watershed ecosystem in New Hampshire, USA was felled. Run-off of soluble nitrogen increased by about 20 times, the loss during the first year after cutting being about equal to the normal annual turnover of nitrogen in the system. The increased loss is related not only to decreased uptake of nitrogen in the system but also to increased water run-off with the reduction in transpiration. The harvesting of an arable crop can produce a similar effect. The removal of the vegetation cover in the autumn, while soil temperatures are still high enough for bacterial activity to produce soluble nitrogen from organic matter, leads to the observed leaching of nitrate into rivers with the heavier rainfall in late autumn. Although it is popular to blame increased fertilizer use for increased leaching of nitrogen, in practice inorganic fertilizer in properly timed applications is less likely to produce leaching than is usage of organic manure to produce an equivalent crop response. The peak nutrient demand of a crop can be met by a properly timed application of the right amount of fertilizer. In order to provide the same available nutrient level at the peak growth period, a much higher level, in terms of total nitrogen, of organic manure would have to be applied, and much of this would remain in the soil at harvest and contribute to the loss in the following winter.

The agricultural crop is essentially an unnatural type of vegetation cover. It is a monoculture, in contrast to the great species diversity of the natural vegetation which it replaces. Even the crop plant itself is often quite arti-ficial, in that such important plants as wheat, barley and maize are very different from their wild ancestors. In the wild vegetation around the edges of the agricultural field, all the pests and diseases of the crop can be found, but in this situation their populations do not, in general, reach epidemic proportions, the speed of spread being limited by the hazards involved in getting from one widely separated host to another. In the crop situation this hazard is removed, spread can be rapid, and an epidemic quickly follows. It is thus unrealistic to expect that the natural biological controls which operate in diverse wild vegetation can operate unaided to control pest, disease and weed spread in the wholly artificial crop situation. Man has produced crop monocultures essentially to aid his own crop harvesting, and it is not surprising that this also aids harvesting by the other competitive forms of life which we call pests and diseases. Biological control methods by themselves seldom work well, except in a relatively few special cases where an introduced pest, proliferating rapidly in the complete absence of its normal parasites and predators, has been controlled by the introduction of the necessary species.

There is no doubt that the properly judged use of chemical pesticides can control pest and disease epidemics. Although the use of chemical pesticides, particularly naturally occurring products such as sulphur or mercury, dates back into antiquity, the era of modern synthetic pesticides can be considered to start in the late 1930s with the discoveries of the insecticidal activity of DDT and of the synthetic plant hormone herbicides MCPA and 2,4D. DDT was put into use rapidly and on a wide scale during the war years, with spectacular success, particularly in the control of malaria, typhus and other insect-borne diseases. Together with other organochlorine insecticides it was used to control agricultural pests during the 1940s. In the general euphoria about the use of this highly successful insecticide, the ecological aspects of the matter were largely forgotten until the inherent long-term problems became apparent. We now appreciate that highly persistent toxic materials distributed widespread in the environment are undesirable, and that the use of pesticides having a toxic effect on a wide range of species can produce effects which are not in our long-term interests. Furthermore, numerous pest species are able, through the operation of the normal processes of biological selection, to become genetically resistant to the action of particular pesticides or groups of pesticides. For all these reasons, there is now a requirement for much information on the ecological safety of pesticides and a movement towards an ecologically based coordination of chemical, biological and cultural methods, referred to as integrated control.

Integrated control systems appear to have been most successful either in plantations or orchards, where the crop is established for a sufficiently long time for a biologically balanced system to develop, or in situations where there is close control of environmental and crop conditions, as in glasshouse culture.

In orchard crops, tetranychid mites have become important pests, the situation having been aggravated by the destruction of predators by the use of nonselective insecticides against various insect pests such as aphids and caterpillars. Tetranychid mites are not affected by many insecticides and appear to possess a genetic plasticity which enables them rapidly to develop resistance to specific acaricides. Wearing *et al.* (1978) have described an integrated control programme against the European red mite (*Panonychus ulmi*) in commercial orchards in New Zealand, using the phytoseiid predatory mite *Typhlodromus pyri*, known to be highly resistant to insecticides. The programme was tested in two seasons between 1975 and 1977. By regular monitoring of the levels and ratios of *T. pyri* and *P. ulmi*, the application of acaricides was restricted to occasions when specific population thresholds were exceeded, and then a selective acaricide, cyhexatin, was used to improve the predator/prey relationship.

The selective pyrimidine carbamate insecticide, pirimicarb, is highly effective against aphids but has little effect upon many other insects or upon mites. It can thus be a useful compound when aphids need to be controlled in situations where other pests are being controlled by biological means. It has been successfully used for aphid control in glasshouse crops, such as cucumbers and ornamentals, where acaricide-resistant tetranychid mites can be controlled only by a biological programme using phytoseiid mites.

It is clear that the development of integrated control methods depends upon the availability of a range of highly selective pesticides. When, however, a crop is subject to attack by a wide range of pests, the farmer finds it more economic, at least in the short term, to apply a single spray of a pesticide having a broad spectrum of activity. There are difficulties also from the point of view of the chemical manufacturer. The costs of research and development in this field, with the necessarily large expenditure on toxicological and environmental investigations, implies that only large potential markets on major crops can justify the investment.

Modern agricultural technology has been subject to criticism in relation to the maintenance of satisfactory soil structure for plant growth. In recent years, the availability of short-persistence herbicides, such as paraquat and glyphosate, has led to the development of no-cultivation (direct drilling) and reduced cultivation techniques. Much research in this area has shown that the most important function of cultivation is weed control and that if this can be accomplished chemically, much cultivation is wasteful of energy and damaging to soil structure. Under a reduced cultivation regime, soil structure progressively improves, with the development of a greater amount of organic matter in the surface layers. Edwards and Lofty (1978) have shown that, under cereal crops, a direct drilling regime favours most soil invertebrates, populations being up to three times larger than in comparable ploughed and cultivated soils. Earthworm populations are particularly favoured, and promote root growth in compacted soils.

References

BORMAN, F. H., LIKENS, G. E., FISHER, D. W. and PIERCE, R. S. (1968). Nutrient loss accelerated by clear-cutting of a forest ecosystem. *Science, N.Y.*, **159**, 882–884

COOKE, G. W. (1967). *The Control of Soil Fertility*. London; Crosby Lockwood & Sons, Ltd.

EDWARDS, C. A. and LOFTY, J. R. (1978). The influence of arthropods and earthworms upon root growth of direct drilled cereals. *J. appl. Ecol.*, **15**, 789–795

OWENS, M., GARLAND, H. J. N., HART, I. C. and WOOD, G. (1972). Nutrient budgets in rivers. *Symp. zool. Soc. Lond.*, **29**, 21–40

TANNER, C. C. (1968). Diet, nitrogen and standard of living. *Outlook Agric.*, **5**, 235–240

WEARING, C. H., WALKER, J. T. S., COLLYER, E. and THOMAS, W. P. (1978). Integrated control of apple pests in New Zealand. 8. Commercial assessment of an integrated control programme against European red mite using an insecticide resistant predator. *N.Z. Jl. Zool.*, **5**, 823–837

9

PRACTICAL PROBLEMS OF ENERGY SAVING AND RECYCLING IN BIOLOGICAL HUSBANDRY

N. W. MARTIN and J. KEABLE
42 Fairview Drive, Hythe, Southampton, UK

At first sight the effect of escalating fuel cost inflation would seem, without question, to favour biological husbandry against conventional agriculture, as 25.5 per cent of energy used in conventional agriculture relates to fertilizer (23.1 per cent) and to chemicals (2.4 per cent) (White, 1976). This represents 4 per cent of UK primary energy use. In terms of the national energy budget, it can be argued that savings in this area have a negligible effect—a probable reason for lack of real support from governmental bodies for research into reducing energy expenditure on agriculture. However, this does not take into account the effect on food prices of escalating energy costs, or the potential problems associated with long-term availability. As rising energy costs appear to be a major factor in world inflation, and present UK Government policy is that gas and other fuel prices should lead inflation by up to 10 per cent, it would appear that an agriculture that avoids, or at least reduces, the use of artificial fertilizers and chemicals must have some economic edge, apart from other benefits.

The argument for a reduction in primary energy consumption must be much stronger in lesser-developed countries and those without indigenous fossil energy sources.

On detailed examination of energy use in agriculture, problems emerge. For example, biological husbandry has to date been used in more labour-intensive situations than conventional agriculture. However, if biological agriculture is to be genuinely 'An Agriculture for the Future', then solutions must be found which are appropriate in whatever agrosystem is best suited for a particular situation, whether climatic, economic or sociopolitical. Herein lies the problem: ideas put forward to save energy may have genuine benefits in a highly developed agrobusiness in the West, but may be totally inappropriate if applied in a less-developed area. The reverse can be equally true.

In this chapter, therefore, we shall try to indicate areas in which there is potential for energy saving and possibilities for recycling; we shall also indicate that there are substantial potential savings in energy consumption to be brought about by more radical changes in our agricultural practices, but that these would need, in the West at least, a more fundamental alteration to our social structures and dietary habits.

Before outlining various possibilities, it is perhaps worth considering the factors which inhibit development of the use of organic materials. The importance of these factors will vary from country to country, but an examination of the particular constraint which retards development can be fairly readily identified for each of the recycling or conservation methods described.

The constraints include the following:

1. Lack of awareness of the possibilities and the need, and lack of interest on the part of society towards the programme;
2. Inadequate or unsuitable infrastructure, for example for collection of wastes in rural and urban areas;
3. Inadequate technological development;
4. Lack of skill and knowledge;
5. Social prejudices;
6. Lack of support from national governments;
7. Financial difficulties concerning initial investment and lack of return on that investment.

It is not claimed that the methods described represent any more than a broad-brush picture of the potential scope for various methods of energy reduction and resource recycling, and in most systems described there is a requirement for research and development in various aspects of production and use.

Farm-scale methane production

Methane production on a small scale is widespread in India and China—China alone is reported to have some 7 million units (Van Buren, 1979). Small units were relatively widespread in Europe too, during periods of wartime fuel shortage. Only recently, however, has attention been given to larger-scale units, adapted to the modern farm scale.

Prototype units are reported from Switzerland, Israel and the US, besides the UK. One such system at the dairy unit at Bore Place, Kent, has been described (Keable, 1979). A number of problems have emerged, but it would appear that successful solutions are now in sight. These include:

1. Reliable and convenient systems for collecting and moving slurry—normally from large animal-rearing units or dairy herds;
2. Ways of regulating the solid/liquid ratio;
3. Methods of anaerobic digestion itself, whether at mesophilic or thermophilic temperatures (30–35 °C and 60–65 °C respectively), that allow constant operation;
4. Reliable units for using the methane gas generated, whether burnt for use as heat, electricity, power or refrigeration—units adapted to biogas, which is likely to be more variable in quality than conventional fuels.

The motive for investigating a farm-scale methane unit may be one of several:

1. Energy production;
2. Pollution control;
3. Recycling of nutrients as fertilizer;
4. Recycling of protein as animal feed.

It is unlikely, however, that the economics of a system will look favourable if only one of these reasons is considered. The number of examples where several are applicable is likely to increase rapidly, however, because:

1. Energy prices are 'leading' inflation, which makes energy-source replacement more attractive each year;
2. Pollution control measures are being enacted, and more firmly administered, in most countries today;
3. Fertilizer prices, linked both to fuel prices and transport costs, are also rising sharply;
4. Protein costs are rising rapidly, not only because of transport costs, but also because the prime producing countries increasingly need the material for their own use.

The scene appears to be set, then, for the rapid growth of well-engineered systems.

It is true that the arguments set out above apply to all farming, and not only to organic farming. It would seem that, to this extent at least, the energy factor is likely to contribute to the growth of organic methods.

Practical problems remaining in this area include:

1. The welding together of the various technical solutions and appropriate equipment into a marketable system, with adequate financing, manufacture and servicing arrangements, together with the necessary demonstration and instruction courses. At a one-off level, systems are bound to be too costly. There are signs that this synthesis may develop much more rapidly abroad than in the UK, despite our lead in the research field.
2. Further assessment of the qualities of postdigested material. For example, there is still too little known about the added value of postdigested slurry as a fertilizer, and almost nothing known, in scientifically accepted terms, of its value as a herbicide or pesticide, although this is claimed by some, e.g. a Mr Hutchinson of Kenya (R. P. King, personal communication), who has used nothing else on his coffee plantation for many years, and 'miraculously' avoided the ubiquitous coffee blight. Until these qualities are better understood, economic assessments will remain too hazy.
3. The ability to store and transport biogas for more than a short time or distance is a limiting factor to the most widespread use of anaerobic digestion. Pressure vessels are expensive, and pressurizing by mechanical means is energy-intensive. Storage of the gas *after* combustion may have more future, whether in batteries for electric vehicles (who will make the first electric tractor?) or as short-term heat or refrigeration. Long-term storage remains an unsolved problem, however.

Direct drilling without herbicides

Direct drilling methods are now well known, but conventionally require one or two applications of herbicide. A reduction in the total energy input is obtained and there is a very significant increase in the humus content of the soil. It is suggested that, if methods could be developed which would not require the use of herbicides, the organic farmer could achieve similar reductions in energy consumption and that other benefits would accrue. The energy budget for winter wheat, shown in *Table 9.1,* illustrates the relative energy inputs of conventional and direct drilling methods, and is of general interest in the context of this chapter. It should be noted that the budget is based on a rigorous application of energy used in the manufacture, for example, of the tractors and combine harvesters used. If we choose to exclude those inputs, then the percentage energy reduction is significantly higher.

Table 9.1 RELATIVE ENERGY INPUTS OF CONVENTIONAL AND DIRECT DRILL METHODS FOR SOWING WINTER WHEAT (COURTESY OF THE INSTITUTE OF MECHANICAL ENGINEERS)

Item	Conventional system MJ/(ha.y)	%	Direct drilling system MJ/(ha.y)	%
Nitrogen	7430		7430	
Phosphate	665		665	
Potash	322		322	
Fertilizer subtotal	8417	47.9	8417	53.9
Seed	782	4.4	695	
Herbicides	139	0.8	278	
Tractor fuel, cultivations	1846		212	
Combine harvester fuel	625		625	
Fuel subtotal	2471	14.1	837	5.4
Cultivation equipment	701		627	
Tractor	489		166	
Combine harvester	1590		1590	
Drying plant	550		550	
Machinery subtotal	3330	19.0	2933	18.8
Grain drying	2436	13.8	2436	15.6
Total	17575	100.0	15596	100.0

Table 9.1 indicates an energy saving of 11.3 per cent by direct drilling. In an attempt to establish whether direct drilling could be used without the application of herbicide, field trials have been carried out with flame weeding using bottled LPG as a fuel; early indications are that the method is potentially useful and is economically viable compared with herbicides, but at the present stage of development it uses a large amount of energy (6700 MJ/ha) compared with the manufacture and use of herbicide. There is the long-term possibility of using bottled farm-produced methane but the economics of this are, at best, doubtful.

It will no doubt be argued that the real potential energy saving shown by *Table 9.1* is in reducing the energy input (approximately 50 per cent of the total) attributable to the fertilizer application, particularly nitrogen.

Table 9.2   ENERGY COST OF USING ORGANIC WASTE MATERIALS (AFTER PIMENTAL 1973)

Organic	*Chemical*	
25 tonnes of cow manure/ha	N: 127 kg; P:35 kg; K: 68 kg/ha	
Handling and Spreading	*Production*	*App⸍*
4183 MJ/ha	14860 MJ/ha	381 ᴍ.

However, it should be recognized that the energy cost of using organic waste materials is not insignificant, as the budget shown in *Table 9.2* indicates.

Municipal sewage utilization

The use of sewage sludge for agricultural purposes is well known. Can we improve the quality and hence the acceptability of the practice and increase the quantities that are available for land application? It is worth noting that most papers on the subject refer to disposal rather than to use: the authorities involved are primarily concerned with the economics of disposal of an unwanted material rather than making best use of a valuable resource! This should in no way be construed as a criticism of those involved in the treatment and subsequent removal of the material, but merely as a comment on the general attitude of society as a whole.

Comparison of the data available indicates that the amount of available sludge used on agricultural land is 20 per cent in Scotland, 50 per cent in England and Wales, 57 per cent (the highest in Europe) in the Netherlands, and an estimated 60 per cent in China. The relatively close correlation between China, and England and Wales, suggests that there is perhaps a 'national' level at which agricultural use becomes more difficult, perhaps because of distance or contamination with industrial effluents.

A major inhibiting factor in the use of sludge is the difficulty associated with undigested sludge. When the first sewage works were built, the original practice was to dry the sludge on draining beds and then apply it to the land. This has become increasingly uneconomic as it is labour-intensive and the use of raw sludge in liquid form has increased. Guidance on the use of sludge in this form on the land is found in *Advisory Paper No. 10* issued by the Agricultural Development and Advisory Service of MAFF, the Water Pollution Research Laboratory's *Note No. 57* and the *Report of the Working Party on Disposal of Sewage Sludge,* published by the Department of the Environment and the National Water Council.

However, in the UK, the *7th Report of the Royal Commission on Environmental Pollution* expressed concern about the practice of using raw sludge and it has been suggested that the provision of sludge digestors in strategic locations could be a sound investment. Research work has indicated that, in digested sludge, much of the original nitrogen is converted to a form more readily available to crops (Greer, 1980).

In areas of high sunshine an alternative approach can be adopted. There are significant cost advantages in the use of waste-stabilization ponds, as opposed to more conventional treatment plants. Estimates vary, but total costs are in the order of one-sixth of those of a conventional plant. In this type of pond, decomposition of the organic matter takes place concurrently with algal photosynthesis and the net effect is to produce organic matter in algal cells (FAO, 1977). The design and construction of such ponds has been described (Mara, 1977). Effluent from the ponds is suitable for quick-growing fish and the final effluent can be used for irrigation purposes.

More direct use of the algae has been achieved by separation using aeration and flotation techniques (Rich, 1978). In California, yields of up to 95 t/ha/y have been achieved using domestic sewage as a nutrient (Oswald, 1973).

A further variation of the sewage treatment theme also has potential benefits in agriculture. A treatment system using water hyacinths as the final stage has recently been commissioned. The hyacinths are to be harvested, for sale as a soil enhancer (Anon, 1980).

In the UK, about 0.5 million tonnes of activated sludge are produced from existing treatment plant. A new process has completed pilot trials and shows considerable promise. The solids in the activated sludge are separated by air flotation and filtration with a heat-treatment stage followed by drying. The resulting material has a protein content of about 50 per cent and preliminary animal feeding trials have been completed. Of particular interest is its potential as a fertilizer with similar characteristics to dried blood or blood meal, and field trials are in progress.

Table 9.3 NUTRIENTS PRESENT IN COMMON WASTE MATERIALS

Material	*Approximate analysis (%)*		
	N	*P*	*K*
Bone meal	4	21	—
Blood, dried	12	1	—
Meat meal	8	1	0.3
Hoof and horn	13	2	—
Fish, blood and bone	5	5	6.5
Fish meal	8	8	—
Meat and bone meal	6	14	
Oil seed cake (average)	6	1	1
Dried chicken manure	3	2	2
Seaweed (dried)	2	0.2	1.5
Coffee grounds (dried)	2	0.3	0.3
Eggshells	1	0.4	0.1
Feathers	15	—	—
Cocoa waste	2	—	—
Leather dust	5/12		
Stable manure	0.5	0.13	0.5
Shoddy (wool waste)	3.5/6	2/4	1/3.5
Comfrey (wilted)	0.7	0.2	1.2
Bracken (fresh)	2	0.2	2.75
Bracken (dry)	1.4	0.2	0.1
Deep litter (wood shavings)	2	1.5	1.1
Mushroom compost (spent)	0.6	0.5	0.9

Industrial and urban wastes

The use of industrial and other urban wastes is well known to gardeners and horticulturists (Hills, 1971) and it is likely that those that are readily available and convenient have a high level of utilization. It is questionable whether, if energy saving were the only criterion, there is any great benefit, because the majority of such wastes are delivered or collected in small quantities by individual users. They are, however, an important source of nutrients and these are indicated in *Table 9.3* for many of the more common materials. Additional data on various organic materials are available (MAFF, 1976; Minnich and Hunt, 1979), as is a detailed analysis of dried activated sludges from food processing plants and pulp and chemical plants (Kurihara, 1978).

It is important to consider the modification and use of more substantial and concentrated quantities of waste arising from major industries such as food processing and paper mills. Technical aspects and the role of composting have been considered elsewhere, but also of interest is the use of worms to convert industrial wastes, combined with digested sewage sludge, to a useable soil conditioner. The ability of earthworms to concentrate nutrients in soil has been established (Lunt and Jacobson, 1944); if the ability can be used in treatment systems, then this is a potentially useful method. Early experiments have been described (Collier, 1978), but it is considered by the authors that a great deal more needs to be learnt before the system can be described as established. However, it is claimed that the system is operating commercially in the US and in Japan (Anon, 1977).

Biological nitrogen fixation

The use of rotations with leguminous crops is outside the scope of this paper, but biological processes for fixing atmospheric nitrogen offer considerable potential and should be exploited, particularly in relation to the secondary treatment of sewage effluent. Recent interest in the production of 'biomass' has been stimulated by the search for alternative fuels, but the potential for use as a biofertilizer should not be ignored.

The importance of algae in the production of wet rice has been described (Russell, 1973) and in China use is made of azolla, which consists of a floating branched rhizome with small leaves and simple roots which hang down into the water. Nitrogen-fixing blue-green algae are found in the cavity of the leaves. It is used as a fertilizer in rice production and under suitable conditions produces large quantities of green matter (150–300 t/ha/y) and also fixes 100–300 kg/ha/y of atmospheric nitrogen (FAO, 1978). It is necessary to provide phosphorus which, in the particular method described, is supplied as superphosphate, but no doubt other sources including sewage effluent could be used.

The use of algae has also been investigated in China, India and Japan. The higher yield obtained when using algae is apparently attributable not only to nitrogen fixation by the algae but also to the release of some vitamins and growth-promoting substances. In field trials in Japan, algae inoculation has resulted in increased yields: 2 per cent in the first year; 8 per cent in the second year; 15 per cent in the third year and 20 per cent in the fourth year.

These examples show that there is a considerable potential for a reduction of the energy input into agriculture and scope for recycling, particularly if some of the inhibiting factors listed earlier are overcome and there is a change in the authorities' attitude to the whole question of waste treatment and utilization. However, we would like to stress that there is not one solution as 'technological fix' but each geographical area should examine its needs and resources in the light of the technical information available, to produce a local solution. One of the great advantages of chemical fertilizer, with which any organic system will have difficulty in competing, is its ease of transport and application: organic materials by their very nature are bulky and thus transport will always be an essential factor in the cost-and-energy equation. Important as energy is, it is frequently difficult to present an overriding argument that 'organic is best' (or lowest in terms of energy) but we must consider the other benefits which accrue in terms of soil structure, reduced pollution and long-term effects on the soil; hence it is difficult to present the case on a technological basis in isolation. For example, the stickiness of soils will be reduced if we repeatedly add compost (and hence humus), to the extent that the power input for ploughing is significantly reduced. However, we can hardly argue that the primary objective of applying compost is to reduce the energy input of cultivation.

In considering the energy input into agriculture, it is impossible not to comment on the changes which have taken place during this century. If we regard agriculture as a system to produce energy to sustain human life and we relate the energy input and output in that system for various crops, we find that changes in dietary habits and the increased industrialization of agriculture has totally changed the energy regime.

Shifting agriculture and the wet-rice culture of South-East Asia produces five times as much in terms of food energy as the energy input needed to produce that food. In contrast, feed-lot beef production produces around one-tenth of the energy input. At the beginning of the century, in the US food system, the energy input and output were approximately equal but, by the 1970s, the energy output was between one-fifth and one-tenth of the input. If we, in the UK, regard this as a US phenomenon, we should remember that in the UK 70 per cent (or 780×10^{15} J/y) of the gross total primary production of plants is fed to animals, yielding as food only 76×10^{15} J/y (Blaxter, 1974) and we should question the long-term viability of a system which produces only 10 per cent of the energy put into it.

These figures illustrate the changes which have taken place since the beginning of the century. Not many years ago it might have been argued that there was little need to change in order to halt or reverse the trend, and that this would probably be impossible anyway because of rising expectations throughout the world. Indeed, Leach *et al.* (1979) in their book, *Low Energy Strategy for the UK,* suggest that the changes that can be made in agricultural energy consumption are relatively minor, and they do not mention any reduced use of energy from a reduction in the use of fertilizer and other chemicals. We have to accept that, at the present rate of *application* of known and potential technology, they are no doubt correct to assume that there will be little change in this area, unless there is a considerable change in attitude of all concerned.

In agriculture traditionally, and until relatively recently, a closed-cycle system was normal within a relatively small geographical area. Plant and animal residues were returned to the soil and the complete system was balanced and sustainable. This has been changed by the way that agriculture and technology have developed, but we believe that a basically natural and sustainable cycle may be rediscovered in the context of modern technology and on a different scale. This will not be done without solving the problems, some of which we have outlined here, associated with a major shift in the emphasis of our existing technology.

References

ABE, R. K., BRAMAN, W. L., ROGERSON, A. C. and SIMPSON, O. C. (1978). Organic recycling and earthworms. *F.A.O. Soils Bulletin No. 36: Organic Recycling in Asia*, pp. 205–211. Rome; F.A.O. (417 pp.)

ANON. (1977). Global report—the worm turns for Japanese Fir. *Wall Street Journal* Aug. 1, 1977

BLAXTER, K. L. (1974). Vermicomposting. *Vermiculture Journal*, **1**, 9–12

F.A.O. (1977). China: recycling of organic wastes in Agriculture. *F.A.O. Soils Bulletin No. 40*. Rome; F.A.O. (107 pp.)

GREER, W. T. (1980). Land disposal—a local authority viewpoint. In *River Pollution Prevention—Aspects of Sewage Sludge Disposal*, 1–19. Institute of Water Pollution Control (51 pp.)

HILLS, L. D. (1971). *Grow Your Own Fruit and Vegetables*. London; Faber and Faber (328 pp.)

KEABLE, J. and DODSON, C. (1979). A modular sytstem for biogas production using farm waste. In *Proceedings of the International Solar Energy Society Silver Jubilee Congress*, pp. 83–87. Paragon & Press (2305 pp.)

KURIHARA, K. (1978). The use of industrial and municipal wastes as organic fertilizers. *F.A.O. Soils Bulletin No. 36: Organic Recycling in Asia*, pp. 248–266. Rome; F.A.O. (417 pp.)

LEACH, G., LEWIS, C., ROMIG, F., VAN BUREN, A. and ROLEY, G. (1979). *A low energy strategy for the United Kingdom*. London; The International Institute of Environment and Development (259 pp.)

LUNT, H. A. and JACOBSON, H. G. M. (1944). *Soil Science*, **58**, 367

M.A.F.F. (1976). *Ministry of Agriculture and Fisheries and Food Bulletin 210*. London; H.M.S.O. (78 pp.)

MARA, D. (1977). *Sewage treatment in hot countries overseas division building research establishment*. London; Department of the Environment (15 pp.)

MINNICK, T. and HUNT, M. (1979). *Rodale Guide to Composting*. Emmaus Pa., USA; Rodale Press (405pp.)

OSWALD, W. J. (1973). *Solar Energy*, **15**, 107

PIMENTAL, D. (1973). Food production and the energy crisis. *Science*, **182**, 443–449

RICK, V. (1978). Israel's place in the sun. *Nature*, **275**, 581–582

RUSSELL, E. W. (1973). *Soil Conditions and Plant Growth*. London; Longman (849 pp.)

SERFLUNG, S. A. and MENDOLA, D. M. (1979). The Solar Aquacell AWT Lagoon System for the City of Hercules, California. In *Proceedings of Water Reuse Symposium*, Vol. 2. Washington D.C.; American Waterworks Association

VAN BUREN, A. (ed.) (1979). *A Chinese Biogas Manual*. London; Intermediate Technology Development Group (135 pp.)

WHITE, D. J. (1976). Energy accounting in agriculture and food. In *Energy Accounting*. London; Institution of Mechanical Engineers

10

THE RELATIONSHIP OF BIOTECHNOLOGY TO BIOLOGICAL HUSBANDRY

L. G. PLASKETT
Biotechnical Processes Ltd, Hillsborough House, Ashley, Tiverton, Devon, UK

Biological husbandry and organic farming are associated with Schumacher's *Small is Beautiful*; one might think of them as unrelated to sophisticated technology of any kind, and depending on small-scale recycling of organic residues for their principal source of materials. Is this a balanced view? This chapter considers whether 'high technology' and biological husbandry are inherently opposed and permanently anathema to each other, or whether, in fact, once biological husbandry has been accepted, technology can be adapted to serve its needs. First, it seems clear that any technology that is to be welcomed into a 'biological husbandry' environment will itself be biological, in other words 'biotechnology': hence the title of this paper.

A definition of 'biotechnology' has been supplied by Schmidt-Kastner (1978): 'The production, isolation, modification and utilization of bio-products from plants, animals, humans or microorganisms on a technical scale' (Definition 1). This is extremely broad, as it includes all processes involving the extraction of natural substances, for example the extraction of sugar from sugar-cane, of quinine from cinchona and of cholesterol from gallstones, even though the process itself is not a biologically mediated conversion. According to this definition, even the thermal processing of biological material to produce oil or gas becomes 'biotechnology', although the techniques used are more in accordance with those used by the non-biological chemical and fuel industries.

A more limited definition is one that confines biotechnology to biologically mediated processes. Such a definition was used by the Joint Working Party on Biotechnology which reported to the UK Government in March 1980; the definition was: 'The application of biological organisms, systems or processes to manufacturing and service industries' (Definition 2). This implies that the conversion process itself must be biological and would have to involve whole organisms, whole cells, subcellular preparations or enzymes as mediators of conversion. This is inherently a narrower definition which I shall use here, although I shall also consider finally what possible processes included within Definition 1, but not within Definition 2, may be of use to biological husbandry. Note that Definition 2 does not include any process, however complex or sophisticated, which comprises a fractionation of biological material by nonbiological methods.

Table 10.1 lists known and existing biotechnical processes that fall within the scope of Definition 2; it indicates the range and scope of such processes.

Table 10.1   A SELECTION OF BIOTECHNOLOGY PROCESSES FALLING
WITHIN THE SCOPE OF DEFINITION 2 (SEE TEXT)

Substrate	Mediating agent	Products
Molasses	Yeast	Single-cell protein
Molasses	Fungi	Citric acid
Molasses	Fungi	Antibiotic drugs
Molasses	Aerobic bacteria	L-Lysine and other amino acids
Molasses	Yeast	Ethanol
Molasses	Yeast	Glycerol
Molasses	Bacteria	n-butanol
Molasses	Bacteria	Gums (e.g. xanthan)
Grape, malt, corn syrup	Yeast	Alcoholic beverages
Corn syrup	Hexose isomerase	Isomerase syrup
Lactose	β-galactosidase	Galactose and glucose
Pig waste	Anaerobic bacteria	Methane gas
Hydrocarbons	Yeast, bacteria	Single-cell protein

Most of these processes and products doubtless appear irrelevant to
biological husbandry, except for single-cell protein, which can be animal
feed, and methane gas, which could doubtless fulfil an important role of
energy supply on biological husbandry farms. It is appropriate, therefore, to
review the potential needs of an agriculture orientated to biological
husbandry, in relation to the range of materials which might, in future, be
supplied by biotechnical processes. At this stage of development of the
subject, lists of needs are likely to vary with personal judgement and
perhaps raise controversy and issues of principle. My list is as follows:

1. Slowly decaying organic carbon as a conditioner for arable soils.
2. Organic nitrogen for addition to all types of soils.
3. Bacterial preparations to enhance natural biological nitrogen fixation.
4. Methods of more efficient use of available organic nitrogen; methods of
   organification of synthetic ammonia.
5. Organic sources of mineral plant nutrients.
6. Preparations of biomolecules or organisms for combating pests, without synthetic chemicals.
7. Preparations of biomolecules for the stimulation or control of plant
   growth and metabolism.

The reasons for these seven requirements and the possible ways of
meeting them in future are considered individually below.

Organic carbon soil conditioners

It is a basic tenet of biological husbandry that organic carbon is to be added
to the soil, yet only rarely does one see a specification of how much organic
carbon, of what type, in what form and from whence it is to be derived.
Organic carbon is, in a very real sense, the main output of agriculture. To
return it to the soil rather than selling or using it does not always appear
viable: growing green-manure crops to plough in, for example, has yet to be
demonstrated clearly to be worth the farmer's while. This makes it essential
to consider quantitative aspects.

Greenland, Rimmer and Payne (1975) found that, for many British arable soil types, 2.5 per cent organic carbon was necessary to give a really stable soil structure: maintainance of organic carbon at this level could be said, therefore, to conform to good biological husbandry practice. Russell (1977) estimated the weight of the ploughed layer of an arable field as about 3000 t/ha; 2.5 per cent therefore corresponds to 75 t/ha of organic carbon, or 129 t/ha of soil organic matter (at the probable level of 58 per cent carbon in soil organic matter). These figures are dry weights, so it is clearly difficult to apply enough organic matter to influence quickly the organic content of a soil. An application of farmyard manure at the rate of 35 t/(ha.y) represents only about 4 t/(ha.y) of organic carbon; the rest is moisture and noncarbon dry components. More than 11 years of such additions would be needed to increase soil organic carbon from 1 per cent to 1.5 per cent, even if none of the added carbon were being lost by oxidation, or carried down by natural processes to lower depths. In practice, soil organic carbon is subject to losses with time, decomposing as a multicomponent system: Jenkinson's three-component model matches the experimental situation fairly closely (Jenkinson, 1965; 1966; 1968). The most rapidly oxidized fraction of fresh plant material is destroyed in a few months, while two other fractions decay with half-lives estimated as 4 and 25 years. The longest-lasting fraction is soil humus: according to Russell (1977) it amounts to about one-sixth of the carbon in straw and one-third of that in farmyard manure. Based on these figures, the 35 t/ha dressing of farmyard manure discussed above contributes only 1.3 t/(ha.y) to this relatively stable soil humus fraction. Under a regime of annual dressings at this rate, soil humus build-up can be only very slow, the modest annual increment in organic carbon being eroded by the annual loss from the soil: the work of Jenkinson and Johnson (1977) showed that at this rate an increase to 3.3 per cent of soil carbon at Hoosfield, Rothamsted, took 100 years. In fact, one hectare of arable land with 2.5 per cent of a 3000 t surface layer as humus, having a 25-year half-life, loses 2.725 per cent of its organic carbon each year, that is 2.04 t/y, which must be replaced with an equal amount of similarly durable humus material. The rapidly decaying fractions of organic carbon may be virtually ignored for this purpose. Such an addition would call for 10.6 t/y of dry matter as farmyard manure (on the basis that one-third of it enters the humus fraction) or 21.1 t/y of dry matter from crops (on the basis that one-sixth of it enters the humus fraction), if there were no additions to soil organic matter from the arable farming operations themselves. In the case quoted, about 4.15 t/(ha.y) of organic carbon was apparently being supplied by the stubble and roots of the crop, and may be taken as supplying 0.69 t/y to the organic carbon of the humus fraction. This has the effect of reducing the calculated farmyard manure requirement to 6.98 t/(ha.y) of dry matter and the calculated crop residue requirement to 13.96 t/(ha.y); the farmyard manure requirement equates with about 35 t/(ha.y), wet weight.

These quantities of organic matter could not be generated by the arable operation itself. By ploughing in an average yield of straw (say 3.7 t/(ha.y), 3.2 t/(ha.y) of solids) and ploughing in a catch crop yielding, say, 3 t/(ha.y) dry matter, a further 0.60 t/(ha.y) of humus organic carbon fraction may be obtained. Nevertheless there remains a deficit equal to 0.75 t/(ha.y) of humus organic carbon, requiring 3.88 t/(ha.y) of farmyard manure dry matter, or 7.76 t/(ha.y) of additional crop residues. In the UK as a whole, if the total tillage area of 4.25×10^6 ha called for the addition of livestock waste dry matter at the rate of 3.88–6.98 t/(ha.y), the demand overall would total between 16.5 Mt/y and 29.7 Mt/y. Either of these levels of addition would be quite impracticable; the lower figure of 16.5 Mt/y represents a fair estimate of total annual cattle, pig and poultry livestock waste production in the UK, well over half of which is in the form of uncollectable cattle waste. The use of much of the remainder as a soil conditioner in arable farming would be precluded economically by long transportation distances between the principal livestock and arable areas, and perhaps also by the danger of nutrient depletion on grassland farms if inorganic fertilizers were not used.

What emerges, therefore, is a complete lack of reconciliation, as long as present patterns of agriculture prevail, between the need for agricultural enterprises to export organic carbon in their produce, and their need to replace it in their own soils in accordance with biological husbandry principles. This need not be interpreted as an argument against the introduction of biological husbandry, but it does indicate that massive changes would be needed to make sufficient organic materials available. The problem arises, clearly, because in agriculture as now practised the primary annual yield of biomass is too low: systems have been optimized to produce maximum yields of the marketable portions of food crops, not total biomass. Two possible solutions are: (1) for agriculture to receive massive inputs of organic matter from outside sources, for example municipal solid waste or seaweed, by composting, although this seems likely to be an expensive option and the resources are open to competition for other purposes; (2) because agriculture exists to produce organic carbon rather than to consume it, the actual production of organic carbon might be markedly increased by combined food and energy farming.

The key factor in this approach is the observation by Callaghan *et al.* (1978) and by Lawson, Callaghan and Scott (1980), that some perennial weeds or exotic species, not normally thought of as crops, notably *Polygonum cuspidatum* and *Polygonum sacchalinense*, are capable of producing annual yields of dry matter of between 20 and 40 dry t/(ha.y) in natural stands, without management or use of fertilizers. These plants are being considered as possible pure-energy crops, the most likely route of utilization being by anaerobic digestion to methane; as pure-energy crops they could only be allocated to nonagricultural or marginal agricultural land. However, Plaskett (1980) has proposed dual utilization of such crops: for example, a 24 t/(ha.y) dry-matter yield of a nonwoody perennial energy species might be composed of 8 t/(ha.y) of a tender-leaf/growing-tip fodder fraction and 16 t/(ha.y) of a fuel fraction, to be passed through an anaerobic digester. The livestock wastes arising after feeding the fodder fraction to ruminants could similarly be passed through an anaerobic digester. That fraction of the original plant feedstock which resists ruminant digestion and anerobic digestion may be equated confidently with the highly resistant fraction which is not susceptible to microbial attack in the soil and which contributes to the long-lived soil humus. This fraction is fibrous and may be separated from the residues by use of a slurry separator. It partitions as the solid fraction of residues, leaving a liquid fraction with fine particles in suspension, from which a useful high-protein animal-feed fraction may be

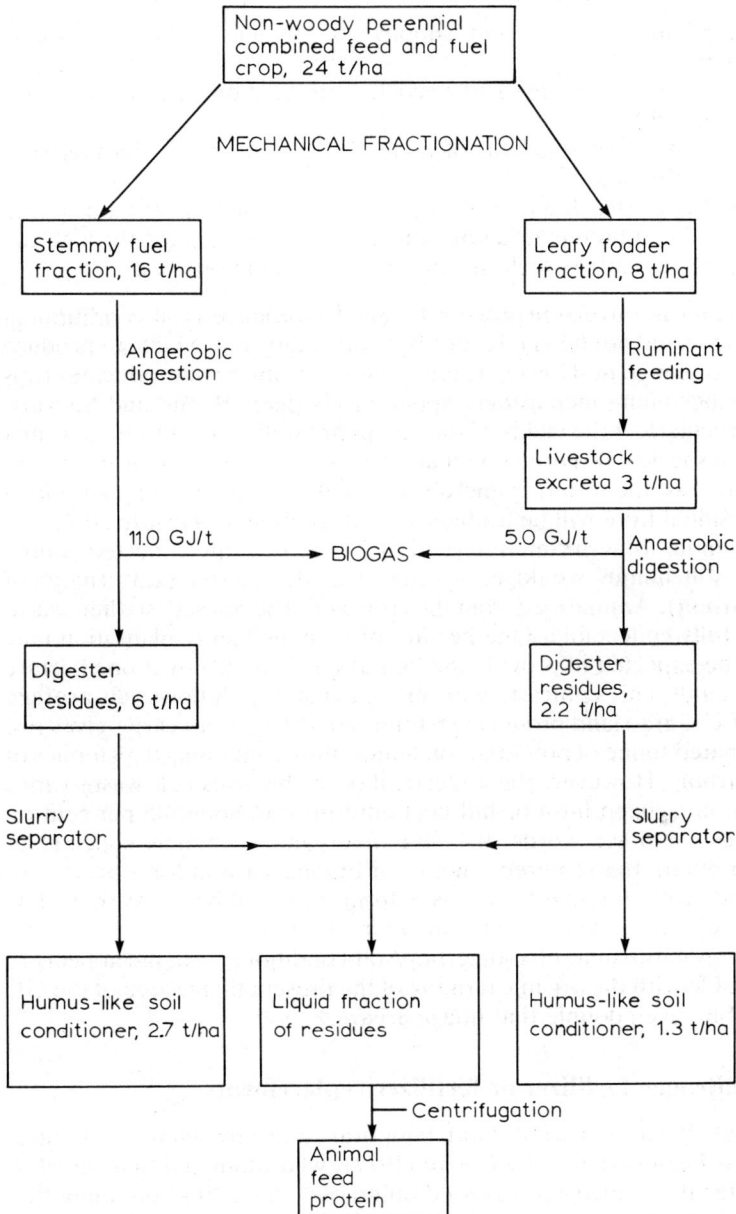

Figure 10.1 Potential impact of energy farming and crop fractionation upon soil humus supply

obtained by centrifuging. This technology can be seen to be relevant to biological husbandry in the light of:

1. The inadequacy of present supplies of organic carbon for arable farming;
2. The current interest in, and need for, production of nonfossil fuels within farming;
3. The likelihood of selection of anaerobic digestion as a biofuel conversion technology;
4. The ability of that fraction of organic carbon which contributes to the long-lived soil humus fraction, similarly to resist passage through the ruminant gut and through the anaerobic digestion process.

A biotechnical conversion process designed to produce a soil-conditioning agent alone would not be viable, nor is it very likely to be viable to produce crops just to plough in. However, the economic viability of producing crops for fuel is becoming increasingly apparent (Badger, Bogue and Stewart, 1979), particularly if the yields of such crops are well above those of normal agricultural species. It is probable that, in combination with such an energy-yielding process, the resulting microbial protein will be salvageable for feed and the residual fibre will be usable as a soil conditioner (*Figure 10.1*).

At an annual total dry-matter yield of 24 t/(ha.y), the expected contribution to soil humus would be 4 t/(ha.y) of dry matter (2.3 t/(ha.y) of organic carbon). Animals fed from the crop would be housed, so their waste would be fully collectable. One hectare of such an energy plantation may therefore be expected to provide the humus needs of just over one hectare of arable land. This contrasts with grassland use as follows: an average hectare of UK grassland producing 6 tonnes of dry matter yearly, gives rise to an estimated tonne of potential soil humus fibre containing 0.58 tonnes of organic carbon. However, the collectability of this livestock waste varies from zero on a sheep farm or hill beef enterprise to about 45 per cent on dairy farms; in other words, 0–0.26 t of organic carbon is collectable. Hence, to obtain the required amount of organic carbon for a hectare of arable land, calls for livestock waste from almost 8 ha of average UK grassland at best, but very often from a far larger area.

The energy-farm/anaerobic-digestion/soil-conditioner scheme appears to be compatible with the organic farming of the present tillage area of the UK and, possibly, even double that tillage area.

Organic nitrogen fertilizer or fertilizer replacements

The nitrogen balance of arable land, temporary and permanent grassland in the UK has been reviewed by Cooke (1977), who estimated that, in 1974, 370 kt of fertilizer nitrogen was used on arable land, 220 kt on temporary grassland and 200 kt on permanent grassland, even though known and estimated nitrogen inputs to the land exceeded known outputs as crops and leaching losses in each case: the estimated excess was 190 kt for arable land, 490 kt for temporary grassland and 565 kt for permanent grassland. Clearly, unmeasured nitrogen-wasting processes are intervening. The strategy of biological husbandry, in seeking to avoid the use of synthetic nitrogen fertilizer, must be either: (1) to find an economically viable way of producing an organic nitrogen fertilizer to replace synthetically fixed nitrogen, which can be added to soil without upsetting the ionic balance; or (2) to stimulate the natural processes of nitrogen fixation; or (3) to identify and prevent, at least in part, the processes that at present cause nitrogen to be lost from the soil economy. New approaches to enhancement of biological nitrogen fixation have been reviewed by Postgate (1977); most of the concepts he discusses should find ready acceptance within biological husbandry if they can be put into practice. Some of these concepts would call for biotechnology production processes: others, notably the introduction of gene sequences related to nitrogen fixation into plant chromosomes, would not, because they are aimed simply at the production of more efficient plant types for use in agriculture.

At present, replacement of the 'bag' nitrogen produced by the chemical industry with acceptable biological alternatives, or the elimination of the need for it, constitutes an unsolved problem. Universal application of biological husbandry methods may become possible only in conjunction with dependence upon the biotechnology processes described below, several of which have not yet been shown to be practical possibilities.

PRODUCTION OF NITROGEN FERTILIZER BY FERMENTATION OR
PHOTOSYNTHETIC PROCESSES

The production of single-cell protein by growing microorganisms upon sugary, starchy or cellulosic substrates has been thoroughly investigated (Matales and Tannenbaum, 1969; Hughes and Rose, 1971). Where the substrate is nitrogen-deficient, inorganic nitrogen must usually be added in the form of ammonium salts to enable protein formation to take place. From the standpoint of biological husbandry, such a process is of inherent interest, even without the introduction of any nitrogen-fixation aspect, because it organifies inorganic nitrogen: the protein is fed to animals and when their excreta are used as fertilizer the soil receives organic nitrogen which has not had to be fixed by biological processes, although no inorganic nitrogen has had to be applied to soil or crops to achieve this (*Figure 10.2*).

If, in addition, the bacteria used were produced in a form having an ability to fix atmospheric nitrogen, the need for externally applied ammonia would disappear. However, yields of protein would probably be reduced, because the energy requirement for ammonia synthesis would have to be

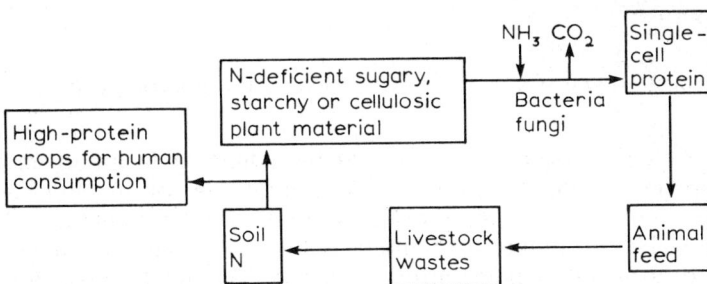

Figure 10.2 The effect of single-cell protein production in organifying inorganic nitrogen: impact upon the nitrogen cycle

subtracted from that otherwise available for biosynthesis of cell biomass. The metabolic requirement for this purpose appears to be in the region of 14.9 MJ/kg of ammonia (Bergersen, 1977) compared with a minimum of 50 MJ for process input to chemical forms of nitrogen (Lockeretz, 1979).

In view of such a major energy requirement for nitrogen fixation, it is interesting to consider the possibility of harnessing biotechnology processes that employ photosynthetic microorganisms for this purpose, giving a direct link between solar energy and the fixation of nitrogen for agricultural purposes. Precisely such a possibility has been emerging from the work of Weissman and Benemann (1977) on the production of hydrogen gas by nitrogen-starved cultures of the blue-green alga *Anabaena cylindrica*: upon introducing nitrogen into the system it is avidly reduced to ammonia, and hydrogen evolution ceases. There are many difficulties standing in the way of harnessing microorganisms commercially for hydrogen generation; however, the nitrogen-rich biomass of photosynthetic blue-green algae could well be incorporated into annual feeding systems such as that shown in *Figure 10.2*, and hence supply organic nitrogen to the soil.

It is also possible that microorganisms capable of fixing nitrogen could be greatly inhibited from incorporating the nitrogenous products into their own cellular protein, so that these would become available as reaction products and usable as nitrogen fertilizer. In theory this might be done by genetic or enzyme manipulation, either with organisms that derive their energy from organic substrates or with photosynthetic organisms, the latter process being more likely to be economically viable.

BACTERIAL PREPARATIONS AND BIOCHEMICALS

Biotechnology processes are also very likely to play a part in the encouragement of natural processes of nitrogen fixation. An obvious case would be rhizobial inoculation of seeds, where supplies of the appropriate strain of *Rhizobium* have to be produced. In time, the direct inoculation of soils, composts and other organic fertilizers may become a desirable and even regular practice to encourage both symbiotic and free-living nitrogen-fixing organisms. The substances added may again be bacterial preparations, perhaps produced in continuous fermenters, or specific biochemicals that may either actively encourage natural nitrogen fixation or inhibit denitrification. This is entirely speculative at the present time, but to envisage these possibilities may be a stimulus to much-needed research in these important areas.

METHODS OF MORE EFFICIENT UTILIZATION OF AVAILABLE ORGANIC NITROGEN

As it appears that nitrogen scarcity would be a limiting factor to any universally applied organic agriculture, at least until some of the biotechnology processes envisaged in the previous sections become available and economically viable, every means should be employed to make the fullest possible use of available organic nitrogen sources. The main sources are livestock waste, plant residues and green manures, which are normally involved in the cycle (*Figure 10.3*).

The direct conversion of livestock-waste nitrogen and green-manure nitrogen into animal-feed nitrogen would be very helpful to overall conservation of nitrogen by avoiding losses that occur through volatilization, denitrification and leaching when these materials are added to the soil. Experimental work undertaken by members of Biotechnical Processes Limited has shown that this is likely to be achieved by recovery of microbial protein after anaerobic digestion.

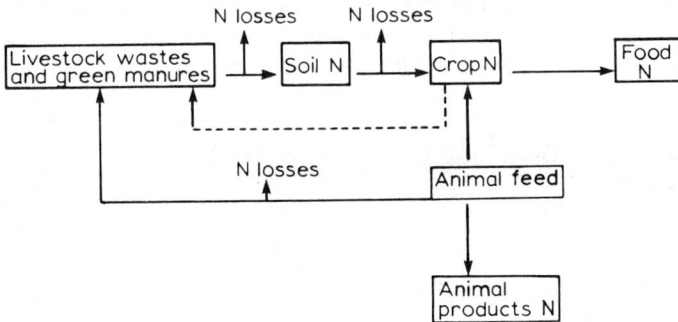

Figure 10.3 Normal role of livestock wastes and green manures in the nitrogen cycle

In this work, anaerobically digested dairy waste, containing 9.1 per cent solids after digestion, was fractionated by means of a slurry separator into a solid fraction, containing 61.8 per cent of the dry matter at 17.15 per cent solids and a liquid fraction containing the remaining 38.2 per cent of the dry matter at 5.07 per cent solids. The solid fraction contained only 35.1 per cent of the total nitrogen of the residues and this fell to only 25.0 per cent after washing. The solids of the liquid fraction were therefore relatively enriched with nitrogen; they were, in fact, 16.47 per cent nitrogen or, allowing for 27.7 per cent of ash in the solids, the remaining, volatile, solids were 22.78 per cent nitrogen. Much of this was shown to be present, as expected, in the form of ammonia, but organic solids were centrifuged from the liquid fraction as a fine, dark material: the crude protein content of this varied from 27.2 per cent in the bottom layer of centrifugate to 55.8 per cent in the upper layer (41.9 per cent crude protein overall). The yield thus obtained represented 16.35 per cent of the solids in the whole digester residue—an estimated 11.77 per cent of the solids of the digester feedstock. The availability of such good yields, from dairy waste, of a material with crude protein contents comparable to that of soya meal, obviously opens up the possibility of basing a substantial animal-feed industry upon such material.

If it is confirmed that the material isolated in our laboratory work performs satisfactorily as an animal-feed concentrate, and if commercial yields were comparable, then the potential production from collectable UK dairy waste alone would amount to the equivalent of about one-third of the UK annual imports of animal-feed protein concentrate. The sequence, as far as the nitrogen cycle was concerned would then be as shown in *Figure 10.4*. A proportion of the digester-residue nitrogen has been converted to animal feed with exceptionally high efficiency and consequent avoidance of losses.

This clearly represents an efficient use and conservation of already-fixed nitrogen that is available now within the agricultural system.

Some plantation energy crops, like cereal straw, may be too nitrogen-poor to be anaerobically digested without added nitrogen nutrients. Ammonium salts may be added, as has already been discussed for aerobic fermentation, in which case the anaerobic bacteria would organify the inorganic nitrogen and make it acceptable as a biological husbandry fertilizer. However, an attractive opportunity exists to fulfil the nitrogen needs of the process photosynthetically if the digestion of the nitrogen-deficient energy crop is accompanied by the growth of nitrogen-fixing blue-green algae, such as *Anabaena,* which is then used as a supplementary

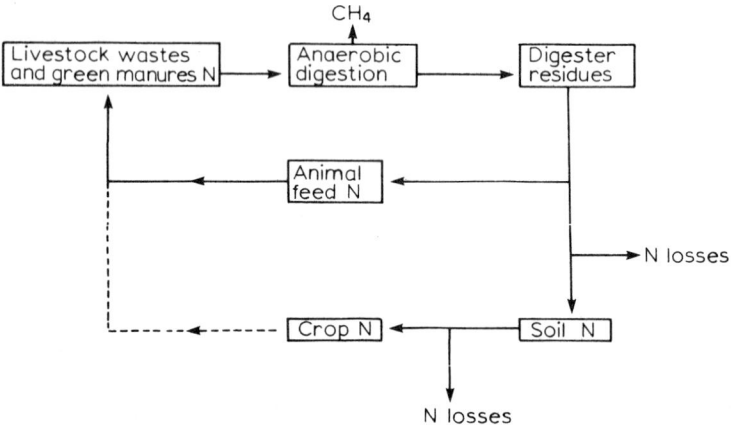

Figure 10.4   Short-circuit of the nitrogen cycle by recovery of protein from digester residues

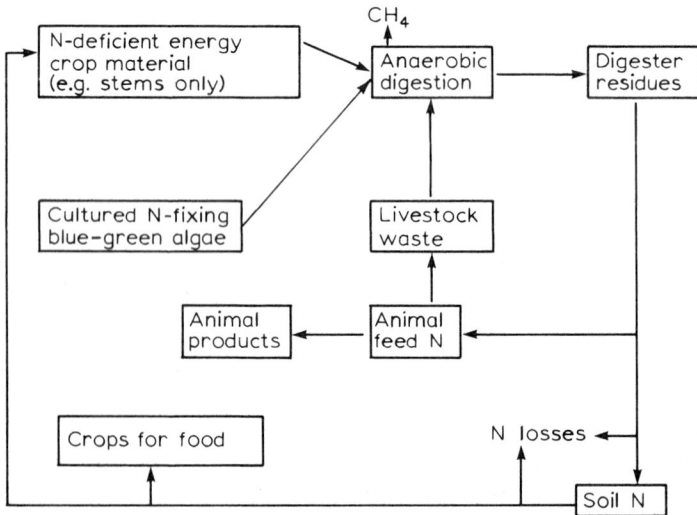

Figure 10.5   Speculative concept of a combined energy/food agriculture supported by biosynthetic nitrogen produced externally to the soil in process plants

feedstock in the anaerobic digestion process (*Figure 10.5*). This process ensures that sufficient nitrogen is always present to permit anaerobic digestion to take place, while the relatively expensive algal biomass is, in effect, used twice: once for energy generation and once for animal feed. Again, the soil receives nitrogen fertilizer in organic form only.

MINERALS IN ORGANIC FORM

In that biotechnology processes will neither form nor destroy minerals, bio-technology has least to offer to biological husbandry in that area. Probably an increasing demand for off-farm organic sources of minerals, such as seaweed, can be foreseen within biological husbandry. However, deep-rooting perennial energy crops can be expected to bring up minerals from lower layers of soil than annual agricultural crops: hence the minerals in their residues after digestion would, in a sense, be a new resource, either for re-application to the energy farm or on to other farmland.

SPECIALIST BIOMOLECULES

This is an area in which biotechnology may well have much to offer, yet its use may well raise issues of principle within biological husbandry. An antibiotic is a case of a natural substance produced in unnatural concen-tration by biotechnical means. If we had an antibiotic against plant diseases, would its use be in accordance with the principles of biological husbandry or not? Its use would not involve a synthetic chemical agent, yet would imply a considerable degree of outside interference over and above the plant's own defensive mechanism. Some such substances might conceivably come from simple extraction of plant tissues and hence fall within Definition 1 of biotechnology, but not within Definition 2.

In time, we may have economically viable options of producing (by bio-logical means rather than synthetic chemical processes, and therefore in biotechnology plants) a range of specialized substances, able to influence plant growth, soil bacteria, plant disease resistance, the behaviour, viru-lence and fertility of pests; all this presupposes a vast amount of research. Before this is undertaken, perhaps proponents of biological husbandry should determine, in principle, to what extent they would like to use the resources of biotechnology—a discipline which potentially, at least, relates to their own in countless different ways. It could constitute for the organic farmer, what the chemical industry is for the so-called 'inorganic farmer', and it could accomplish such a working relationship without contravening any vital principles of biological husbandry, because biology is the corner-stone of both disciplines. Moreover, seen in this light, organic husbandry does not have to spell disaster to the fertilizer and chemical industry, provided that it is prepared to become orientated to biotechnology.

References

BADGER, D. M., BOGUE, M. J. and STEWART,D. J. (1979). Biogas production from crops and organic wastes. *N.Z. Jl. Sci. Technol.*, **22**, 11
BERGERSEN, F. (1977). Recent developments in nitrogen fixation. In *Proceedings of the Second International Symposium on Nitrogen Fixation,*

Salamanca, 1976. Eds W. E. Newton, J. R. Postgate and C. Rodrigues-Barrveco. London; Academic Press

CALLAGHAN, T. V., MILLAR, A., POWELL, D. and LAWSON, G. J. (1978). *Carbon as a Renewable Energy Resource in the U.K. – a Conceptual Approach.* Study undertaken for Energy Technology Support Unit by Institute of Terrestrial Ecology

COOKE, G. W. (1977). Waste of fertilizers. *Phil. Trans. Roy. Soc. Ser. B*, **281**, 231–241

GREENLAND, D. J., RIMMER, D. and PAYNE, D. (1975). Determination of the structural stability class of English and Welsh soils using a water coherence test. *J. Soil Sci.,* **26**, 294

HUGHES, D. E. and ROSE, A. H. (1971). *Microbes and Biological Productivity.* Cambridge; Cambridge University Press

JENKINSON, D. S. (1965). Studies in the decomposition of plant material in Soil, I. *J. Soil Sci.,* **16**, 104

JENKINSON, D. S. (1966). Studies in the decomposition of plant material in soil, II. *J. Soil Sci.,* **17**, 280

JENKINSON, D. S. (1968). Studies in the decomposition of plant material in soil, III. *J. Soil Sci.,* **19**, 25

JENKINSON, D. S. and JOHNSON, A. E. (1977). Soil organic matter in the Hoosfield barley experiment. In *Rothamsted Exp. Stn. Rep. for 1976,* Part 2, p. 87

JOINT WORKING PARTY ON BIOTECHNOLOGY (1980). Advisory Council for Applied Research and Development; Advisory Board for the Research Councils, The Royal Society. London; HMSO

LAWSON, G. J., CALLAGHAN, T. V. and SCOTT, R. (1980). *Natural Vegetation as a Renewable Energy Resource in the UK.* Study undertaken for Energy Technology Support Unit by Institute of Terrestrial Ecology

LOCKERETZ, W. (1979). Energy inputs for nitrogen, phosphorus and potash fertilizers. In *Energy in Agriculture.* Ed. D. Pimentel. CRC Handbook Series. West Palm Beach, Florida; CRC Press

MATALES, R. I. and TANNENBAUM, S. R. (1969). *Single-Cell Protein.* Cambridge, Mass; MIT Press

PLASKETT, L. G. (1980). The potential role of crop fractionation in the production of energy from biomass. Paper at the *Energy from Biomass Conference, Brighton,* 1980, Commission of the European Communities, in preparation

POSTGATE, J. R. (1977). Possibilities for enhancement of biological nitrogen fixation in agricultural efficiency. *Phil. Trans. Roy. Soc. Ser. B*, **281**, 249–260

RUSSELL, E. W. (1977). The role of organic matter in soil fertility in agricultural efficiency. *Phil. Trans. Roy. Soc. Ser. B*, **281**, 209–219

SCHMIDT-KASTNER, G. (1978). Methods for isolation and purification of enzymes. In *Proceedings of the 1st European Congress on Biotechnology, Interlaken, September 1978*

SCHUMACHER, E. F. (1973). Small is Beautiful. London; Blond & Briggs

WEISSMAN, J. G. and BENEMANN, J. R. (1977). Hydrogen production by nitrogen-starved cultures of *Anabaena cylindrica. Appl. Environ. Microbiol.,* **33** (1), 123

11

LIMITING FACTORS IN BIOLOGICAL HUSBANDRY—
A BIOCHEMICAL APPROACH

J. COOMBS
Winchcombe Cottage, Bucklebury Common, Reading, UK

Introduction

The actual yield (expressed as weight of dry matter produced per area of land cultivated) for a given crop reflects the interaction between the genetic characteristics of the species concerned and environmental factors. In general, the amount of material harvested will fall short of the potential yield which could have been achieved under optimal conditions, because of the various forms of stress shown in *Figure 11.1.* These stresses may be grouped as physical or biological—the former arising from extremes of such factors as temperature, light, carbon dioxide, oxygen and water, and the

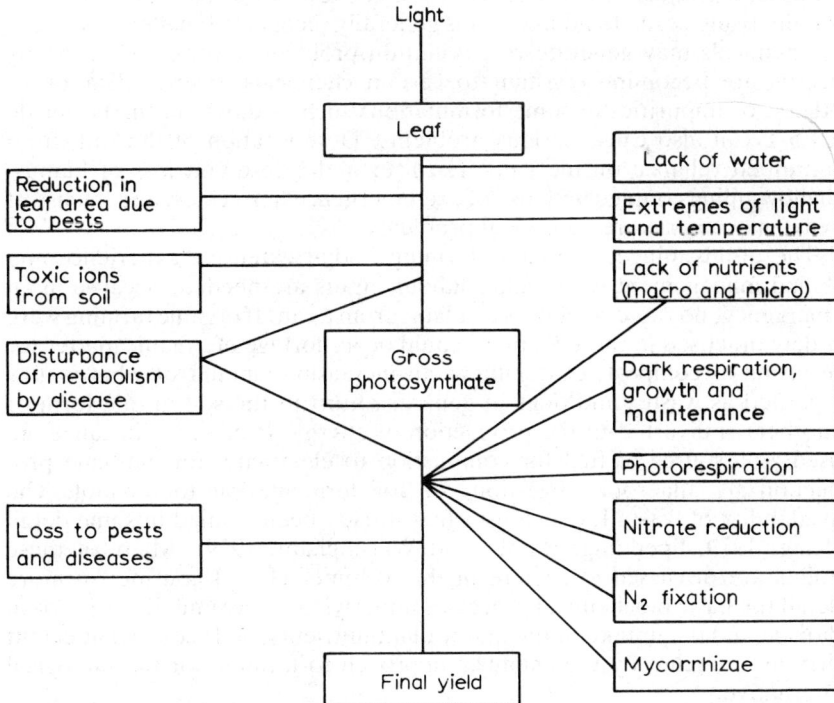

Figure 11.1 Factors contributing to loss of photosynthate during crop growth

157

latter arising from damage to the plants through pests and diseases. In addition, stress may be induced in the plants by lack, or assimilation problems, both of major nutrients such as nitrogen, phosphate, potassium and sulphur, and of a range of essential microelements. Of particular importance is the nature of the soil, where extremes of pH can affect the uptake of both essential and potentially toxic metal ions. The aim of agriculture, whether based on chemical application or biological husbandry, is to improve yields by decreasing, where possible, the effect of these stress factors.

Over the last 30–40 years, agriculture in the developed world has evolved scientifically based methods of overcoming these stress factors, resulting in dramatic increases in yield per unit area. This marked improvement has been brought about partly by breeding of new varieties, and partly by large increases in energy inputs in the form of tractor fuel, fertilizer and pesticides.

The use of inorganic fertilizers is based on extensive trials on most of the major crops. Similarly, most herbicides and pesticides have been developed and tested extensively before marketing. As a result, detailed instructions are available to the chemical farmer, and any problem of nutritional deficiency or attack by pests can be tackled rapidly by opening a sack or making up a spray. In contrast biological husbandry represents an art, in which good management of land and organic resources can often result in similar yields, dispensing with the need for high levels of chemical input. The main disadvantage of the chemical approach is that it relies on oil and oil-based chemicals, and hence over the last few years has been faced with rapidly rising costs. In addition, it is generally recognized that excessive use of chemicals may generate real pollution problems (Anon, 1979). Many insects are becoming resistant to certain chemicals; over-application of others, or impurities in some formulations such as dioxin in the herbicide 2,4,5-T can also cause serious problems. Deterioration of the soil, from continued reliance on inorganic fertilizer and a resultant loss of humus, shallow tilling coupled with use of pre-emergence herbicides, and the use of heavy machinery, are additional problems.

The disadvantage of organic farming is that extensive experience and skilled management with higher labour inputs are needed, because in an emergency, no rapid remedy is available from a can. If organic farming were widely practised in the UK, there could be a shortage of organic matter for recycling as compost, especially as an increasing amount of what is now regarded as waste (and does in general return to the soil in one form or another) is diverted to the production of energy. It may, for instance, be used as a source of fuel for combustion to electricity, for methane production by anaerobic digestion, or for fermentation to ethanol. The availability of waste for such uses has already been studied in some detail (King, 1979; Spedding, Bather and Walsingham, 1979). Many of these points are discussed elsewhere in this volume. Here I consider in more detail the basic biochemical processes underlying the assimilation of carbon dioxide and the uptake of the major plant nutrients, and see to what extent they are affected by the chemical approach to farming, or the biological alternative.

Photosynthetic potential

The potential for photosynthesis in a given area is determined mainly by geographical and climatic characteristics—factors which cannot be avoided except by use of controlled-environment enclosures. Such control is used in the horticultural production of tomatoes, especially in using the nutrient film technique, and by the animal farmer producing pigs or eggs intensively.

As shown in *Table 11.1* and discussed in detail below, the theoretical conversion of solar energy into organic material is 5–6 per cent. At the annual mean global irradiance level which occurs in the UK (around 105 W/m^2), and allowing for the cold winter months when plant growth is precluded, one might theoretically expect yearly yields approaching 100 tonnes of dry matter per hectare. Actual yields are nothing like this, nor are they achieved even in the most productive crops, such as sugar-cane growing for most of the year under optimal conditions in the subtropics. The reasons for the limited yield are indicated in *Table 11.1*. This information may be considered step by step.

Table 11.1 LIMITATION IN THE CONVERSION OF SOLAR ENERGY TO DRY MATTER

Type of limitation	%
Photochemical	
Available solar energy	100
Photosynthetically active radiation	43
Less loss from light scattering	38
After correction for quantum efficiency	11
Biochemical	
After correction for respiratory loss	6
After correction for photorespiration	4
Environmental stress	
Lack of water and nutrients reduces to	1
Pests and diseases	
Result in loss of about half	0.5
Harvest losses	
Only part of crop harvested, varies with crop	0.1 to 0.4
Result of postharvest losses	0.05 to 0.3

Not all the solar energy which reaches the surface of the earth can be used by green plants. In fact, plants use radiation between 400–700 nm, the so-called photosynthetically active radiation (PAR) which has been established as about 43 per cent of the total solar irradiance. Absorption of this radiation by chlorophyll in the chloroplast results in a flow of electrons through the membranes or thylakoids comprising the grana of the chloroplasts. This electron flow in turn results in the splitting of water to form a reductant-reduced NADPH (β-nicotinamide adenine dinucleotide phosphate) and a form of complexed chemical energy, ATP (adenosine-5′-triphosphate). These products are the result of two separate photoreactions working in series. The ATP and NADPH are used as the energy source and reductant to reduce carbon dioxide (CO_2) from the atmosphere to the level of carbohydrate. Fixed CO_2, in the form of carbohydrate, has the energy content of 0.47 MJ/mol, whereas the energy content of a mole quantum (einstein) of red light at 680 nm (the least energetic light able to perform photosynthesis) is 0.176 MJ. Reduction of one molecule of CO_2 requires the flow of four or more electrons through both photosystems, in total at least eight photons. Because of limitations in the efficiency of the photochemical reactions of photosynthesis, this number is, in fact, probably nearer to ten. Hence, 0.47 MJ are trapped as a result of the expenditure of 10×0.176 MJ, and the efficiency will be equivalent to the 43 per cent PAR multiplied by the fraction 0.47/1.76, or 11–12 per cent. However, the overall practical maximum efficiency of conversion of solar energy into fixed carbon compounds over a prolonged period of time is approximately 5–6 per cent. This figure is derived from current knowledge of the process of carbon fixation and of the physiological losses which occur during dark respiration and photorespiration. Dark respiration is required to produce the energy and reducing power needed to convert the primary photo-synthetic product (carbohydrate) to other more reduced or more complex carbon compounds (proteins, fats, nucleic acids and pigments). In addition, the energy released from carbohydrates during respiration is used for assimilation—and where necessary for reduction and incorporation—of inorganic nitrogen, phosphorus and sulphur, as well as the uptake of micro-nutrients in energy-dependent active-transport reactions.

It is now well established that all green plants and algae that are capable of splitting water posses the photosynthetic carbon reduction (PCR) cycle (Bassham and Calvin, 1957) as the major route of net carbon assimilation. In this pathway the initial reaction is the carboxylation of ribulose-bis-phosphate (RBP) resulting in the formation of two molecules of phospho-glyceric acid (PGA). The ATP and NADPH formed in the light reactions of photosynthesis are used to drive this cycle in the direction of net synthesis. Plants which have only the primary PCR cycle are known as C3 plants, because the first-formed product contains three carbon atoms. The ability of C3 plants to assimilate and retain atmospheric carbon as organic matter is limited by the fact that carbon fixation may be inhibited by oxygen. This inhibition is associated with a loss of some of the carbon which has already been fixed, in the process known as photorespiration (Tolbert, 1971). Oxygen inhibits photosynthesis by competing with CO_2 at the binding site on the enzyme which catalyzes the initial carboxylation reaction (Ogren and Bowes, 1971) so that the two following reactions occur in competition with one another:—

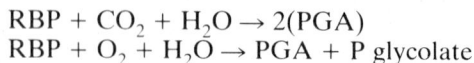

$$RBP + CO_2 + H_2O \rightarrow 2(PGA)$$
$$RBP + O_2 + H_2O \rightarrow PGA + P \text{ glycolate}$$

In the C2 or photorespiratory pathway (Tolbert, 1963) two molecules of glycollic acid are used to form one of PGA (thus conserving at least some of the fixed carbon) and the fourth carbon is lost as CO_2. It can be estimated that flow through this pathway results in a loss of at least 15 per cent of the carbon fixed. However, because photorespiration is favoured by high light intensities, high partial pressures of oxygen, low CO_2 concentrations and high temperatures, on hot still days crop losses in large stands—fields of wheat for example—may reach 30–40 per cent. Furthermore, although one might expect that yields of all crops would increase as one moved towards the equator, this may in fact not be true for C3 plants, as photorespiratory losses increase.

However, some species known as C4 plants have evolved a second pathway (the C4 pathway) for assimilating CO_2 from the atmosphere, involving the carboxylation of a second acceptor molecule—phospho-enol-pyruvate (PEP). Carbon dioxide is fixed in the cytoplasm and transported to the chloroplasts in the form of a 4C organic acid (malate or aspartate) where the CO_2 is released and reassimilated through the normal PCR cycle. The C4 cycle is similar to the PCR cycle in that it consists of a carboxylation reaction followed by a reduction. However, in the C4 cycle there is no net gain of CO_2. In other words, the C4 cycle does not act as a synthetic pathway in the true sense, but rather as a pump (Hatch and Slack, 1970; Coombs, 1976), moving CO_2 from the atmosphere to the chloroplast. As a result the CO_2 concentration at the carboxylation site of the PCR cycle is increased, and hence the inhibitory effect of oxygen is decreased.

As a result of the addition of the C4 cycle, the C4 plants are more efficient than C3 plants at fixing CO_2, and thus produce higher yields of dry matter as long as climatic conditions are right. However, the extra cycle requires more light energy to drive it. Hence, C4 plants are, in general, restricted to warmer climates with higher light intensities. The major C4 crops of importance are maize, sorghum and sugar-cane. Advances in plant breeding have extended the range of maize into southern England. Another advantage of C4 plants is that they are more efficient users of water. For instance, maize may lose (transpire) only 370 g of water from the leaves per gram of dry matter produced, whereas temperate cereals may require over 500 g of water per gram (Woolhouse, 1978). Attempts have been made to incorporate characteristics of C4 plants into C3 species by hybridization of plants from such genera as *Atriplex,* which contain both types (Bjorkman, 1976). However, so far no commercially useful strains have emerged from this work.

There is probably little that can be done to improve the actual photochemical efficiency of carbon assimilation. Yields could be improved in both C3 and C4 plants if the dark-respiratory losses could be reduced (McCree, 1976), or if plants with low rates of photorespiration could be identified. Already there is some indication that it is possible to select strains with lower rates of dark respiration, but in spite of considerable work, especially in the United States (Ogren, 1976), no such plants have yet been identified. Hence, at present it would appear that for all practical purposes, irrespective of the farming approach adopted, yields of around 4 per cent depending on available solar radiation may be regarded as the optimum for C3 plants, with C4 plants showing a slightly higher potential.

Nitrogen

Where nutritional stress is concerned, the most pressing problems are related to lack of nitrogen. The reason for this is that most of the nitrogen sources available to plants are soluble in water, and are thus leached from the soil by rain or irrigation. This fact is reflected in the dramatic increase in use of nitrogen fertilizer since the Second World War. Although the annual use of P_2O_5 has remained at about 400 thousand tonnes, and the use of K_2O has almost doubled to reach the same level, during the same period the use of nitrogen fertilizer has increased seven-fold, so that now about 1.2 million tonnes are being applied each year (Anon, 1979). However, much of this does not, in fact, end as plant constituents, but is lost by leaching to rivers and lakes or degraded by microorganisms. Considerable research has been undertaken to investigate the possibility of replacing a proportion of this with biologically fixed nitrogen. In leguminous plants the property of fixing nitrogen from the atmosphere depends on the presence of symbiotic *Rhizobium* bacteria. A number of free-living blue-green algae and bacteria, for instance *Azotobacter*, *Beijerinckia* and *Clostridium* species, are also capable of fixing atmospheric nitrogen, and some plants—tropical grasses in

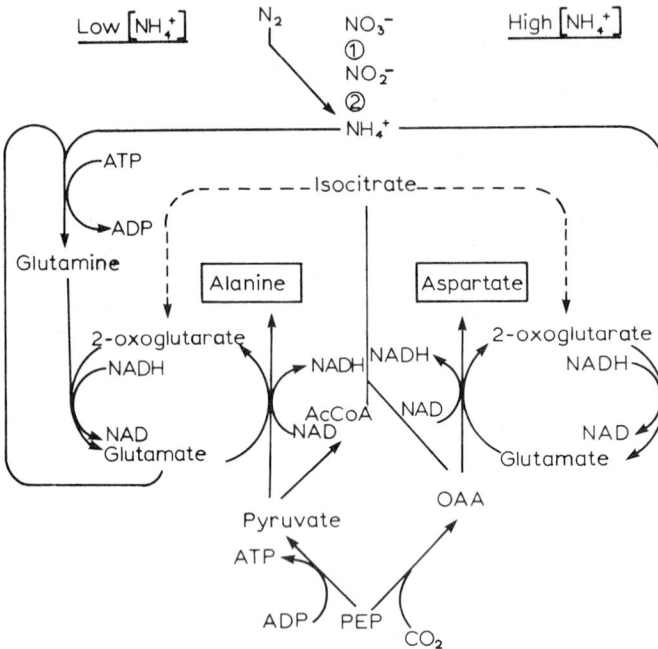

Figure 11.2   Pathways of assimilation of atmospheric and combined nitrogen

particular—have been shown to live in association with such nitrogen-fixing microorganisms as *Spirillum lipoferum*. This subject is covered in some detail in Granhall (1976). Although it has been widely suggested that genetic material coding for N_2 assimilation could be introduced into other plant species, this possibility is still remote. Thus for most plants nitrogen is available only in the form of inorganic combined nitrogen sources such as nitrate, urea and ammonium salts. The biochemistry of its assimilation, following reduction where necessary, is now well established (*Figure 11.2*).

For most plants, nitrate is the immediate source of nitrogen. Assimilation of nitrate occurs as follows:

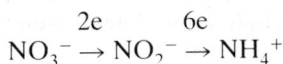

$$\overset{2e}{NO_3^-} \rightarrow \overset{6e}{NO_2^-} \rightarrow NH_4^+$$

This sequence is catalyzed by two metalloproteins, nitrate and nitrite reductase. The sources of electrons (indicated as e in the above equation) are reduced NAD (β-nicotinamide adenine dinucleotide), NADP or ferredoxin, depending on the source and type of enzyme. These reductants may in turn be produced by photosynthesis or respiratory activity. In either case there is direct or indirect competition for energy, to be used in carbon fixation, cellular metabolism or nitrogen reduction.

Nitrate reductase is a complex multicomponent enzyme containing flavin, haem and Mo prosthetic groups. Electrons provided by NAD(P)H are transferred to nitrate through the electron-transport chain of the enzyme:

$$H^+ + NAD(P)H \rightarrow \{FAD - Cytochrome_b - Mo\} \overset{NO_3^-}{\rightarrow} NO_2^- + H_2O$$

Of particular significance is the fact that assimilatory nitrate reductase from most organisms has a fast turnover rate, being present at high levels only when the plants are fed with nitrate, but becoming inactive if the plants are supplied with ammonium ions. For detailed references to this and other aspects of nitrogen metabolism *see* Hewitt and Cutting (1979) and Miflin (1980).

Most higher plants contain a chloroplastic nitrite reductase which accepts electrons from photosynthetically reduced ferredoxin via a tetranuclear iron–sulphur centre, the electron-transport chain being as follows:

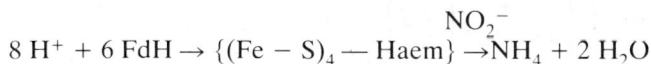

$$8\,H^+ + 6\,FdH \rightarrow \{(Fe - S)_4 - Haem\} \overset{NO_2^-}{\rightarrow} NH_4 + 2\,H_2O$$

Traditionally it was assumed that ammonium ions were assimilated into glutamic acid in the reaction catalyzed by glutamate dehydrogenase:

$$NH_3 + 2\ oxoglutarate + NAD(P)H \rightarrow glutamate + NAD(P) + H_2O$$

The mystery surrounding this enzyme involved its high K_m, or low affinity for ammonia, inconsistent with the ability of plants to assimilate low ammonia concentrations. However, an additional and more efficient pathway has now been discovered in higher plants (Miflin and Lea, 1976), the overall reactions being as follows:

1. Glutamine synthetase (GS)
   $$NH_3 + glutamate + ATP \rightarrow glutamine + H_2O + ADP + P_i$$
2. Glutamate synthase (GOGAT)
   $$Glutamine + 2\cdot oxoglutarate + NAD(P)H \rightarrow 2(glutamate) + NAD(P)$$

Glutamine synthetase is localized in the cytoplasm and chloroplasts of leaves, whereas GOGAT is situated mainly in the chloroplasts. Once again there will be competition for ATP and reductant between these enzymes and the fixation and reduction of CO_2, whereas glutamic dehydrogenase requires only reductant.

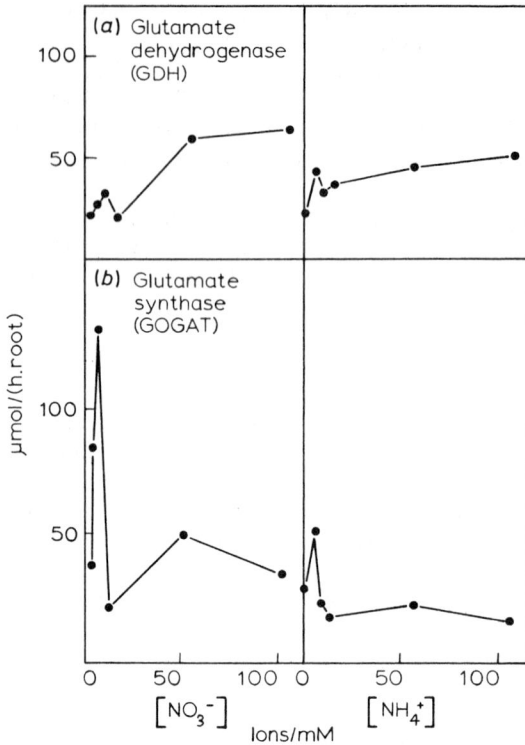

Figure 11.3   Effect of growing *Zea mays* plants, irrigated with a complete nutrient supply containing varying levels of nitrate and ammonium ions, on levels of (*a*) glutamate dehydrogenase (GHD) and (*b*) glutamate synthase (GOGAT), detectable in partially purified root extracts

Although it is now accepted that the GS/GOGAT pathway is of major importance in assimilation of combined nitrogen, we have some evidence (*Figure 11.3*) to suggest that the relative activities of the two pathways of ammonia assimilation may, in fact, vary according to the level of externally fed nitrate or ammonium ions. This might indicate that application of high levels of inorganic nitrogen fertilizer could suppress the pathway responsible for assimilating low levels of nitrogen.

Fixation of nitrogen gas

As already mentioned, only certain blue-green algae and bacteria can fix atmospheric nitrogen. These organisms may be free-living or symbiotic.

The basic metabolic requirements for atmospheric nitrogen reduction are (1) a strong reductant; (2) a source of energy—ATP; (3) low local oxygen concentration. The enzyme complex known as nitrogenase (for reviews *see* Hardy and Silver, 1977; Newton, Postgate and Rodriguez-Barrueco, 1977; Rawsthorne *et al.*, 1980), catalyzes the overall reaction:

$$N_2 + 3\,XH_2 + 6\,ATP \rightarrow 2\,NH_3 + 3\,X + 6\,ADP + 6\,P_i$$

This enzyme, again a multi-subunit protein, has usually been found to consist of two major components. The larger component, molybdoferredoxin, has four subunits and contains Mo, nonhaem iron and sulphide, whereas the smaller component, azoferredoxin, has two subunits containing iron and sulphide. Its biological nitrogen-fixation energy requirement of 355 KJ/mol NH_4^+ is roughly half that of the industrial process, which has an energy requirement of roughly 680 KJ/mol. The energy input, both biological and industrial, is needed to overcome the activation energy for the reduction of nitrogen: in the industrial process energy is needed also for the production of hydrogen.

The growth of legumes as a ley is obviously attractive as a biological source of nitrogen. However, the position for grain legumes is more complex. During the nitrogen-fixation process, of the photosynthate transported to the roots the largest proportion (46–76 per cent) representing

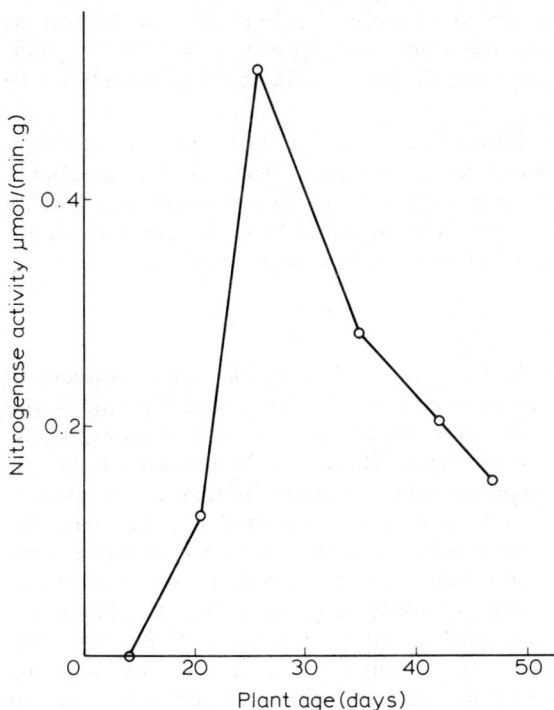

Figure 11.4 Changes in the level of nitrogenase (determined as ability to reduce acetylene) during the growth of dwarf french beans (*Phaseolus vulgaris* L.) up to the onset of flowering and early stages of pod-filling

from 19 per cent to over 40 per cent of net shoot photosynthesis, is lost as respired CO_2. Hence there is significant competition between the use of carbon for growth, and as an energy source for fixation of nitrogen. This competition becomes especially significant during the period of seed set and pod filling, and may result in a rapid drop in the rate of nitrogen fixation during the very period that the plant has the greatest need for nitrogen (*Figure 11.4*). A second problem is that the application of combined nitrogen reduces the ability of the *Rhizobium* to fix nitrogen (Kamberger, 1977; Latimore, Giddens and Ashley, 1977) and may also alter the pattern of incorporation of nitrogen into translocated metabolites (Cookson, Hughes and Coombs, 1980).

A similar reduction in the ability to fix nitrogen is seen in such free-living species as *Azotobacter* when treated with combined nitrogen. Coombs and Baldey (unpublished work) We have made an extensive study of the possibilities of using cells of *A. vinlandii* as a bacterial fertilizer. In a number of trials an increase in productivity of treated horticultural crops was observed. However, this could not be associated with nitrogen-fixation alone. In common with other workers in this area (*for example* Brown and Burlingham, 1968; Azcon and Barea, 1975) we concluded that part of the action of these organisms resulted from the production of chemical compounds with activities similar to those of 'plant growth substance'. However, in addition there was evidence that a compound produced by the bacteria was antagonistic to the growth of pathogenic fungi. The presence of *Azotobacter* on the root surface could suppress colonization by deleterious organisms: at the same time the stimulated growth caused by the plant growth hormones could strengthen the plant, making it less susceptible to infection.

However, in general we found that there was little point in adding a bacterial fertilizer to a soil which had insufficient organic content to support a natural population of the microorganism, or to one which was heavily fertilized with combined nitrogen. On the other hand there is little doubt that plant growth and bacterial growth are interdependent.

Phosphate and potassium

It is now fairly well established that most plant species are dependent to varying degrees on their association with certain specialized fungi—the mycorrhizae—which assist the plant in the assimilation of phosphorus, potassium and other essential nutrients (Mosse, 1973). In addition there is evidence that these fungi can protect the host plants from attack by disease-producing organisms. The problem here is again similar to that with the nitrogen-fixing bacteria. In the absence of applied fertilizer, improvements are seen on inoculation of sterile soils. However, application of phosphate, for instance, reduces the ability of these fungi to assimilate phosphate. Again, in parallel with nitrogen-fixation by symbiotic *Rhizobium*, the symbiotic fungi undoubtedly exact a toll from their hosts by diverting carbon compounds to their own use: in fact over 50 per cent of the carbon translocated to the roots may be used by the fungi. This may have a significant effect in decreasing root growth, so the compensatory benefits, if essential mineral nutrients are not supplied, have to be large to make up for it.

Conclusions

Deficiencies in yields of crops, brought about by physical factors and climatic extremes, lie outside the control of the farmer, whether orthodox or organic. Nutrient deficiencies may be compensated for, either by application of inorganic fertilizers, or by provision of a suitably enriched organic soil, favouring the proliferation of soil bacteria and fungi. However, these microorganisms, especially the symbiotic ones, will require considerable amounts of organic carbon compounds as an energy source for assimilation and reduction of applied nutrients. Competition between host and symbiote may result in yields lower than those obtained by application of inorganic fertilizers. The biochemistry of the plant and soil microorganisms is such that little distinction will ultimately be made between the source of the major nutrients. In general, heavy application of combined sources of such nutrients will reduce the activity of soil and symbiotic microorganisms.

Although not discussed in detail here, some of the most significant losses of available energy are those resulting from pests and diseases. Hence a careful and systematic programme of chemical pest control may in some instances show a greater effect on yields than fertilizer application. However, a combination of good agricultural practice, crop rotation, and selection of resistant varieties can and should be used to reduce the need for pesticide applications to a minimum. There is now firm evidence that both symbiotic and free-living microorganisms can produce natural chemical compounds antagonistic to pathogenic organisms.

References

ANON (1979). *Royal Commission on Environmental Pollution, Seventh Report on Agriculture and Pollution*. London; HMSO

AZCON, R. and BAREA, J. M. (1975). Synthesis of auxin, giberellins and cytokinins by *Azotobacter vinelandii* and *Azobacter beijerinkii* related to effects produced on tomato plants. *Pl. Soil*, **43**, 609–619

BASSHAM, J. A. and CALVIN, M. (1957). *The Path of Carbon in Photosynthesis*. Englewood Cliffs, NJ; Prentice-Hall

BJORKMAN, O. (1976). Adaptive and genetic aspects of C4 photosynthesis. In *CO$_2$ Metabolism and Plant Productivity*, pp. 287–310. Eds R. H. Burris and C. C. Black. Baltimore and London; University Park Press

BROWN, M. E. and BURLINGHAM, S. K. (1968). Production of plant growth substances by *Azotobacter chroococcum*. *J. gen. Microbiol.*, **53**, 135–144

COOKSON, C., HUGHES, H. and COOMBS, J. (1980). Effects of combined nitrogen on anapleurotic carbon assimilation and bleeding sap composition in *Phaseolus vulgaris* L. *Planta*, **148**, 338–345

COOMBS, J. (1976). Interactions between chloroplasts and cytoplasm in C4 plants. In *The Intact Chloroplast*, pp. 279–314. Ed. J. Barber. Amsterdam, New York, London; Elsevier Scientific Publishers

GRANHALL, U. (1976). Environmental role of nitrogen-fixing blue-green algae and asymbiotic bacteria. *Ecological Bulletin No. 26*. Uppsala, Sweden; Swedish Natural Science Research Council

HARDY, R. W. F. and SILVER, N. S. (1977). *A Treatise on Dinitrogen Fixation.* New York; John Wiley

HATCH, M. D. and SLACK, C. R. (1970). Photosynthetic fixation pathways. *A. Rev. Pl. Physiol.*, **21**, 141–162

HEWITT, E. J. and CUTTING, C. V. (1979). Nitrogen assimilation of plants. In *Proceedings of 6th Long Ashton Symposium.* Ed. London; Academic Press

KAMBERGER, W. (1977). Regulation of symbiotic nitrogen fixation in root nodules of alfalfa (*Medicago sativa*) infected with *Rhizobium meliloti. Arch. Microbiol.*, **115**, 103–108

KING, G. H. (1979). Department of Energy, solar biological programme: biofuels. In *Energy from the Biomass*, pp. 70–75. Report number 5. London; The Watt Committee on Energy Ltd.

LATIMORE, M., GIDDENS, J. and ASHLEY, D. A. (1977). Effect of ammonium and nitrate nitrogen upon photosynthate supply and nitrogen fixation by soybeans. *Crop Sci.*, **17**, 399–404

McCREE, K. J. (1976). The role of dark respiration in the carbon economy of a plant. In CO_2 *Metabolism and Plant Productivity*, pp. 177–184. Eds R. H. Burris and C. C. Black. Baltimore and London; University Park Press

MIFLIN, B. J. (1980). *The Biochemistry of Plants*, vol. 5. New York; Academic Press

MIFLIN, B. J. and LEA, P. J. (1976). The pathway of nitrogen assimilation in plants. *Phytochemistry*, **15**, 873–885

MOSSE, B. (1973). Advances in the study of vesicular–arbuscular mycorrhizae. *A. Rev. Phytopathol.*, **11**, 171–196

NEWTON, W., POSTGATE, J. R. and RODRIGUEZ-BARRUECO, C. (1977). *Recent Developments in Nitrogen Fixation.* London and New York; Academic Press

OGREN, W. L. (1976). Search for higher plants with modifications of the reductive pentose phosphate pathway of CO_2 assimilation. In CO_2 *Metabolism and Plant Productivity*, pp. 19–30. Eds R. H. Burris and C. C. Black. Baltimore and London; University Park Press

OGREN, W. L. and BOWES, G. (1971). Ribulose diphosphate carboxylase regulates soybean photorespiration. *Nature New Biology*, **230**, 607–614

RAWSTHORNE, S., MINCHIN, F. R., SUMMERFIELD, R. J., COOKSON, C. and COOMBS, J. (1980). Carbon and nitrogen metabolism in legume root nodules. *Phytochemistry*, **19**, 341–355

SPEDDING, C. R. W., BATHER, D. M. and WALSINGHAM, J. M. (1979). Fuel crops—an assessment of the UK potential. In *Energy from the Biomass*, pp. 17–23, Report number 5. London; The Watt Committee on Energy Ltd

TOLBERT, N. E. (1963). Glycollate pathway. In *Photosynthetic Mechanisms in Green Plants*, pp. 648–662. National Research Council Publication 1145. Washington, DC; National Academy of Science

TOLBERT, N. E. (1971). Microbodies, peroxisomes and glyoxysomes. *A. Rev. Pl. Physiol.*, **22**, 45–74

WOOLHOUSE, H. W. (1978). Light gathering and carbon assimilation processes on photosynthesis, their adaptive modifications and significance for agriculture. *Endeavour*, **2**, 35–46

12

ALLELOCHEMICALS IN A FUTURE AGRICULTURE

J. V. LOVETT and J. LEVITT
*Department of Agronomy and Soil Science, University of New England,
Armidale, N.S.W., Australia*

Introduction

The paradox of the growing plant in agriculture is that, with few exceptions, it combines the advantages of autotrophy with the disadvantages of lack of mobility. Plants are also devoid of the special adaptations for communication which are found in many other organisms. Swain (1977) points out that communication between organisms of all levels of complexity depends largely upon an array of chemical signals. Even man, in many ways the most mobile and communicative of organisms, depends upon chemical stimuli for much of his sensory perception. Given the twin disabilities of the growing plant, noted above, it would seem logical to expect that chemical stimuli would play a relatively major role in communication by plants with other organisms, thus also offsetting immobility.

More than 10 000 organic compounds of low molecular weight are known to exist in plants (Swain, 1977). They have been termed 'secondary compounds' by Levin (1976) because they are produced as byproducts of primary metabolic pathways. Some are associated with particular families of plants, for example the glucosinolates with the Cruciferae (Kjaer, 1976). Others, like the alkaloids, are more cosmopolitan. Swain (1977) considers that the function of such secondary compounds is as chemical signals in ecosystems and there are documented examples of the significance of such chemicals in communication between plants and other organisms, including the performance of defensive and even offensive functions for the plant which produces them. Many of the defences of plants against insect attack are chemically based (Maxwell, Jenkins and Parrott, 1972) and Levin (1976) discusses resistance to fungi, bacteria, viruses and nematodes which may also be based upon the presence of secondary compounds, particularly the phenolics. There remain, however, deficiencies in documentation of the part played by plant-derived chemicals in interplant relationships.

The nature of allelopathy and allelochemicals

Molisch (1937) uses the term 'allelopathy' to describe all biochemical interactions, whether positive or negative, among plants of all levels of complexity, including microorganisms. This definition is more useful than the strict linguistic interpretation of the term, that is, 'mutual harm'.

Allelochemicals belong to a few major groups of the secondary compounds, including phenolic acids, flavonoids and other aromatic compounds, terpenoids, steroids, alkaloids and organic cyanides (Whittaker and Feeny, 1971). Such chemicals are a normal part of the environment of land plants (Whittaker, 1970) and are released in a number of ways, including the leaching of water-soluble toxins from the above-ground parts by the action of rain, fog or dew, the exudation of volatile chemicals, the exudation of water-soluble toxins from plant roots, and leaching from plant litter, with or without decomposition involving the action of microorganisms (Putnam and Duke, 1978). While the effects of allelochemicals have been investigated in some agricultural systems, their role in natural plant communities (*see,* for example, the work of Muller, 1966) is the best documented.

In natural plant communities, allelochemicals are part of the checks and balances that maintain relative stability. Many responses to allelochemicals are therefore subtle, having developed during evolution of the community or, in some agricultural examples, during coevolution of weed and crop species. Whittaker (1970) states that '. . . the known examples of allelopathic effects are only the most conspicuous cases, which have been observed because they stand out from a matrix of more general and less obvious influences.' It might be expected that some examples of allelopathy involving weeds will be relatively conspicuous because the species which promote them are not, generally, in equilibrium with other components of the system in which they occur.

Harper (1977) uses the term 'interference' to describe changes in the environment of plants resulting from influences of neighbouring plants upon the environment. Interference thus includes the accepted competition which occurs for environmental resources, together with any allelopathic effects which may occur. In this chapter, three examples of allelopathy in weed/crop associations have been selected for discussion. They indicate part of a wide range of plant-with-plant interactions which have applications for the future agriculture.

The role of allelochemicals in weed/crop associations

CAMELINA SATIVA AND *LINUM USITATISSIMUM:* INTERMEDIARY ORGANISMS AND ALLELOPATHY

The crop plant *Linum usitatissimum* L. (linseed) and the weed *Camelina sativa* L. Crantz (Cruciferae) have been associated since prehistoric times (Grigson, 1955). There has thus been the opportunity for coevolution between the crop and the weed.

Lovett and Sagar (1978) noted that *C. sativa* leaf washings consistently stimulated the growth of the radicles of germinating *L. usitatissimum* seeds. The presence of bacteria in the *C. sativa* phyllosphere was necessary for the stimulatory phenomenon. Lovett and Duffield (1981) have demonstrated that washings of *C. sativa* leaves contain organic acids which sustain rapid bacterial growth. During this growth, complex organic compounds present in the leaves are broken down to simpler compounds, at least one of which, benzylamine, is allelopathic in activity. Lovett (1979) has shown that, in Petri dishes under controlled conditions, benzylamine at 10 and 100 ppm invokes stimulatory responses in germinating seedlings of *L. usitatissimum* which are similar to those produced by *C. sativa* leaf washings. However, recent data (*Table 12.1*) suggest that, in soil under controlled conditions, benzylamine at concentrations greater than 10 ppm has inhibitory rather than stimulatory effects upon germinating *L. usitatissimum*.

Table 12.1 EFFECTS OF BENZYLAMINE ON EARLY GROWTH IN SOIL OF *LINUM USITATISSIMUM* (MEANS OF TEN REPLICATES)

Variables	Days after sowing	Benzylamine (ppm)				P
		0	10	100	1000	
Radicle length (mm)	8	167.0[a]	148.8[ab]	137.9[b]	108.0[c]	< 0.001
Shoot height (mm)	7	23.4[f]	23.1[f]	21.4[g]	20.7[g]	< 0.01

Means which share a common letter are not signicantly different at the 5% level, Studentized Range Test

While, 8 days after sowing, the emergence of *L. usitatissimum* was not impaired by benzylamine in concentrations up to 1000 ppm, 10 000 ppm of this chemical completely inhibited germination. At the same stage of growth, 100 and 1000 ppm benzylamine reduced radicle length and shoot height. Detrimental effects of *Camelina* species upon the growth of *L. usitatissimum* in the field have previously been noted by Grümmer and Beyer (1960).

The data presented in *Table 12.1* were obtained using lower concentrations of solution, in terms of ml/g of medium to which they were applied, than were used in Petri dish experiments (Lovett, 1979). Two phenomena may account for this difference: first, the more complete and sustained contact between seed and soil compared with seed and filter paper, leading to greater uptake of benzylamine; secondly, the induction by high concentrations of benzylamine of hydrophobic conditions in soil, inhibiting uptake of water by seeds.

Concentration of the allelochemical is clearly critical, and it remains to be determined whether benzylamine can evoke, in the field, the type of response to which Grümmer and Beyer (1960) refer.

Ruinen (1974) argues that the phyllosphere is essentially an aquatic milieu. In the work of Grümmer and Beyer (1960) and of Lovett and Sagar (1978) application of water to the foliage of *C. sativa* was essential for the occurrence of allelopathic phenomena. For Grümmer and Beyer (1960), the time of occurrence of rainfall was important for allelopathic manifestations.

The involvement of bacteria as intermediary organisms in promoting allelopathy in the *C. sativa/L. usitatissimum* association appears to be unique in the literature. However, Trenbath and Fox (1976) have suggested that leaf-feeding insects concentrate allelochemicals produced by leaves of *Eucalyptus globulus* ssp *bicostata* (Maiden *et al.*) Kirkp. In their experiments, inhibition of germination of a phytometer species by insect frass was greater than that due to any other solid preparation derived from the leaves of that species. Aqueous leachates of *Eucalyptus* leaves slightly encouraged germination in these experiments, again emphasizing the significance of concentration of allelochemicals. Trenbath and Silander (1978) subsequently found strong inhibition of early root-growth of grassland species by such insect frass. In the field these species were absent beneath stands of *E. globulus* ssp *bicostata* upon which insects were feeding. The allelochemicals were selective in their inhibitory effects.

SALVIA REFLEXA HORNEM AND *TRITICUM AESTIVUM:* MORPHOLOGICAL
ADAPTATIONS AND ALLELOPATHY

The production of volatile growth inhibitors by aromatic shrubs has been investigated in detail. In particular, *Salvia leucophylla* Greene, *S. apiana* Jeps and *S. mellifera* Greene (Labiatae), which grow naturally in the soft chaparral of Southern California, contain phytotoxic terpenes which volatilize and inhibit the establishment of seedlings of a wide variety of plants (Muller, 1966). The zone of inhibition may extend 8–10 m beyond either the lateral spread of the *Salvia* root system or the shrub crown, favouring the hypothesis that the toxic material released by these *Salvia* plants is volatile.

Bioassay techniques confirmed that *S. leucophylla* leaves were the site of production of phytotoxic terpenes. Gas chromatographic analysis revealed the presence of α-pinene, β-pinene, camphene, cineole and camphor, all of which inhibited the germination and radicle growth of phytometer species. It was shown that the soil, particularly when dry, acted as a reservoir for the terpenes, accounting for inhibitory effects observed in the field.

Preliminary experiments with *S. reflexa* Hornem, a significant weed of cropping areas in north-western New South Wales and south-eastern Queensland, showed that aromatic substances liberated by whole plants slightly stimulated early growth of wheat. Wetting the foliage greatly enhanced the scent of *S. reflexa,* a phenomenon noted by Muller (1966) in his work with aromatic species. Experiments with *S. reflexa* failed to demonstrate strong competitive effects (Lovett and Lynch, 1979) but aqueous leachates had highly significant allelopathic effects upon germination and early growth of wheat (*Triticum aestivum* L.), *Table 12.2.*

Table 12.2   EFFECT OF LEAF WASHINGS OF *SALVIA REFLEXA* ON GERMINATION AND EARLY GROWTH OF *TRITICUM AESTIVUM* IN SOIL (MEANS OF FOUR REPLICATES)

| | | Days after sowing | | | | | |
		5	6	7	8	9	10
Emergence	Sterile water (control)	42.5	70.0	77.5	80.0	82.5	87.5
	Salvia washings	20.0	32.5	45.0	47.5	50.0	50.0
	*LSD (t) 5%	2.1	2.1	2.0	1.8	2.3	1.8
	P	<0.05	<0.01	<0.01	<0.01	<0.05	<0.01
Height of coleoptile (mm)	Sterile water (control)	20.8	29.5	54.8	81.8	102.0	111.7
	Salvia washings	9.5	18.7	35.1	66.3	86.0	108.1
	LSD (t) 5%	7.1	4.8	16.4	—	—	—
	P	<0.01	<0.001	<0.05	N.S.	N.S.	N.S.

*Least significant difference—t test

At the end of the experiment (Day 10), seeds and seedlings were recovered from soil. The emergence counts represent all seeds which had germinated, indicating that washings of *S. reflexa* inhibited germination of wheat. Those wheat seeds which germinated were retarded in growth early in the experiment, as indicated by height of the coleoptile, but subsequently recovered. In the field, such a check in early growth could place surviving crop plants at a disadvantage. Thus, while *S. reflexa* may not interfere strongly with crop plants through competition for environmental resources, it has the potential to interfere through allelopathy.

Examination by scanning electron microscopy of leaf surfaces of *S. reflexa* revealed trichomes (glandular hairs or 'bladders' *Figure 12.1*) of mean diameter 50 μm, evenly distributed at intervals of about 200 μm (Lovett and Speak, 1979). Immersion in water caused rapid collapse of the bladders. The observation of strengthening odour when water contacts *S. reflexa* leaves, coupled with the collapse of the bladders when immersed in water, suggests that osmotic absorption of water swells and ruptures the bladders.

It is possible that the bladders on the leaves of *S. reflexa* function as a means of sequestering volatile terpenes (possibly α-pinene) pending release, through the action of rain or dew, into the environment; the

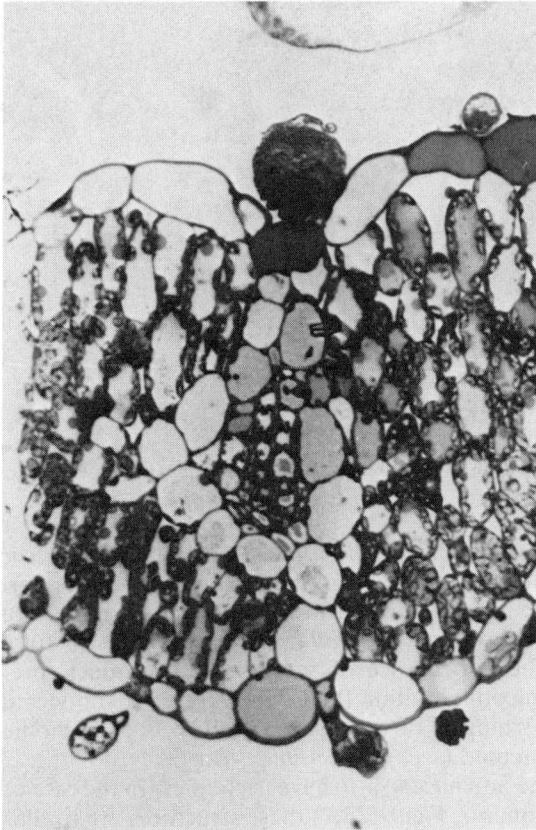

Figure 12.1 Transverse section of leaf of *Salvia reflexa*. Trichomes on upper and lower surfaces are apparently linked to the vascular system. (Diameter of upper bladder cell = ± 50 μm)

importance of water in promoting allelopathic activities is again apparent. Muller (1966) demonstrated that terpenes, including α-pinene, produced by *S. leucophylla* were retained in the soil and inhibited subsequent germination of herbaceous species. Washings of *S. reflexa*, both in controlled conditions and in soil, are potent germination inhibitors, raising the possibility of a longer-term effect, as described by Muller (1966), as a further possible means by which allelochemicals might affect growth of subsequent crops.

That trichomes have a role in plant defence is well documented (Levin, 1973). Not only may they act as a physical deterrent (for example in *Urtica* species) but trichomes also function in the resistance of plants to insect attack. Alkaloids secreted by trichomes of several species of *Nicotiana* (Solanaceae) have been shown to be specifically insecticidal towards the green peach aphid *Myzus persicae* (Sulzer). Because aphids are killed by

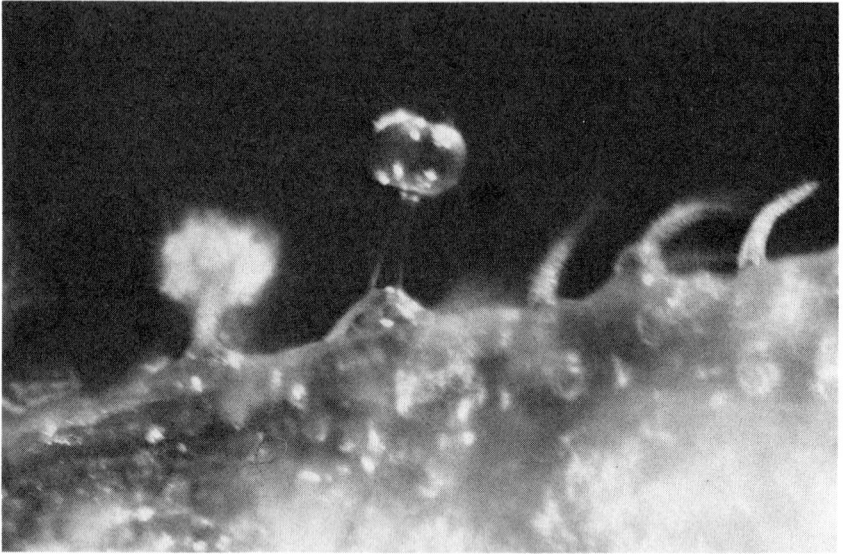

Figure 12.2   Trichome on upper leaf surface of *Lycopersicon esculentum*. (Diameter of bladder cell = ±40 μm)

contact with these secretions, resistance of *Nicotiana* species to *M. persicae* is attributable to the trichome secretion (Thurston, Smith and Cooper, 1966). Electron-microscope studies have shown that the trichomes of *Nicotiana tabacum* L. (tobacco) are glandular and undergo structural changes concurrent with the appearance of the secretory product (predominantly the alkalkoid, nicotine) within the cells (Akers, Weybrew and Long, 1978). Many other examples of the toxicity to insects of trichome secretions have been documented (*see, for example,* Whittaker, 1970).

Other crop members of the Solanaceae also have trichomes. In the case of *Lycopersicon esculentum* (tomato, *Figure 12.2*) these structures are similar to those of weed members of the family, such as *Datura stramonium*, for which they may have a role in allelopathy.

DATURA STRAMONIUM AND *HELIANTHUS ANNUUS*: WEED SEED DORMANCY
AND ALLELOPATHY

Datura stramonium L., an erect summer-growing annual of the family
Solanaceae, is an important weed of many irrigated summer crops (Felton,
1979). In common with many of the Solanaceae, *D. stramonium* contains
many alkaloids, principally scopolamine and hyoscyamine (Mothes, 1955),
which are translocated from the site of synthesis in the roots to accumulate
in the leaves and seeds (Dawson, 1948).

. The alkaloids of *D. stramonium* have a certain protective function in that
they make the plant taste bitter and hence unpalatable to stock (Parsons,
1973). Lovett *et al.* (1981) established that aqueous leachates of *D.
stramonium* seeds and leaves contained scopolamine and hyoscyamine,
seed leachates having a significantly higher concentration of total alkaloid
than leaf leachates. The application of various concentrations of leachate
and alkaloid solutions to seeds of a phytometer species *Linum usitatissimum*
L. on sterile filter paper indicated that both alkaloids were major active
principles in the depression of germination and radicle elongation.

One *D. stramonium* plant may produce up to 24 000 seeds in a season
(Parsons, 1973) with each seed capsule containing about 250–700 seeds.
These fall to the ground below the parent plant, providing a high con-
centration of alkaloid-containing material in the soil.

Dormancy is an important factor in dispersal in time of many weed
species. In *D. stramonium*, dormancy is broken when alkaloids are leached
from the seed coat. Rain is vital to this process, and is known also to remove
significant quantities of alkaloid from leaves of *D. stramonium* (Mothes,
1955). Crop seeds adjacent to *D. stramonium* seeds in the soil will,
therefore, be exposed to varying amounts of alkaloid. Because alkaloids in
the soil behave as cations and are adsorbed on to the exchange complex,
they are relatively resistant both to leaching from the profile and to bacterial
degradation (Winter, 1961).

Observations by Winter (1961) suggested that plants take up alkaloids
from the soil by processes similar to those involved in nutrient uptake.
Thus, soil factors which influence plant nutrient uptake (such as particle
size, and clay and organic matter contents) may also influence the uptake of
alkaloids. Data resulting from the application of *D. stramonium* seeds,
seed leachate and a 0.5 per cent scopolamine solution to *Helianthus*

Table 12.3   EFFECT OF *D. STRAMONIUM* SEEDS, SEED LEACHATE AND A
0.5% SCOPOLAMINE SOLUTION ON *H. ANNUUS* RADICLE ELONGATION IN
TWO MEDIA (MEANS OF FIVE REPLICATES)

Medium	H. annuus *radicle length after 72 h (mm)*				
	50 ml *sterile water*	50 ml *sterile water* + 550 *D.* stramonium *seeds*	50 ml 0.5% *scopolamine solution*	50 ml *sterile water* + 5500 *D.* stramonium *seeds*	*P*
Sterile coarse sand	50.3 a	34.6 b	29.4 bc	25.9 c	< 0.001
Sterile 4 : 1 soil + peat	53.2 g	40.6 h	38.8 h	15.1 i	< 0.001

Means which share a common letter are not significantly different at the 5% level, Studentized Range Test *annuus* L. seeds sown either in sterile coarse sand or in a sterile 4 : 1 mixture of soil and peat, are presented in *Table 12.3*.

In either medium the response of *H. annuus* to the various treatments was similar; however, the seed leachate treatment produced a greater inhibition of *H. annuus* radicle elongation in the soil and peat medium than in the coarse sand medium. This indicates that the *H. annuus* seedlings may be taking up more alkaloid at relatively high concentration in the former medium, possibly because of the greater contact provided between soil and seed.

These data suggest that there may be an allelopathic component involved in interference between *D. stramonium* and summer crops such as *H. annuus,* as well as the factors of competition for light, moisture and nutrients which are cited in the literature (Parsons, 1973). If alkaloids in a weed such as *D. stramonium* can complement competitive ability with a crop, the reverse may be possible for an alkaloid-containing crop plant. It is, for example, a relatively simple genetic procedure to increase the absolute amounts of alkaloid produced by *Nicotiana tabacum* L., and this applies to most commercially important Solanaceous species (James, 1949). Any benefit in terms of weed control resulting from such genetic manipulation must, however, be weighed against possible detrimental effects to the commercial value of the crop.

Some roles of allelochemicals in a future agriculture

It has been estimated by a number of authorities (*see, for example,* Russell, 1978) that as much as 50 per cent of the yield potential of crop species is sacrificed annually to the depredations of weeds, insects and diseases. In the past half-century the agriculturalist has come to rely increasingly upon synthetic chemicals to offset these losses, but they continue relatively unabated.

If allelochemicals can be harnessed to the advantage of crop plants, it would be reasonable to expect both direct benefits, in terms of reduced dependence upon herbicides, and indirect benefits resulting from the alleviation of stress by interference from weeds. The cumulative effect of enhancing plant defences against insects and diseases, as well as through allelopathy, could contribute to a reduction in the need for synthetic chemical inputs to agriculture.

In the examples discussed in this chapter it is the weeds—relatively aggressive, colonizing plants—that have been identified as producing allelochemicals. Natural selection might be expected to favour the development of allelochemicals as a part of plant defence or offence. Because many crops cannot survive in plant communities without the aid of man, it may be that artificial selection, as applied to the majority of crop species to date, has reduced natural chemical protection against weeds as well as against pests and diseases, although the evidence for this assertion is largely circumstantial. In our own work, two members of the Cruciferae, *Camelina sativa* and the crop species *Brassica napus* L. (rape), contain glucosinolates, but only the weed appears able to support the bacterial population which is a necessary intermediary to allelopathy. Rice (1974) has documented allelopathic activity by sunflowers indigenous to the United States. Our work with open-pollinated cultivars of sunflower has failed to demonstrate such activity although some hybrids may have allelopathic potential. In most instances any reduction in potential allelochemical content of crop plants has probably been unintentional, but in others the presence of secondary compounds may be linked with features which adversely affect economic yield. A good example occurs with cotton (*Gossypium* spp) where the principal pigment in the seed, gossypol, is a poisonous phenolic. The quantity of gossypol varies with cotton species and appears to be susceptible to genetic manipulation (Maxwell, Jenkins and Parrott, 1972). Most gossypol is rendered harmless when the seed is processed, but monogastrics, e.g. pigs, poultry, are sensitive to it and this may militate against enhancement of gossypol content for plant defence. Analogous situations might be expected to occur with allelochemicals.

Given that the allelochemical content of weed species is relatively high, it may be possible to capitalize upon this content in a number of ways. As more information is gleaned about the identity and mode of action of the chemicals concerned, certain of these compounds may be harvested for agricultural use. As they are naturally occurring and often selective, it seems reasonable to expect that their use as 'natural herbicides' would have fewer detrimental side-effects than is the case with many synthetic compounds. A second possibility is to produce synthetic analogues of allelochemicals, as has already been done—for example with the insecticidal synthetic pyrethroids (Leahey, 1979). Some allelochemicals are selective in action and it may therefore prove possible to introduce them through companion planting of chemically compatible species. In the short term, because of their relatively high allelochemical content, deliberate introduction of weeds into certain crops might be indicated. A longer-term objective would be to use companion crops offering mutual defence. A few such combinations may already exist (Putnam and Duke, 1978).

Crops themselves offer at least two other avenues for development in relation to allelochemicals. The work of Kimber (1967; 1973) shows that crop residues produce phytotoxins which remain active in soil for significant periods after harvest. Research into minimal cultivation techniques suggests that, with their adoption, it may become more common for residues to remain on soil surfaces than has recently been the case. Timing of cultivation and sowing could be manipulated to take best advantage of any benefits of these substances in terms of weed control. The long-term effect described by Muller (1966) (*see above*) may also be important in natural plant communities.

The most exciting possibility, however, is the genetic enhancement of plant defences. Enhancing resistance to disease is a frequent objective of plant breeding programmes, and Chapman (1974) states that: '. . . plant varieties which are resistant to insect activity by virtue of their chemical components may be produced by selective breeding programmes' As has already been indicated, it is common for crop plants to contain potential allelochemicals but, apparently, at too low a concentration to be effective.

Before genetic manipulation can be contemplated, it is necessary to understand the synthesis of a potential allelochemical within the plant. The major factors which influence the concentration of the chemical present must also be investigated in relation to possible autotoxicity. It is also necessary to know the concentration critical in promoting allelopathy because this may be at variance with commercial objectives and may preclude an allelopathic approach in some species. In addition, critical growth stages for both donor and receiver plants must be defined.

The effects of allelochemicals need not be direct. For example, Rice (1974) discusses data which suggest that the grass *Aristida oligantha* Michx. inhibits nitrogen fixation by bacteria and by blue-green algae, thereby maintaining a low nitrogen status in the soil in which it grows, and effectively retarding further stages of plant succession. The example of *C. sativa* with *L. usitatissimum* and phyllosphere bacteria is also indirect, in the sense of allelopathy, and it may be possible to devise manipulations of agricultural systems which exploit these relationships rather than direct plant-to-plant transfers of allelochemicals. The involvement of intermediary organisms indicates the complexity of interactions which contribute to the relative stability of natural ecosystems. This complexity contrasts with the dangerously simplistic approaches taken to counter apparent imbalances of some organisms in contemporary agricultural practice.

Conclusions

Research into allelopathy has, in part, been placed at a disadvantage by some of the techniques used to obtain allelochemicals from plants, and by the fact that many dramatic effects, documented under controlled conditions, have not been demonstrated in the field. Stowe (1979) discusses these and other reservations regarding allelopathic phenomena. In the examples discussed in this chapter the chemicals involved have been obtained from whole plants by means consistent with occurrences in the field. The use of aqueous leachates emphasizes the essential role of rainfall in plant/plant or plant/soil/plant transfer of allelochemicals. All of the effects cited have been confirmed in soil, as well as under laboratory conditions; identification of the chemicals involved, including estimates of concentrations necessary to promote allelopathic activity in Petri dishes and in soil, have been documented in two of the three examples. The subtlety of allelopathic effects, for example in inhibiting rather than killing a competitor, is emphasized by these data.

The possibilities for exploiting allelopathy in a future agriculture parallel much of what has already been achieved in investigations of plant and insect associations. For example, the role of biochemistry in such relations in discussed by Feeny (1975), while Chapman (1974) comments on plant breeding for improved chemical resistance of plants to insects, and Levin (1973) reviews the role of trichomes in plant defence, with special reference to their deterrence of feeding by phytophagous insects.

As emphasized by Whittaker (1970), chemical defence of plants is relative and very seldom complete: however, it may be sufficient to reduce stress to the point where a crop is less vulnerable to pest, disease or weed influence, thereby reducing or even eliminating the need for pesticide applications. With regard to weeds, it is now widely accepted that cosmetic ('total kill') weed control is both unattainable and undesirable. Management of weeds at tolerable levels by manipulation of natural crop defences is a logical corollary of this acceptance. This concept aligns with a view of coevolution in which weed and crop species coexist at levels which are tolerable to both, the balance between them being maintained by biochemical and other means.

The objective of future research into allelochemicals must be to permit emphasis to be placed upon chemicals which are of ecosystems, thus reducing perturbation of such ecosystems and countering the instability of agriculture which, ecologically, is its ultimate weakness.

References

AKERS, C. P., WEYBREW, J. A. and LONG, R. C. (1978). Ultrastructure of glandular trichomes of leaves of *Nicotiana tabacum* L. cv. Xanthi. *Am. J. Bot.*, **65**, 282–292

CHAPMAN, R. F. (1974). The chemical inhibition of feeding by phytophagous insects. *Bull. ent. Res.*, **64**, 339–363

DAWSON, R. F. (1948). Alkaloid biogenesis. *Adv. Enzymol.*, **8**, 203–251

FEENY, P. (1975). Biochemical coevolution between plants and their insect herbivores. In *Coevolution of Animals and Plants*, pp. 3–19. Eds L. E. Gilbert and P. H. Raven. Austin; University of Texas Press

FELTON, W. L. (1979). The competitive effect of *Datura* species in five irrigated summer crops. In *Proceedings of the 7th Asian-Pacific Weed Science Society Conference,* pp. 99–104. Haymarket, N.S.W., Council of Australian Weed Science Societies

GRIGSON, G. (1955). *The Englishman's Flora.* London; Phoenix House

GRÜMMER, G. and BEYER, H. (1960). The influence exerted by species of *Camelina* on flax by means of toxic substances. In *The Biology of Weeds,* pp. 153–157. Ed. J. L. Harper. Oxford; Blackwell

HARPER, J. L. (1977). *Population Biology of Plants.* London; Academic Press

JAMES, W. O. (1949). Alkaloids in the plant. In *The Alkaloids*, vol. 1, pp. 16–86. Eds R. H. F. Manske and H. L. Holmes. New York; Academic Press

KIMBER, R. W. L. (1967). Phytotoxicity from plant residues. I. The influence of rotted wheat straw on seedling growth. *Aust. J. agric. Res.*, **18**, 361 374

KIMBER, R. W. L. (1973). Phytotoxicity from plant residues. II. The effect of time of rotting of straw from some grasses and legumes on the growth of wheat seedlings. *Pl. Soil*, **38**, 347–361

KJAER, A. (1976). Glucosinolates in the Cruciferae. In *The Biology and Chemistry of the Cruciferae,* pp. 207–219. Eds J. G. Vaughan, A. J. Macleod and B. M. G. James. London; Academic Press

LEAHEY, J. P. (1979). The metabolism and environmental degradation of the pyrethroid insecticides. *Outlook*, **10**, 135–142

LEVIN, D. A. (1973). The role of trichomes in plant defence. *Q. Rev. Biol.*, **48**, 3–15

LEVIN, D. A. (1976). The chemical defences of plants to pathogens and herbivores. *A. Rev. Ecol. Systematics*, **7**, 121–159

LOVETT, J. V. (1979). The ecological significance of odour in weeds. *Proceedings of the 7th Asian-Pacific Weed Science Society Conference,* pp. 335–338

LOVETT, J. V. and DUFFIELD, A. M. (1981). Allelochemicals of *Camelina sativa* (L.) Crantz. *J. appl. Ecol.*, **18** (in press)

LOVETT, J. V. and LYNCH, J. A. (1979). Studies of *Salvia reflexa* Hornem. I. Possible competitive mechanisms. *Weed Res.*, **19**, 352–358

LOVETT, J. V. and SPEAK, M. D. (1979). Studies of *Salvia reflexa* Hornem. II. Examination of specialized leaf surface structures. *Weed Res.*, **19**, 359–362

LOVETT, J. V. and SAGAR, G. R. (1978). Influence of bacteria in the phyllosphere of *Camelina sativa* (L.) Crantz on germination of *Linum usitatissimum* L. *New Phytol.*, **81**, 617–625

LOVETT, J. V., LEVITT, J., DUFFIELD, A. M. and SMITH, N. G. (1981). Allelopathic potential of *Datura stamonium* (Thorn apple). *Weed Res.*, **21**, (in press)

MAXWELL, F. G., JENKINS, J. N. and PARROTT, W. L. (1972). Resistance of plants to insects. *Adv. Agron.*, **24**, 187–265

MOLISCH, H. (1937). *Der Einfluss einer Pflanze auf die andere-Allelopathie*. Jena; Fischer

MOTHES, K. (1955). Physiology of alkaloids. *A. Rev. Pl. Physiol.*, **6**, 393–432

MULLER, C. H. (1966). The role of chemical inhibition (allelopathy) in vegetational composition. *Bull. Torrey bot. Club*, **93**, 332–351

PARSONS, W. T. (1973). *Noxious Weeds of Victoria*. Sydney; Inkata

PURSEGLOVE, J. W. (1968). *Tropical Crops, Dicotyledons*. London; Longmans

PUTNAM, A. R. and DUKE, W. B. (1978). Allelopathy in agroecosystems. *A. Rev. Phytopathol.*, **16**, 431–451

RICE, E. L. (1974). *Allelopathy*. New York; Academic Press

RUINEN, J. (1974). Nitrogen fixation in the phyllosphere. In *The Biology of Nitrogen Fixation*, pp. 121–167. Ed. A. Quispel. Amsterdam; North-Holland

RUSSELL, G. E. (1978). *Plant Breeding for Pest and Disease Resistance*. London; Butterworths

STOWE, L. G. (1979). Allelopathy and its influence on the distribution of plants in an Illinois old-field. *J. Ecol.*, **67**, 1065–1085

SWAIN, T. (1977). Secondary compounds as protective agents. *A. Rev. Pl. Physiol.*, **28**, 479–501

THURSTON, R., SMITH, W. T. and COOPER, B. P. (1966). Alkaloid secretion by trichomes of *Nicotiana* species and resistance to aphids. *Entomologia exp. appl.*, **9**, 428–432

TRENBATH, B. R. and FOX, L. R. (1976). Insect frass and leaves from *Eucalyptus bicostata* as germination inhibitors. *Austr. Seed Sci. Newsl.*, **2**, 34–39

TRENBATH, B.R. and SILANDER, J. A. (1978). Bare zones under eucalyptus and the toxic effects of frass of insects feeding on the leaves. *Aust. Seed Sci. Newsl.*, **4**, 13–18

WHITTAKER, R. H. (1970). The biochemical ecology of higher plants. In *Chemical Ecology*, pp. 43–70. Eds E. Sondheimer and J. B. Simeone. New York; Academic Press

WHITTAKER, R. H. and FEENY, P. P. (1971). Allelochemics: chemical interactions between species. *Science*, **171**, 757–770

WINTER, A. G. (1961). New physiological and biological aspects in the interrelationships between higher plants. In *Mechanisms in Biological Competition*, pp. 228–244. Symposium of the Society for Experimental Biology No. 15. Cambridge; Cambridge University Press

13

THE ROLE OF SEAWEED IN 'CLOSED CYCLE' AGRICULTURE

J. L. H. CHASE
*C.C. 353, 8400 Bariloche, Argentina and Chase Organics Ltd,
Shepperton TW17 8AQ, UK*

Seaweed recycles plant nutrients eroded from the land and contains every trace element known to be essential to animal and plant growth. It also contains the whole range of vitamins and several groups of hormones.

In a closed cycle, any element deficient in the soil cannot be replaced except from an outside source. The use of seaweed ensures that there will never be a complete absence of any element.

Seaweed is usually applied to crops in the form of a liquid extract which acts both indirectly through the soil and directly through the leaf. Its value lies mainly in improving the quality of the crop and in giving resistance to pests and diseases and marginal frosts.

Seaweed is fed to animals in the form of a meal, traditionally at about 10 per cent of the ration. During the last 20 years, however, balanced blends of a number of different varieties have been fed to cattle, pigs, sheep, horses, goats and poultry as a feed supplement at a rate of less than 1 per cent. This has resulted in improved health, resistance to disease and inclement weather, and improved fertility.

III

BIOLOGICAL HUSBANDRY IN THE TROPICS

Introduction

Good tropical soils, warm and well-watered, are second to none in their capacity for heavy crop production. Traditional methods, often involving multiple cropping, have for generations produced a well-balanced abundance of carbohydrates, oils and proteins for local populations; good farming practices have been as rewarding in the tropics as in any other part of the world. However, the balance is all too easily upset. Economic pressures to grow large-scale monocultured cash crops, and the needs of rapidly expanding human populations, have overtaxed tropical, no less than temperate soils: tropical soils have proved just as vulnerable to damage by overcropping as temperate ones, and rather more likely to be lost for good once they have been abandoned to the elements.

Biological husbandry has much to offer in tropical climates: as a natural extension of traditional practice, requiring low inputs of energy and chemical fertilizers, it is often better suited to local pace and local pockets than imported techniques, and local agricultural colleges and research institutes are currently examining its potential. The seven papers in this section deal with various aspects of biological husbandry practice in the tropics.

Dr O. A. Ojomo's introductory paper sets the scene for Nigeria, pointing out some of the problems in a country that seeks to develop its agriculture using little more than its own natural resources; pilot experiments involving manures and nitrogen-fixing bacteria give promise for the future. Dr G. F. Wilson and Mr B. T. Kang report on experimental work to develop and improve upon traditional bush fallow cultivation, intercropping food species with selected shrubs—mostly legumes—that accelerate soil nutrient regeneration and permit more intensive use of the land. Mr H. W. Dalzell reviews current problems besetting agriculture in India, and suggests practical measures for dealing with them based on his studies in Medak; his emphasis on small-scale, low-cost, labour-intensive solutions is entirely in keeping with the main theme of this symposium. Dr O. A. Ochwoh and Dr J. Y. Kitungulu-Zake present their studies on the effects of adding organic materials to soils in Uganda, underlining the importance of added humus in enhancing cation exchange capacity. Professor R. K. Jana reports on erosion control experiments in Tanzania, in which techniques of mulching and minimum tillage under intercropping systems showed up favourably. Dr I. Carruthers and Dr R. Palmer-Jones discuss the work of Sir Albert Howard, a tropical botanist and agricultural scientist whose pioneering work on composting of half a century ago may be relevant in many tropical countries today. Finally Dr A. Abidogun sounds a note of warning: the scope for biological agriculture in tropical countries may be limited by short-term factors—hunger is a compelling example—that cannot be overridden.

14

THE SCOPE FOR BIOLOGICAL AGRICULTURE IN NIGERIA

O. A. OJOMO
National Cereals Research Institute, PMB 5042, Ibadan, Nigeria

Introduction

Before the commercial exploitation of oil in Nigeria, the agricultural sector contributed most both to her gross domestic product and to foreign exchange earning. Agriculture has since yielded place to oil. Despite this, it still provides employment for about 70–75 per cent of the populace, who adopt three main types of farming systems.

MIXED FARMING

This is practised mainly in the northern parts of Nigeria where cattle, sheep- and goat-keeping is practised together with cropping. The animals graze and clean up the crop stubbles and at the same time they deposit their dung on the farmland.

MIXED CROPPING

Two or more crops are grown together on the same piece of land. In actual practice, the land is used for 2–4 years depending on the initial fertility of the soil. It is then abandoned and allowed to grow back to bush. This is usually referred to as shifting cultivation.

Research results in Nigeria by Andrews (1972), Remison (1978), Olafare and Ojomo (1980) and others have shown that yields obtained from such mixtures exceed yields obtained if those crops had each been sown alone on the same piece of land.

SOLE CROPPING

In this system, only one crop is grown at a time on usually large stretches of land. The various research institutes in Nigeria have been developing the necessary inputs (seeds, fertilizers, herbicides) and technology (mechanization, growing methods) to suit this system of farming. The problems of this system have not been fully solved.

So far, some of the problems which have been identified from these types of farming systems may be summarized as follows.

187

Population pressure

The rate of population growth cannot allow for an indefinite system of shifting cultivation. Something has got to be done to develop a more sedentary farming system.

Social changes (shortage of farm labour)

Because of a high concurrent spate of industrial development, youths are continuously lured from farms. Only the old are left behind to continue farming (Tiffen, 1971).

Availability of inputs

Technological.   Monocropping is highly technical, requiring a more-or-less sophisticated range of farm energy and tools. These are usually not easily available and maintenance is difficult. In addition, their effects on soil degradation are not known.

Fertilizers, chemicals, herbicides and other chemicals.   These are mainly developed elsewhere and imported into the country at great expense. Reports from countries which produce them show that caution must be exercised in view of side-effects and environmental pollution.

Capital.   Mechanized tools, energy, seeds, chemicals, land and labour are inevitable annual requirements in a highly intensive agriculture. Costs go up yearly. It will be highly desirable to keep down the costs of these inputs.

Energy squeeze

In the manufacture of nitrogen fertilizers etc., a large quantity of energy (electric or oil) is consumed. This could be more profitably used in other areas.

Weather conditions

The northern parts of the country are dry in contrast to the south, where torrential rainfall occurs. In all cases temperatures are high. The result is that the nutrient status of the soils and soil organic matter are low generally. The problem is more acute as one goes northwards.

   It becomes evident that working alternatives must be evolved in the face of the facts set out above. One of these approaches is to turn towards adapting what is naturally available. Some efforts have been made in this direction in Nigeria and the results obtained indicate very good possibilities.

Manures

The status of soil fertility and its maintenance have been investigated for over 30 years. Reports by Vine (1953), Watson and Goldsworthy (1964), Dransfield and McDonald (1967), Amon and Adetunji (1969), Bache and Heathcote (1969), Jones (1971, 1973) and Baker, Lombin and Abdullahi (1977) indicate quite conclusively that Nigerian soils are generally low in organic carbon (range 0.08–4.70 per cent, mean 0.62–0.75 per cent). In addition, the level of carbon decreases as one travels northwards.

Their results show that inorganic fertilizers are of great help, but that they do not lead to any permanent improvement of the soil structure. Their results also show that an application of 3–7.5 tonnes of manure per hectare improved soil fertility and structure tremendously and led to significant increases in crop yield. Addition of supplementary inorganic fertilizers (nitrogen, phosphorus and potassium) at various levels and combinations provided further increases. However, manure in this quantity is not readily available on a national scale.

Jones (1976) and others have suggested that proper crop or fallow residue management can partly solve the problem. The slashing and burning of the thrash *in situ* gave maximum immediate benefits. At the National Cereals Research Institute, Ibadan, the possibilities of utilizing agricultural and some industrial wastes as manures are being investigated.

The possibility of adopting the use of manures is very high in Nigeria. Peasant farmers in northern Nigeria are already widely using farmyard manure: any limitation has been due to its scarcity.

Nitrogen-fixing bacteria

Very early fertilizer trials with cowpea often showed significant yield response to low (20–40 kg) nitrogen application. Excessive lush vegetative growth and poor yields were obtained at higher levels. This has been confirmed by more recent reports by Ezedinma (1964a) and Osimame (1978).

The later studies showed that nitrogen was chiefly required in the first few weeks of growth before effective nodulation is fully established. Thus, in practice only 20–30 kg nitrogen is routinely applied to cowpea before, or at, planting. Experience in Britain, New Zealand and elsewhere suggests that between 300–600 lb nitrogen per acre (135–270 kg/4047 m^2) per year can be fixed by these soil organisms. The amounts of nitrogen fixed under Nigerian conditions have not been determined.

Ezedinma (1964a, b) stated that cowpea rhizobia are commonly present in all agricultural soils of Nigeria. This means that natural inoculation of cowpea occurs on the field without extra effort and cost. However, he cautioned that *Rhizobium* inoculation alone would not provide all the nitrogen necessary for rapid early growth.

Soybean has been found to be a good source of oil, protein and livestock cake in Nigeria. Its cultivation is thus becoming increasingly popular. The major problem is the lack of effective nodulating rhizobia for the crop in Nigerian soils. It is conjectured that by inoculating the seeds with rhizobia before sowing, farmers could reduce the cost of fertilizer applied, as with the cowpea crop.

Lack of trained personnel and facilities are major constraints of this programme. It is reliably learnt that FAO/UNCP plan to be actively involved in a programme to promote nitrogen fixation in West Africa.

AZOLLA FILICULOIDES

This aquatic organism is capable of fixing high levels of nitrogen. A report by Talley and Rains (1980) showed that *Azolla pinnata* is being widely grown as a fertilizer crop in Vietnam. Currently it occupies 400 000 ha of fallow rice fields. In their trials in the Sacramento Valley, California, Talley and Rains reported that when 60 kg nitrogen/ha as decomposing *A. filiculoides* was pressed into the top 18 cm of soil, rice yields increased by 1.6 t/ha compared with controls.

The potential of this organism in Nigeria, where swamp, mangrove and deep-flooded rice cultivation is widely practised, is good. These areas must occupy several thousand hectares. Routine fertilizer application is difficult because of the flooded conditions. Not much work has been done so far, but the West African Rice Development Agency (WARDA) is proposing to conduct trials on the application of *Azolla* in the West African zone.

VESICULAR–ARBUSCULAR MYCORRHIZA (VA)

The potential of these organisms for enhancing phosphorus nutrition and the water economy of plants is widely known, especially on low-fertility soils. The work in this area in Nigeria is rudimentary. Nevertheless, the results obtained have been encouraging.

Redhead (1968) and Sanni (1976a) have shown that Nigerian soils are commonly infested with endogenous spores forming VA mycorrhiza. They each described many species related to economic tree and food crops.

Sanni (1976a, b) also reported that the addition of VA to the soil encourages phosphorus absorption and utilization by cowpea, tomato, maize and rice. These crops are widely grown in Nigeria.

Natural predators

Large sugar-cane plantations are being established throughout the wetlands of Nigeria to reduce the current trend of sugar importation. Among other problems, nematodes have been found to reduce cane yields drastically. Some of the more commonly available nematocides are harmful to man. Thus, the control of nematodes poses a great problem.

Salawu (unpublished work) recently found an unknown endozoic fungus parasite associated with the sugar-cane cyst nematode in one of our trial sites. Further work is in progress to study and identify the organism to see whether it can be exploited economically.

Conclusion

Vast areas of agriculturally utilizable land exist in Nigeria. The present method of farming is a mixture of small- and large-scale farming. In all cases the cost of various inputs in terms of capital and/or energy is prohibitive. In addition, the problem of irreversible environmental pollution is imminent. This cost and risk must be reduced.

It is very important to investigate the use of alternative natural resources. Reports obtained in other countries have indicated that this possibility is feasible. In Nigeria, there is evidence that useful microflora abound in the soils throughout its length and breadth. These need to be pressed into service: so far, research results are rudimentary but encouraging. Reports show that effective use is already being made of farmyard manure and nitrogen-fixing bacteria for crop production, the latter at no additional cost to the farmers. The uses of VA, *Azolla* and predatory organisms are being explored.

The work so far, except for that involving farmyard manure, has been done by individual scientific teams in different laboratories. There must be central coordination if much is to be achieved. There are additional constraints, such as the lack of trained manpower and of facilities successfully to pursue and implement the projects. Virtually nothing is known about the organisms themselves, their population dynamics and optimum environment, and these must be studied in order that the organisms may be cultured and made available on a large scale for general use.

Lastly, what will be the influence of biological agriculture on global politics and economy *vis-à-vis* the developed, developing and underdeveloped nations of the world?

References

AMON, B. O. E. and ADETUNJI, S. A. (1969). Review of soil fertility investigations in Western Nigeria. *Min. of Agric. Res. Report No. 55*

ANDREWS, D. J. (1972). Intercropping with sorghum in Nigeria. *J. agric. Sci., Camb.*, **79**, 531–540

BACHE, B. W. and HEATHCOTE, R. G. (1969). Long term effects of fertilizers and manure on soil and leaves of cotton in Nigeria. *Expl. Agric.*, **5**, 241–247

BAKER, E. F. I., LOMBIN, G. and ABDULLAHI, A. (1977). Long term fertility studies at Samaru. 1. Direct and residual effects of single super phosphate and FYM on yields of cotton, sorghum and groundnuts grown in a rotation. *Samaru Misc. Paper 67*

DRANSFIELD, M. and McDONALD, D. (1967). The influence of cropping practice and application of FYM on the soil microflora at Samaru, Northern Nigeria. *Samaru Res. Bull. 81*

EZEDINMA, F. O. C. (1964a). Notes on the distibution and effectiveness of cowpea *Rhizobium* in Nigeria soils. *Pl. Soil*, **21**, 134

EZEDINMA, F. O. C. (1964b). Effects of inoculation with local isolates of cowpea *Rhizobium* and application of nitrate-nitrogen on the development of cowpea. *Trop. Agric., Trinidad*, **41**, 243–249

JONES, M. J. (1971). The maintenance of soil organic matter under continuous cultivation in Samaru, Nigeria. *J. agric. Sci, Camb.*, **77**, 473–482

JONES, M. J. (1973). The organic matter content of the savanna soils of West Africa. *J. Soil Sci.*, **24**, 42–53

JONES, M. J. (1976). The significance of crop residues to the maintenance of fertility under continuous cropping at Samaru, Nigeria. *J. agric. Sci., Camb.*, **86**, 117–125

OSIMAME, O. A. (1978). The fertilizer requirement of Ife Brown cowpea. *Grain Legume Bull. (IITA)*, **11** & **12** 13–15

OLAFARE, S. O. O. and OJOMO, O. A. (1980). The response of cowpea intercropped with maize having different leaf canopies—a preliminary report. *Paper presented at the 16th Annual Conf. Agric. Soc. Nigeria. August 17–22, 1980*

REDHEAD, J. F. (1968). Mycorrhizal association in some Nigerian forest trees. *Trans. Br. mycol. Soc.*, **51**, 377

REMISON, S. U. (1978). Neighbour effects between maize and cowpea at various levels of N and P. *Expl. Agric.*, **14**, 205–212

SANNI, S. O. (1976a). Vesicular–arbuscular mycorrhiza in some Nigerian soils and their effect on the growth of cowpea (*Vigna unguiculata*), tomato (*Lycopersicon esculentum*) and maize (*Zea mays*). *New Phytol.*, **77**, 667–671

SANNI, S. O. (1976b). Vesicular–arbuscular mycorrhiza in some Nigerian soils. The effect of *Gigaspora gigantea* on the growth of rice. *New Phytol.*, **77**, 673–674

TALLEY, S. N. and RAINS, D. W. (1980). *Azolla filiculoides* Lam. as a fallow-season green manure for rice in a temperate climate. *Agron. J.*, **72**, 11–18

TIFFEN, MARY (1971). Changing pattern of farming in Gombe Emirate, North-Eastern Nigeria. *Samaru Misc. Paper 32*

VINE, H. (1953). Experiment on the maintenance of soil fertility at Ibadan, Nigeria. 1922–1951. *Emp. J. exp. Agric.*, **21**, 65–85

WATSON, K. A. and GOLDSWORTHY, P. R. (1964). Soil fertility investigations in the middle belt of Nigeria. *Emp. J. exp. Agric.*, **32**, 290–302

15

DEVELOPING STABLE AND PRODUCTIVE BIOLOGICAL CROPPING SYSTEMS FOR THE HUMID TROPICS

G. F. WILSON and B. T. KANG
International Institute of Tropical Agriculture, Ibadan, Nigeria

Introduction

Population growth rates indicate that by the year 2000, 60 per cent more food will be required to meet the requirement of the world population (FAO, 1977). Increasing agricultural productivity to meet this demand poses a special problem, as the greater proportion of the population increase is expected in developing tropical countries where food production is relatively low, and in many cases falls below the demands of the present population. The alternative solutions to the above problem are: (1) expanding the area of production; (2) increasing the productivity of the area under production; or (3) a combination of (1) and (2).

Expanding the area under food production will be possible only in areas where suitable lands are available. Unfortunately many developing nations are already fully using the area regarded as arable, and further expansion would bring into production lands of marginal productivity under existing local practices.

Increasing productivity of unit area can be approached by:

1. Introducing high input and management technologies based mainly on fossil-fuel energy and inorganic fertilizers; or by
2. Developing a more efficient low-input system based on biological recycling of energy and chemical nutrients.

Although approach (1) offers increased yields (Kumm, 1976) it may not be acceptable in all developing tropical countries, as anticipated responses are unlikely to justify high-input expenditures, especially in subsistence economies. It is also feared that uncontrolled use of such technologies in the tropics may cause serious degradation of the environment (Clehara, 1976).

Achieving desired results through approach (2) depends on the identification or development of effective soil nutrient-restoring technologies, as all agricultural systems pivot on their soil-fertility maintenance or restoration techniques (Benneh, 1972). The search for, or development of, these new systems should be based on some existing systems from which basic principles are extractable. Because the 'bush fallow' has been the mainstay of nearly all stable biological tropical cropping systems, it has been chosen as the model or basic system for research orientated towards developing improved land productivity in the tropics.

The bush fallow and shifting cultivation

Shifting cultivation and bush fallow systems *(Figure 15.1)* are still the dominant traditional agricultural system practised for crop production in humid tropical Africa. These systems involve a few years of cultivation alternating with several years of fallow for soil-fertility regeneration (Ruthenberg, 1971). With increasing population pressure and development of more

Figure 15.1   A natural bush fallow, about 7 years old, in the humid tropics

sedentary agriculture, much of the traditional shifting cultivation, in which the farmers move from place to place as farms are changed from year to year, has given way to a range of traditional farming systems or varying cropping intensities and periods of bush fallow. According to Okigbo (1974), in order of cropping intensities this may include:

1. Long-term bush fallow in areas of low population density;
2. Short-term natural bush fallow or planted fallow of 2–5 or more years' duration;
3. Rudimentary sedentary of often not more than 2 years; and
4. Permanent or continuous cropping systems or compound farming which rely heavily on farm or household refuse and ample rotation of crops for soil-fertility maintenance.

Shifting cultivation and the bush fallow systems have often been criticized as being wasteful and inefficient, causing degradation of soils and vegetation, rapid decline in soil fertility and crop yields (FAO, 1957; FAO-SIDA, 1974). However, where land is abundant it is generally agreed that, considering the environmental constraints and availability of tools and resources, the traditional farmers have evolved an efficient and stable system, which has sustained them and their families for many generations. The problem arises only when there is not enough land to meet the need for the increasing population, so that the fallow period becomes progressively shorter and fertility restoration during the fallow period correspondingly becomes less effective.

Although generally there is a random regeneration of species during the bush fallow period, in certain parts of the humid tropics e.g. in south-eastern Nigeria, farmers intentionally retain certain species. During land clearing, species such as *Acioa baterii, Anthonata macrophylla, Alchornia* sp are retained; the small trees are only trimmed and the big branches retained for staking yams and other climbers. The tops are spread on the soils and burned. In this system after the cropping cycle, there is rapid regeneration of the fallow. The bush fallow has thus a double function of (1) providing staking materials and (2) nutrient recycling *(Figure 15.2)*.

Figure 15.2 Crops growing in field recovered from natural bush fallow. The fire-killed *in situ* stems of tree and shrubs are used for supporting yams and beans

During the fallow period, large quantities of plant nutrients are accumulated in the vegetation (biomass). These nutrients are released during the burning of the fallow vegetation following land clearing, to prepare the land for cropping. Nye and Greenland (1960) indicated that all the nutrients in the fallow, except for nitrogen and sulphur, are preserved and returned to the soil in the ash. Seubert, Sanchez and Valverde (1977) however, showed at Yarimaqua, Peru, that not all the nitrogen was lost during the process of burning.

Despite the fact that burning is the most common practice for seedbed preparation, the benefit from this practice may vary depending on the soil type. Studies on Alfisols (Kang, unpublished work) have shown that the main benefit from burning is to hasten the release of the phosphorus in the fallow biomass to the crop. On the acid Ultisols, the plant ash besides being an important source of plant nutrients is also an important source of liming material. Very few quantitative and qualitative data regarding the amount of nutrient immobilized during the fallow period and also of the mechanisms of their release after burning are available for the humid tropics. Seubert, Sanchez and Valverde (1977), reporting data from a burnt 17-year-old secondary forest in an Ultisol of Yarimaquas showed an average ash of 4 t/ha. The true ash and partly burned or charred plant material added approximately 70 kg/ha nitrogen, 6 kg/ha of phosphorus, and 37 kg/ha potassium, plus 240 kg/ha of dolomitic lime and substantial amounts of micronutrients. Okoro (unpublished data) observed in a study on an Ultisol in Eastern Nigeria that burning of a 6-year-old fallow yielded only 563 kg/ha of ash, containing (in kg/ha) 5 nitrogen, 6 phosphorus, 20 potassium, 33 calcium, 9 magnesium, 1.4 manganese, 0.3 zinc and 3.8 iron. Although addition of the above amount of ash raised the pH of the surface soil (0–7.5 cm) from 4.30 to 5.00, and reduced the total acidity from 50 to 11 per cent, the effect on maize yield was less than expected and also of short duration. Use of the fallow vegetation as mulch appears to be more beneficial than burning on this low-nitrogen soil (*Table 15.1*).

Table 15.1   EFFECT OF MULCHING AND BURNING OF THE FALLOWS RESIDUE AND FERTILIZER APPLICATION ON MAIZE GRAIN YIELD (OKORO, UNPUBLISHED)

Treatment	Maize grain yield (kg/ha)	
	Residue burned	Residue as mulch
No fertilizer	1190	1740
NPK, Mg, Zn + CaCO$_3$ (1 t/ha)	3495	3405

Burning of the crop residue, besides the indirect effect on soil properties, also has a direct effect. Flash burning at low temperature may have a beneficial effect on soil nutrient status; however, heating the soil at high temperature and for long duration, as may take place in concentrated burning, can have a deleterious effect on soil properties (Kang and Sajjapongse, 1980). Thus, burning the crop residue without care can reduce the more general beneficial effects.

Planted fallow

The efficiency of the bush fallow, as measured by the time required for effective soil-fertility restoration, is associated with the plant species comprising the fallow. It is well known that some plant species are more efficient than others in improving plant nutrient levels in soil. Thus, leguminous species are known for their nitrogen-fixing ability, because of their symbiotic association with *Rhizobium* bacteria, and they form the major group of plants intentionally grown for soil-fertility restoration. This practice is common in temperate zones where mostly herbaceous legumes serve the dual purpose of providing forage and regenerating soil fertility in the rotation.

Many previous attempts to introduce planted leguminous species for soil-fertility restoration in tropical cropping systems (Webster and Wilson, 1966) have now become popular, possibly because herbaceous-type legumes similar to those used in temperate regions were being emphasized in an environment where trees are the favoured dominant plant types in the ecology. Tropical farmers have recognized the plant-nutrient regenerating superiority of certain tree species, most of them legumes, and have used them in improving the efficiency of the bush fallow (Benneh, 1972; NAS, 1977). Invariably these are trees or upright vigorous shrubs. In certain regions of eastern Nigeria, as indicated earlier, farmers have recognized the soil-fertility regenerating properties of certain plant species, namely *Acioa baterii* (a non-legume) and *Anthonata macrophylla,* and have encouraged them in the bush fallow (Benneh, 1972; Okigbo, 1976). Although actual

Figure 15.3   Yam vines supported by chemical-killed *Leucaena* stems from 18 months' planted fallow

planting is seldom done, preferential protection and care warrant the claims that these are cultivated fallow species. In the area around Ibadan, glyricidia (*Glyricidia sepium*) established from stakes used for yam support has become the recognized fallow species. Farmers in the area claim that only 2 years resting under glyricidia fallow is required for restoring soil productivity to economic levels *(Figure 15.3).*

In Asia, leucaena (*Leucaena leucocephala*) and sesbania (*Sesbania grandiflora*) are among the legumes recognized as efficient soil restorers (NAS, 1979; Benge, [no date]; Guevarra, 1976). The contribution from these tree species is not by the direct transfer of nutrient from nodules to the soil but through material released from decomposing leaves or small twigs resulting from natural leaf-fall or slashing during land clearing. On the usually poor tropical soils, the tree-type legumes with deep roots are able to recycle nutrient from the deeper soil layer, thus contributing (in addition to nitrogen of symbiotic fixation) other essential plant nutrients.

The ability of plants to contribute to soil regeneration depends on the soil, climatic conditions and management. In Hawaii, leaves of giant leucaena cut to a height of 1 metre every 3 months yield 500–600 kg/ha of nitrogen annually (Guevarra, 1976). Such a high yield may not be obtainable on many of the poorer tropical soils, but even on these, substantial contributions are obtainable.

Alley cropping

The term 'alley cropping' was coined by the authors to describe a cropping system in which crops, especially food crops, are grown in alleys formed by trees or shrubs, established mainly to hasten soil-fertility restoration and enhance soil productivity. Depending on the state of the crop or the soil-restoring species, pruning or other methods of suppression may be used to prevent the trees or shrubs from competing with the crop *(Figure 15.4)*.

Systems of growing crops between rows or hedgerows of soil-regenerating or fallow species in farmers' fields have been described by Benge [no date] and have been observed in Africa by the authors. Guavarra (1976) experimented with the system in Hawaii and determined the nutrient

Figure 15.4 Planted fallow of leguminous species, *Trephosia candida* (foreground) and *Leucaena leucocephala* (background) after 1 year regrowth. The alleys are being planted by using a rolling injector no-till planter

contribution from pruned leaves of leucaena placed on, or incorporated into, the soil on maize yield in alleys one metre wide. He found that, where nitrogen was limited, leucaena leaf nitrogen contributed to a significant 23 per cent yield increase over the control. The total leucaena nitrogen added to the maize plant from single and double leucaena rows average 100 and 162 kg N/ha respectively.

Table 15.2   NITROGEN YIELD OF *LEUCAENA LEUCOCEPHALA* TOPS 'ALLEY CROPPED' WITH MAIZE ON APOMU SOIL (PSAMMENTIC USTHORTHENT)

Yield	Nitrogen rate (kg N/ha)[1]			
	0	60	120	180
First season[2]				
Nitrogen yield (kg N/ha), sum of three cuttings	95.7	107.7	125.9	153.8
Second season[2]				
Nitrogen yield (kg N/ha), sum of two cuttings	111.2	89.6	93.2	104.1
Total/year (kg/ha)	206.9	197.3	219.1	257.9

[1] Nitrogen applied to associated maize crop
[2] First season trimming April–June and second season trimming September–October. Leucaena row trimmed to height of ± 150 cm

In trials at IITA, the dry-matter yield and potential nitrogen contribution from leucaena rows of alleys 4 m wide substantiate Guavarra's findings (*Table 15.2*). Nitrogen applied to maize in the first season resulted in an increased leucaena leaf nitrogen contribution but had a slightly depressing effect in the second season when the quantity of nitrogen applied was less. Rapid leaching during the heavy rains of the second season, plus suppression of nodule development by high nitrogen levels in the first season, could have influenced the trend. The maize yield over the two seasons (*Figure 15.5*) was significantly affected by added urea nitrogen but leucaena nitrogen gave acceptable yields. As leucaena leaves were not incorporated into the soil, high nitrogen loss through volatilization was partly responsible for these low yields.

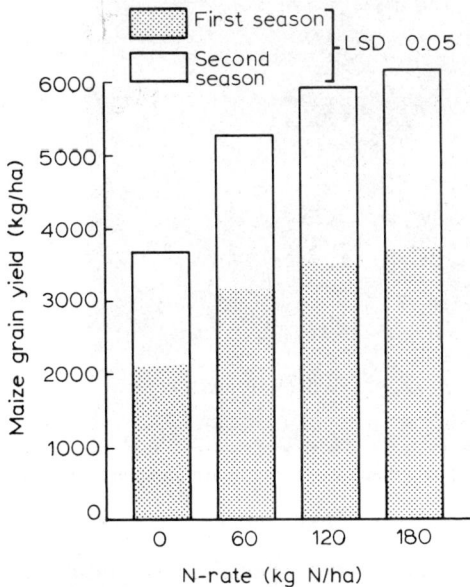

Figure 15.5   Effect of nitrogen rates of grain yield of maize 'alley cropped' with *Leucaena leucocephala* on Apomu soil (Psammentic usthorthent). (First season, maize variety TZPB; second season, maize variety TZE. Two-thirds nitrogen applied to first-season crop)

The potential nutrient contributions of glyricidia (*Glyricidia sepium*), pigeon pea (*Cajanus cajan*) and tephrosia (*Tephrosia candida*), bordering alleys 2 m wide to interplanted maize (*Table 15.3*) are indicative of the potential of other legumes to contribute substantially to crop production in the tropics.

Table 15.3 NITROGEN YIELD FROM VARIOUS LEGUMINOUS CROPS

Shrubs		Leaf yield	
Shrubs	Dry weight (kg/ha)	N (%)	(kg/ha)
Pigeon pea	4100	3.6	151
Tephrosia	3067	3.8	118
Glyricidia	2300	3.7	84

Alley cropping, which may be regarded as an organized form of bush fallow in which selected species are planted in orderly patterns, was designed to facilitate crop growth and easy execution of activities associated with crop management in systems based on nutrient recycling by plants (*Figure 15.6*). The alley width, the fallow species and spacing along the rows should be selected with respect to the tools to be used, the crops and the agricultural operations anticipated. Observations include certain advantages of alleys 2 and 4 m wide, for hand tools and for tractor-oriented maize production respectively.

Figure 15.6   Leaves and twigs from shrub-planted fallow serve as mulch and plant nutrient source in 'alley cropping'

Establishment of the fallow species deserves special consideration, as tropical peasants with limited labour would not willingly devote themselves to establishing fallow plants from which there are no direct returns. Development of a simple and economic establishment procedure is there-

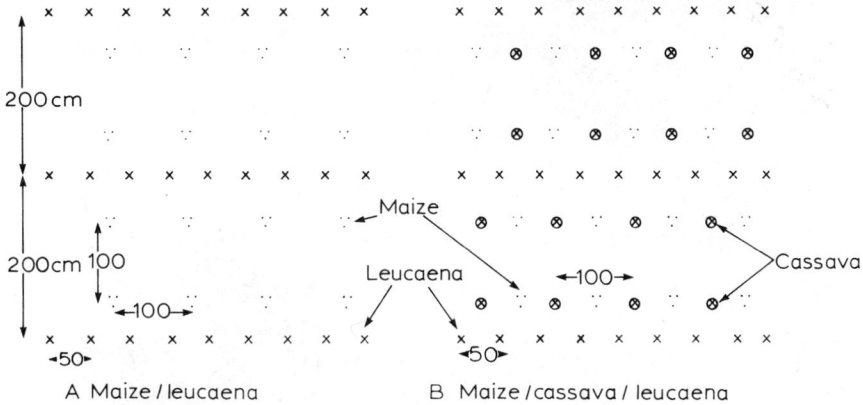

Figure 15.7 Diagram of intercropping for establishment of leucaena

fore receiving priority attention. *Figure 15.7* shows the layout for establishing leucaena in association with maize and maize/cassava. The initial phase of a 2-year rotation involving maize/leucaena followed by yam/leucaena was established in a similar pattern. Crop yield and leucaena development (*Table 15.4*) indicated no serious disadvantage to the intercropping establishment of leucaena. Where soil nutrient levels are low, fertilizer applications may be justified until the fallow species are large enough to regenerate sufficient nutrient to maintain crop production. As fallow species are not destroyed during cropping there is an automatic reversion to fallow, under an efficient species, immediately cropping pressures are removed. In addition to nutrient recycling, the systems also contribute wood for fuel and other uses. The rapid regeneration of trees suppresses weed growth, prevents erosion and prevents the environmental deterioration which is often associated with absence of trees (*Figure 15.8*).

As with all systems there are disadvantages. The major fears are that certain fallow species could become hard-to-control weeds, or provide favourable conditions for multiplication of pests and other undesirable organisms.

Table 15.4 YIELDS OF MAIZE, CASSAVA AND YAM AND DEVELOPMENT OF LEUCAENA IN INTERCROPPING ESTABLISHMENT OF LEUCAENA

Crop			Year 1		Year 2
	Maize	Cassava	Leucaena Height	Girth	Yam
	(kg/ha)		(cm)		(kg/ha)
Maize	2512	—	—	—	—
Maize/cassava	3058	20 187	—	—	—
Maize/leucaena	2857	—	375	2.72	14 905
Maize/cassava/leucaena	1840	29 229	370	2.62	12 333
Leucaena	—	—	445	3.14	13 690
LSD 0.05	NS	NS	74	0.25	NS

Figure 15.8 'Alley cropping': young maize growing in alley formed by *Leucaena leucocephala*. The *Leucaena* leaves and twigs were used for the mulch while the wood piled in the background is from the stem

Conclusion

Alley cropping is an improved bush fallow system in which selected shrub or tree species, usually legumes, are planted in association with food crops to accelerate soil-nutrient regeneration, thus shortening the fallow period. The ultimate aim is the development of a more efficient food production system for continuous cropping based on biological nutrient recycling for the tropics.

The research is directed to the lesser-developed countries where financial and infrastructural constraints may limit the use of fossil-fuel and inorganic fertilizer-based technologies, or in areas where a biological alternative is required to control pollution and other possible forms of environmental degradation associated with excessive chemical input technologies.

References

BENGE, M. P. [No date]. Bayani, Giont ipil-ipil (*Leucaena leucocephala*) a source of fertilizer feed and energy for the Philippines. *USAID Development Series*

BENNEH, G. (1972). Systems of agriculture in Tropical Africa. *Econ Geog.*, **48**, 244–257

CLEHARA, G. (1976). An overview of the soils of the arable tropics. In *Proc. of a Workshop on Exploiting Legume–Rhizobium Symbiosis in Tropical Agriculture. Aug. 23–28, 1976* Hawaii; University of Hawaii

FAO (1957). Shifting cultivation. *Trop. Agric. (Trinidad)*, **34**, 159–164

FAO (1977). *State of World Agriculture*. Rome; FAO

FAO-SIDA (1974). *Shifting Cultivation and Soil Conservation in Africa*. Soils Bulletin 24. Rome; FAO

GUEVARRA, A. B. (1976). *Management of* Leucaena leucocephala *(Lam) de Wit for maximum yield and nitrogen contribution to intercropped corn.* PhD Thesis, University of Hawaii

KANG, B. T. and SAJJAPONGSE, A. (1980). Effect of heating on properties of some soils from southern Nigeria and growth of rice. *Pl. Soil* (in press)

KUMM, K. I. (1976). An economic analysis of nitrogen leaching caused by agricultural activities. *Ecol. Bull (Stockholm)*, **22**, 169–183

NAS (1977). *Leucaena: Processing Forage and Tree Crop for the Tropics.* Washington, DC; National Academy of Sciences

NAS (1979). *Tropical Legumes: Resources for the Future.* Washington, DC; National Academy of Sciences

NYE, P. H. and GREENLAND, D. J. (1960). *The Soil under Shifting Cultivation.* Tech. Comm. 51. Harpenden, Herts; Comm. Bureau of Soils

OKIGBO, B. N. (1974). Fitting research to farming systems. In *Proc. Second Int. Seminar on Change in Agriculture, Reading, England*

OKIGBO, B. N. (1976). Role of legumes in small holdings of the humid tropics of Africa. In *Proc. of a Workshop on Exploiting the Legume– Rhizobium Symbiosis in Tropical Agriculture,* University of Hawaii, College of Tropical Agriculture. Misc. Publications 145

RUTHENBERG, H. (1971). *Farming Systems in the Tropics.* London; Oxford University Press

SEUBERT, C. E., SANCHEZ, P. A. and VALVERDE, C. (1977). Effect of land clearing methods on soil properties of an Ultisols and crop performance in the amazone jungle of Peru. *Trop. Agric. (Trin)*, **54**, 307–321

WEBSTER, C. C. and WILSON, P. N. (1966). *Agriculture in the Tropics.* London; Longman Green and Co.

16

AN APPROPRIATE TECHNOLOGY FOR INDIAN AGRICULTURE

H. W. DALZELL
Agricultural Centre, Medak 502 110 A.P. India

The goals to be achieved by Indian agriculture in the next quarter-century are to increase and stabilize food production for a population which will expand to that of present-day China, and to generate employment and income, either directly from the production process or indirectly in post-harvest industry and processing. These goals must be achieved while preserving or improving the ecological infrastructure essential to sustained agricultural progress. It would be unwise to rely on technology which depends primarily on massive inputs of nonrenewable energy and nutrients, or on continuing widespread and often indiscriminate use of highly toxic pesticides.

The international trend to nonecological, high-energy, inorganic nutrient and synthetic pesticide input agriculture, often inaccurately referred to as orthodox, has been facilitated by:

1. Ignorance or contempt of ecological concepts. Continued ignorance is inexcusable and contempt antisocial.
2. Economic advantages to energy and agrochemical suppliers in centralized decision making, in easily repeatable chemical processing and in international marketing. The propriety of such an economic order is doubtful and politically it is unlikely to be allowed in future.
3. Acceptance of ready-made 'miracle' inputs because of their spectacular short-term advantages. Alternative technologies have the disadvantage of greater educational and motivational requirements.

As surely as the high-energy, agrochemical revolution took place in a favourable social, political and economic infrastructure, so any change in technical emphasis will require a shift in that infrastructure consciously dictated by man, the unique central figure in the ecosystem.

The organization and strategies to be adopted are open to discussion, but without the integration of natural, social, political and economic sciences, the development of highly productive ecological agriculture along the lines discussed below is unlikely. Natural science includes chemical and physical science as well as biological science. To neglect the potential contribution of physical and chemical inputs to increased production would be as immature as to disregard the dangers of concentrating on such inputs alone.

The present agricultural output of India is estimated to use 25 per cent of theoretically available resources (Swaminathan, 1979). A consideration of

205

land, climate, nutrient and energy resources indicates the potential for development.

Land

Production can be increased by bringing more land under cultivation, or by improving yields, either through better husbandry and/or converting rain-fed land to irrigated land. Judged by traditional cost/output analysis, it would be uneconomic to bring the available poor-quality uncultivated lands into production. Most Chinese peasants, however, would not agree with this analysis: although the labour cost would be high, it would be productive labour which could be paid for out of existing governmental Food for Work Schemes. The implications of improving yields and of increasing the area of irrigated land are quantified in *Tables 16.1 and 16.2.*

Table 16.1   EFFECT ON ANNUAL PRODUCTION OF INCREASED YIELDS AT PRESENT IRRIGATION AREA

	Cultivated area (10^6 ha)	Production levels (t/ha.y) Current (CPL)	Potential (PPL)	Annual production (10^6 t grain equivalents) CPL	PPL
Irrigated	34	2.14	4.0	72.8	136.0
Rainfed	108.8	0.48	1.8	52.3	195.8
Total	142.8	—	—	125.1	331.8

Table 16.2   EFFECT ON ANNUAL PRODUCTION OF INCREASED IRRIGATION AREA AT CURRENT AND POTENTIAL YIELDS

	Potential area (10^6 ha)	Annual output (10^6 t grain equivalents) CPL	PPL
Irrigated	77.6	166.0	310.4
Rainfed	65.2	31.3	117.36
Total	142.8	197.3	427.76

Production varies from year to year, but the relative size of the figures is more important for the purpose of the argument than the accuracy of the totals. Land areas quoted are based on figures calculated by Swaminathan (1979) and production figures on Government of India statistics for 1978.

The potential increase from irrigation by itself is less than that from improving production levels and can be maintained only if backed by adequate engineering and husbandry to prevent problems of erosion, silting, soil alkalinity and salinity. It is reasonable to conclude that improvement of husbandry is a prerequisite for sustained development. Because husbandry is the responsibility of a large number of independent farmers, it cannot be improved without motivation and education. Farmer-level studies near Hyderabad have shown an input management efficiency of 60 per cent for fertilizer and pesticide (Krishnamurthy, Choudhury and Spratt, 1977). Transfer of technology, or in many cases the efficient application of technology already in the hands of farmers, is one of the major problems to be overcome in increasing production.

Climate

Receipts of solar energy are such that very high crop growth rates are possible at most times of the year throughout India. However, in that part of the country south of latitude 25 degrees North, the temperature does not fall below 15 °C for long enough to promote tillering of wheat, which is widely grown. Safflower grown as an alternative has consistently shown a yield double that of wheat (Krishnamurthy, Choudhury and Spratt, 1977).

Rainfall is the other important climatic factor and its effect on cropping is modified by amount, seasonal distribution and intensity. A vast country like India will inevitably have drought or flood affecting a large area of land in any one year. Planning should be to maximize production in good years and to stabilize it at a moderate level in normal years, thus building reserves to carry over poor years. Adequate soil moisture is critical to production, and hence improved management of soil moisture is essential to improved yields. It has been estimated that efficient management of soil moisture could increase grain production on nonirrigated land from the present levels of 19 kg/(ha.cm) rain to 63 kg/(ha.cm) rain (ICRISAT, 1977).

The problems arising directly from rainfall pattern are soil erosion and either too much or too little soil moisture. Fortunately, measures available to the farmer to increase the absorptive capacity of soils while preventing waterlogging will also control erosion.

The first step in control is to ensure optimum layout of plots. Ideally, this can best be achieved by developing large watersheds consisting of many farmers' lands, and by shaping, contouring and grading so that field boundaries are contiguous with property boundaries, infiltration is maximized, and run-off along gentle slopes is canalized to efficient storage for supplementary irrigation of crops during periods of moisture stress. Political and social problems prevent this in most areas, but surely the time has come to weigh the technical cost of such considerations. Second best is for individual farmers to lay out their own plots scientifically.

It is not necessary to have a dumpy level and bulldozer to do so. A simple 'A' frame, constructed locally, is accurate enough for surveying which can be done by the farmer himself, and bullock-drawn equipment is adequate for most land-shaping and bund construction. Large contour bunds can be a disadvantage in view of their cost and the fact that, if soils are shallow, most topsoil will be scraped off for the bunds. Graded ploughing and small graded bunds are often adequate, cost less to build and can be easily realigned if experience shows a need for adjustment.

Tillage offers a second method of moisture control. Ploughing after pre-monsoon showers opens the soil to maximize infiltration on the first monsoon showers and the effects can be as dramatic as a doubling of wetting depth, which not only facilitates earlier sowing and a longer growing season, but also extends the planting season. Ploughing may have to be done within a few hours of rain falling, and with slow-moving ploughs the acreage covered before the soil dries and becomes unploughable is low.

Table 16.3   EFFECT ON YIELD OF TILLAGE DEPTH AT ANANTAPUR (AFTER KRISHNAMURTHY, 1973, WITH SUPPLEMENTARY DATA)

Crop	Yield (100 kg/ha)			
	1973–74		*1974–75*	
	Shallow	*Deep*	*Shallow*	*Deep*
Redgram	9.08	11.78	8.10	11.90
Bajra	7.73	10.29	5.80	9.60
Groundnuts	8.55	11.68	—	—

In many areas deep tillage has been shown to give greatly increased yields, as detailed in *Table 16.3*. Depth of tillage is often limited by the bullock power available. Improved implements can offset this difficulty, as can repeated ploughing along the line to be seeded. It may be that much of the effect of deep tillage is due to the associated suppression of weeds, and provided that weed control can be achieved by other means, shallow tillage would be adequate to improve moisture control. With its lower energy requirement and less soil disturbance, shallow tillage makes more eco-logical sense.

Table 16.4   MOISTURE USE AND FERTILIZER APPLICATION (AFTER KRISHNAMURTHY, CHOUDHURY AND SPRATT, 1977)

Nitrogen	Wheat crop 1972–73, Hoshiapur	
	Yields	*Moisture use*
(kg/ha)	(q/ha)	(cm)
0	15.8	26.3
40	29.8	30.6
80	38.5	31.6
120	46.0	33.1

Nutrient availability, as shown in *Table 16.4*, has an important effect on the ability of a crop to withstand moisture stress. The reason seems to be that a quick-growing crop develops a strong rooting system with a better ability to draw moisture from the soil. A fertilized crop may yield 100–150 per cent more, requiring only 20–25 per cent more water (Krishnamurthy, Choudhury and Spratt, 1977). Putting fertilizer, organic or inorganic, in the seeding line can enhance this increase and requires little more than a simple adaptation of traditional bullock-drawn ploughs.

Crop selection is an important factor in managing moisture. Crops should be selected in relationship to the probable length of the growing season and ability of the variety to cope with moisture stress or waterlogging. The farming system must be adaptable enough to change the crop to be sown if rains are delayed, as in the example in *Table 16.5*, applicable to the Hyderabad region.

Some crops can withstand drought better than others; for example pigeon pea is more resistant than maize. One reason for this is deep rooting. Within individual crops there may be genotypic differences in rooting pattern, leading to variation in drought resistance. An example is that GAUCH I castor has a strong rooting system which can penetrate compact soil layers,

Table 16.5 APPROPRIATE CROP IN RELATION TO
ONSET OF RAINY SEASON

Monsoon onset	Most appropriate major crop
June 3rd week	Sorghum or castor
July 3rd week	Pearl millet or sunflower
Mid-August	Foxtail millet
September 1st week	Field beans, cowpea
Early onset but long gap in middle	Cut sorghum for fodder and ratoon with September rains

whereas Aruna, a high-yielding castor variety comparable with GAUCH I in loose soils, dries up more quickly in compact soils as its roots cannot penetrate compact layers.

It has also been observed that vigorous high-yielding hybrid crops have better tolerance to moisture stress than slow-growing local varieties. While there are obvious dangers in specializing in a restricted number of genotypes, a balance must be found of a range of hybrids with location-specific advantages or preferably careful selections of true-breeding vigorous genotypes.

A good method of reducing surface moisture loss is a mulch, organic matter being better than soil. In heavy compact black soils, a vertical mulch in trenches 50 cm deep, 15 cm wide and several metres apart, mulched to 10–15 cm above soil level with sorghum stalks, has shown substantial yield benefits which are detailed in *Table 16.6*.

Table 16.6 PERFORMANCE OF SORGHUM IN SOILS WITH VERTICAL MULCHING AT DIFFERENT INTERVALS (AFTER KRISHNAMURTHY, 1973, PLUS ADDITIONAL DATA)

Spacing of mulch trenches	Yield (q/ha) (rabi season 1973–74) Grain	Straw
Control (0)	11.2	26.5
2	16.41	30.33
4	16.92	32.51
8	16.14	28.59
Soil cracks filled with sorghum stalks	13.10	26.98

Under traditional farming methods, rainfed lands are cropped once a year. If the soil is light enough to crop in the monsoon, it will be too dry for good establishment of a second crop after the rain, and the heavy soils which can produce a crop sown with late rains are too sticky to crop in the monsoon. Improved moisture management opens up the possibility of increasing cropping intensity, either by ratooning or sequential cropping on heavy soils, and by ensuring better production from late-harvested crops sown as part of an intercrop on light soils. Intercropping is an important tool in exploiting soil moisture efficiently and has the added advantage that it can provide soil cover during heavy September rains after the cereal component has been cut.

With the exception of inorganic fertilizer, all the above methods of optimizing the use of sunlight and rainfall and of minimizing erosion, are consistent with maintaining or improving the ecological infrastructure. They use locally available renewable resources and either have a low cash requirement or depend on labour, farming skill in the field and management knowledge. For land surveying and shaping the skill is undoubtedly inherent and the major problem is how to mobilize it. For all methods, meaningful practical education is required; this may prove costly and requires close liaison between farmers and research workers. The general picture emerging from research and field trials is that good soil and water management can lead to a doubling of crop yields for a given level of nutrient input.

Plant nutrition

Increased use of inorganic fertilizers has been responsible for much of the expansion which has taken place in food production in recent years. This undoubtedly gives rise to ecological dangers and problems of renewability, but underlines the central importance of plant nutrients in the production process. It would be unwise to expect a self-imposed decrease in usage and unrealistic to attempt to decrease consumption at this time. What is required is the development of organic manuring in its many forms, to a level where the need further to increase inorganic fertilizer usage is removed.

Some systems of organic manuring, such as *Azolla* production, are producers of nutrient, while others, such as composting, are processors of wastes. Each system has advantages and disadvantages and hence cannot be a sole answer. Rather, the appropriate course is to integrate the various methods. The effects of some organic manures on soil fertility are described in detail in a report from China (FAO, 1977a). The principles, practice and economics of composting in tropical conditions are described by Dalzell, Gray and Biddlestone, 1979.

One of the reasons for very low average dryland yields is that much of the production takes place on marginal land. The use of high levels of nitrogen and phosphorus on these lands, in conjunction with hybrid seeds, definitely increases yields substantially in the short term. However the economy of trace elements in these soils, which is stable under low crop yields, becomes drained rapidly under intensive production. Organic manuring prevents this drainage and stabilizes increased production.

Most Indian farmers appreciate the benefits of organic manures, but adopt the attitude of using only what comes from their own animals, or of carting small quantities from the bottom of irrigation tanks in the dry season, using low-capacity transport systems. As a result, application rates are of the order of quintals and not normally more than 2–3 t/ha.

This compares very unfavourably with the rates quoted in *Learning from China* (FAO, 1977b) and shown in *Table 16.7*. Potential rates of use for India are shown in *Table 16.8*.

While not all farmers can find the entire supply of nutrients for high yields from organic sources, there will be many farmers who have access to

Table 16.7   USE OF ORGANIC MANURES IN CHINA (AFTER FAO, 1976b)

Crop	Manure (t/ha)
Rice (first crop)	60– 70
Rice (second crop)	60– 70
Wheat	75–105
Corn	75– 90
Sorghum	60– 70
Millet	70– 75
Soybean	30– 45
Cotton	70– 75
Sugar-cane (first crop)	75–135
Sugar-cane (ratoon)	75–135 plus 675 t/ha of mud from bottom of fishponds

Table 16.8   POTENTIAL AVAILABILITY OF NUTRIENTS IN INDIA

Source			10^6 t/y
Night soil (50% recovery of 0.75 kg per person per day)			82
Animal manures:			
226 million cattle and buffaloes @ 10 kg recovered per day			861
105 million goats and sheep @ 0.6 kg recovered per day			23
Poultry, pigs and other stock @ 10% of cattle, sheep and goats			88
Trees, hedgeplants, green manures and crop residues @ 2 t/ha of cultivated land			244
Recovery of eroded soil @ 50% of annual controlled erosion rate of 2 t/ha			142
All other sources @ 3% of above			144
Total availability			1584
	N (kg)	P (kg)	K (kg)
Approximate availability (11.9 t/ha)	43.2	11.6	41.6
Normal recommendation for dryland	40	40	—

sufficient manure and need not use inorganic fertilizers. For India to reach these levels, much more production of raw materials will have to take place. Production of these quantities of manure need not reduce cropping intensity. Every village has some waste land suitable for tree planting which would yield valuable litter. Planting of *Glyricidia* species on bunds round paddy fields can yield 6–7 tonnes of green leaves per hectare per year, which would compost to 3 tonnes organic manure, representing an important nutrient source.

Improvement of village grazing lands would reduce the dependence of animals on low-grade cereal byproducts, which would then become available for composting. Development of biogas production would result in the safe processing of human waste for organic manuring.

Production of plant nutrients from *Azolla* species in symbiosis with blue-green algae, or by blue-green algae alone, is the subject of much interest in India. (For a description of *Azolla* production methods and biology *see*

FAO, 1977a). A directorate has been established by the Central Government to stimulate and coordinate research, and many farmers have attempted to fertilize their land by these methods. Generally speaking, results to date have been variable, but indicate that the practices have an important contribution to make.

Several conclusions can be drawn from this outline of nutrient availability. It is difficult to envisage a system at present which will provide all the nutrients for very high crop yields from organic sources. This is especially true of phosphorus, which is the element most limited in predictions of availability from inorganic sources. Accordingly, research effort is called for to improve the biological production of manures, as distinct from biological processing of waste, and to devise farming systems with least demand on phosphorus. Given the total quantity of phosphorus physically present in most soils, greater knowledge is required of how to incorporate it into the farming cycle. Organic manuring on the scales calculated cannot be achieved without substantial changes in attitude and technology in relation to night soil handling, tree production, erosion control and irrigation tank cleaning. None of these depends on nonrenewable energy.

Organic manuring is not a single factor within the farming system and need not be seen as an alternative to inorganic manuring: in fact, a real possibility exists to use inorganic manuring as a supplement to organic production. Undoubtedly, organic manuring offers a nonpolluting, self-sustaining contribution which will be of increasing importance to Indian agriculture.

Energy

The predominant sources of domestic energy in Indian village life are firewood and dried cattle dung. Traditionally, agriculture has relied on human energy and animal power for cultivation, harvesting, transport and irrigation. For many farmers, diesel or electrically powered pumpsets have now replaced animals as irrigation power sources; however, although there has been a steady increase in the acreage cultivated by tractors, most farmers still rely on animals.

Organic manure requirements and the dangers of deforestation argue against the energy appropriation of dung and firewood. The nonrenewability of oil and the associated high foreign exchange cost argue against a dependence on oil energy. Coal, of which India has reasonable reserves, is an obvious major energy source, but coal is finite. Clearly, there is a need for alternative sources of energy.

Wind power is a possibility but, because of the capital cost, is unlikely to be used much, other than for limited water-pumping. Wind patterns, however, do not always result in power availability at times of maximum irrigation demand. Solar pumps with small-scale irrigation applications are now being manufactured in India and, given time for research and total adaptation to local conditions, should become more cost effective and capable of larger-scale applications. It is argued that solar energy does not have much to offer in domestic use, as storage systems are expensive and cooking is usually done at times of low energy receipts. Ultimately, technical possibilities may prevail over social traditions.

Fermentation of organic wastes to biogas is important from the energy viewpoint and as a means of safely processing night soil for use as organic manures. Much experience has been gained in various biogas technologies (Sathianathan, 1975) and the process is becoming widespread. The gas can be used for cooking, lighting, and fuelling piston engines. The cost of biogas production systems and the requirement for steel in some designs are disadvantages: now that alternative energy costs are rising and nonsteel designs are available, the future of biogas for domestic and irrigation purposes looks secure. This is an area where, for technical and social reasons, small-scale technology seems more satisfactory. The sludge remaining after biogas production can be spread directly on to fields, although ultimately this may create problems of acidity: processing it with wastes containing a high percentage of carbon in a compost heap provides a good alternative. The impact of biogas production is interrelated with other aspects of ecological agriculture. The ultimate limiting factor to the quantity of gas produced is likely to be the availability of wastes. A strong positive correlation exists between the quantity of fodder produced in India and the quantity of waste available. Thus, improvement of fodder production would be necessary to maximize gas production. Increased use of gas would lessen the pressure on forests, increasing the availability of timber for nonfuel purposes and of litter for manuring. The importance of preventing deforestation and so reducing erosion cannot be overstressed.

Although possibilities do exist, solar energy and bioenergy are not readily harnessed for cultivation. Energy is, however, a secondary factor to the cultivation system and implements used. Evidence from surveys of the economics of tractor cultivation in the semiarid tropics of India indicates that tractor use does not improve cropping intensity or yields, but does displace labour (Bingswanger, Ghodake and Thierstein, 1979).

There can be no universally typical case, but often production in India is limited not so much by lack of cultivation energy as by lack of adequate implements. The development of bullock-drawn tool carriers emphasizes the under-utilization of bullock power in traditional Indian agriculture, and offers great scope for the development of bullock power as an efficient energy source for improved cultivation and output. Such tool carriers are unlikely to be economic unless they are used to improve husbandry, rather than merely to speed up operations (Thierstein, 1979). A tool carrier consists of a basic frame mounted on wheels and a tool bar which can be lifted in and out of work and on which various tools, such as ridgers, ploughs and weeders can easily be interchanged. The tool carrier can be fitted with a cart frame for transport purposes. The facility for multirow weeding at rates much faster than hand-weeding at equivalent cost offers a definite possibility for intensive cropping without using herbicides.

A pair of bullocks pulling a tool carrier has a work output at least twice that of a pair with traditional equipment. The range of tools available equips the farmer to practise with precision the improved agronomic practices for guaranteed high yields. Experience shows that four or five tool carriers can replace a medium horsepower tractor or 16 pairs of bulls with traditional equipment. This raises a major issue, as ownership of bulls confers agricultural independence and status, whereas many landholdings are too small to justify individual ownership of tool bars. A reduction in draught animal numbers would improve animal health, and total food production could be increased either by putting surplus pasture under grain crops or, preferably, by replacing redundant animals with sheep and milch stock, thus maintaining the availability of manures.

The possibility of producing energy for industry by processing maize, sugar-cane and other crops to alcohol, is being given serious attention at government level.

Summing up the energy situation, it is clear that, although Indian agriculture is less dependent on nonrenewable energy than most First World agriculture, the oil crisis has created pressure to consider alternatives. The development of solar energy technology, biogas production and animal-powered equipment can go a long way towards meeting requirements and to integrating other ecological aspects of the productive process.

The brief description given above of land, climate, plant nutrients and energy resources indicates that considerable scope exists in India to increase production through self-sustaining processes and to enhance the ecological infrastructure.

Weed, insect and disease problems are likely to provide a major constraint to this increase. Pest attacks have become an increasing problem in the last 20 years, and are positively correlated with monoculture, especially when limited to a narrow genetic base.

A wide range of pesticides are marketed and used in India and, while skilled technical advice is available through government services, there is a lot of misuse of pesticides caused by ignorance on the part of the users. This not only causes economic waste in outlay and lost production, but has serious ecological implications. When one considers the enormous potential which exists for fish production in the network of canals and irrigation tanks, it is clear that pest control by nonchemical methods is preferable. In the event of major pest attacks, the use of ultra-low volume sprayers seems to be advisable.

Biological methods of control are used in certain areas, and success has been reported on cotton, sugar-cane and rice crops. The importance of breeding for genetic resistance cannot be overstressed. Examples are the greater susceptibility, compared with local varieties, of hybrid pearl millet to ergot, and of hybrid sorghum to *Striga.*

One of the best methods of control is to adopt farming methods which prevent the development of crisis-scale pest occurrence. An example is crop rotation: not only does this prevent pest build-up, but it also balances nutrient demands, minimizes risk due to rainfall irregularity, irons out labour and energy peaks and spreads economic risk. A typical example of the dangers of not using rotation is that continued cropping with hybrid sorghum can easily result in a build-up of shoot fly and *Striga.* On black soils, rotation with cotton and black gram, and on red soils rotation with castor, green gram and groundnut, will prevent the problem. Rotations can, of course, be used in conjunction with other control measures: thus, breaking continuous rice cropping with irrigated maize or oil seeds can prevent serious outbreaks of stem borer, while cutting stubble high and burning can minimize the carryover from crop to crop.

Perhaps the most appropriate long-term technology for Indian agriculture is the development of intercropping systems which, in addition to the benefits of rotation, can optimize use of root zones and light and moisture receipts. By growing adequate populations of cereal and legumes as intercrops, yields per hectare can be obtained of up to 90 per cent of the cereal, and 60 per cent of the companion crop, when grown as sole crops. Thus, two hectares of land intercropped can yield 1.8 times the cereal and 1.2 times the companion crop, compared with cropping one hectare of each crop as sole crops. Good management and moderate nutrient inputs have helped to produce total yields of up to 5 t/(ha.y) with cereal legume intercrops on dry land.

Much interest has been stimulated in this area of farming and many combinations of crops, including vertical intercropping, are under trial in various parts of the country. Among the main combinations are those shown in *Table 16.9.*

Table 16.9 COMBINATIONS OF CROPS

Cereal–cereal	Sorghum + pearl millet
Cereal–grain legume	Maize + red gram
Cereal–oilseed	Maize + groundnut
Cereal–fodder	Sorghum + cowpea
Plantation crop–fodder–fruit	Cocoa + napier + banana

Traditionally, villagers intercrop several crops such as maize, vegetable gourd, black gram and coriander. Understanding the experience which has led to the various combinations used, could well indicate directions for progress. Undoubtedly, part of the explanation is the insurance value of several crops, but as insurance can be gained by rotations, this cannot be the only reason.

Intercropping practices almost always include provision for animal fodder and, given the importance of draught animals, must continue to do so; however, much could be done to improve village grazing lands. A major obstacle is the commons system of grazing, with the result that no individual can expect much return from private improvement measures. Another obstacle is overstocking, leading to grazing-out of all but the hardiest species, and to widespread erosion. Broadcasting *Styloxanthes* spp has led to marked improvement of pastures where grazing can be controlled. A combination of pasture improvement, grazing control and reduction of livestock numbers is critical to marked improvement of animal production.

Conclusions

A brief consideration of the available resources has indicated that various husbandry practices offer great scope for enhanced production. Agriculture is, of course, more than a series of independent husbandry practices. Animal production is closely integrated with manure and energy production and utilization. Energy production from renewable resources, such as the sun and animal manure, has definite positive implications for forest development, which will undoubtedly help to prevent erosion and to provide litter for manuring. Improved methods of erosion control and moisture management, such as timely and appropriate cultivation with improved implements, will lead to increased yields which can be further enhanced and stabilized by drawing on the benefits of rotations and intercropping. The resultant reduction in pest problems will encourage the development of fish farming, which is otherwise limited by pesticide spraying. Experience has shown that the processing of animal manures through fish ponds can enhance fish growth rates, and the pond mud can later be returned to the land. Increasing the quantity scale of organic manuring will improve soil fertility in the long term and relieve ecological and economic difficulties.

Clearly, each factor has a contributory part to play in increasing food production in India. Equally clearly, the relationships which exist between factors result in positive interactions when several factors are combined in a farming system. The benefits of such interactions can be gained only if the improved technology outlined is practised by a very great number of smallholders, who must be given access to adequate working capital and equitable marketing and pricing arrangements. Almost all the practices described are inherently nonpolluting, are renewable, and help to improve the production environment: with the further qualification that many of them are labour-intensive, it is justifiable to conclude that they combine to offer an appropriate technology for the sustained development of Indian agriculture. The realization of this development depends on effective education and motivation, which can present the technology in acceptable social, political and economic terms.

References

BINSWANGER, H. P., GHODAKE, R. D. and THIERSTEIN, G. E. (1979). Observations on the economics of tractors, bullocks and wheeled tool carriers in the semi-arid tropics of India. Paper presented at a *Workshop on Socio-Economic Constraints to Development of Semi-Arid Tropical Agriculture*. ICRISAT—19–23 February, 1979

DALZELL, H. W., GRAY, K. R. and BIDDLESTONE, J. A. (1979). *Composting in Tropical Agriculture*. Review Paper Series No. 2. Ed. R. D. Hodges. Suffolk, England; International Institute of Biological Husbandry

FAO (1977a). *China—Recycling of Organic Wastes in Agriculture*. Soils Bulletin 40. Rome; Food and Agriculture Organization

FAO (1977b). *Learning from China*. Regional Office for Asia and the Far East, Bangkok; Food and Agriculture Organization

ICRISAT (1977). *Annual Report, 1976–77*. Hyderabad, India; International Crops Research Institute for the Semi-Arid Tropics

KRISHNAMURTHY, C. (1973). Water management in rainfed agriculture. Paper presented at the *Symposium on Water Resources of India and their Utilisation in Agriculture*, Indian Agricultural Research Institute, New Delhi, March 21

KRISHNAMURTHY, C., CHOWDHURY, S. L. and SPRATT, E. D. (1977). New horizons for dry farming in the rainfed tropics. *Results of the All India*

Coordinated Research Project for Dry Land Agriculture. Hyderabad; Indian Council of Agricultural Research

SATHIANATHAN, M. A. (1975). *Biogas Achievements and Challenges.* New Delhi; Association of Voluntary Agencies for Rural Development

SWAMINATHAN, M. S. (1979). Role of national programmes in linking research and development. *ICRISAT Symposium on Development and Transfer of Technology for Rainfed Agriculture and the SAT Farmer, 28 August–1 September.* Hyderabad, India; International Crops Research Institute for the Semi-Arid Tropics

THIERSTEIN, G. E. (1979). Possibilities for mechanising rainfed agriculture in the semi-arid tropics. *IX International Agricultural Engineering Congress, East Lancing, Michigan, U.S.A., July 8–13, 1979.* Hyderabad, India; International Crops Research Institute for the Semi-Arid Tropics

17

THE EFFECTS OF THREE ORGANIC MATERIALS ON THE CATION EXCHANGE CAPACITY OF THREE UGANDAN SOILS

V. O. A. OCHWOH and J. Y. KITUNGULU-ZAKE
Department of Soil Science, Makerere University, Uganda

Introduction

The effects of organic matter in the soil originate from its contribution to the cation exchange capacity (CEC), its supply of nutrients, and its effects on the physical properties. Because most tropical soils contain much iron and aluminium which contribute to the soil structure, the most important contributions of organic matter to these soils are the negative charges and the nutrients (Russell, 1963). With little organic matter in the soil, the CEC is reduced to the extent that it is very easy to unbalance the proportions of different cations in the soil. The disproportionate supply of nutrients in Ugandan soils, especially with limited application of fertilizers, may be one of the major problems facing Uganda's agriculture. The maintenance of fertility of soils in Uganda has often been achieved by resting the land under a natural fallow dominated by grasses (Mills, 1954). Under this practice, the nutrients are retained in the soil by being recycled by plants. In many cases it has not been possible to attain as high yields in Uganda using commercial fertilizers, as when organic matter was used (Mills, 1952).

There are abundant sources of organic matter that could be used by Ugandan farmers to increase the soil CEC and to supply nutrients. The list includes coffee husks, cow manure, cotton-seed husks, and leaves of some plants. Not much work has been done on the ability of these organic materials to effect changes in CEC and plant nutrients. Their effectiveness was tested on yield responses by indicator crops, but yields depend on various factors. Therefore, the objective of this research was to evaluate the effects of the three readily available organic matter sources (coffee husks, cow manure and elephant-grass leaves) at various rates on the CEC of three different Ugandan soils.

Materials and methods

Coffee husks were collected from a nearby pulping factory at Kawempe. Cow manure was collected from a farm near Kawanda Research Station. For elephant grass, obtained from Kabanyolo Farm, the leaves were cut and mixed well to minimize the variations due to age and/or varieties. These materials were then dried and ground to pass through a 2 mm sieve before application to soils for incubation.

220   *Cation exchange capacity of three Ugandan soils*

Soils A and B (Kitende and Kabanyolo East Valley soils respectively) were highly leached ferrallitic soils, and were thus expected to have low CEC. Soil B differed from soil A only in that it had more organic matter. These two soils represent a major portion of Ugandan soils. Soil C (Lina soil) was a topographic Vertisol. It was a young soil, of which the mineral contribution of CEC was expected to be high. This soil type represents major agricultural areas of Uganda, such as the coffee areas of Bugisu, the tobacco areas of West Nile and the cotton areas of Busoga.

Incubation was carried out in plastic pots of 1500 ml where 1 kg of each soil was used for rates of 0, 25 and 50 t/ha of each organic material. The soils were kept at field moisture capacity by determining the field capacity of each soil, then the soil moisture level was restored to field capacity by adding water after every 4 days. Soil samples were taken at 4-weekly intervals throughout the 24 weeks of incubation. A 'Hoffer' soil tube was used for sampling, which enabled the whole length, from top to bottom of the pots, to be sampled. The soils sampled were air-dried and ground to pass through a 2 mm sieve before they were analyzed. Cation exchange capacity was determined by sodium saturation, a method adopted by Kawanda Research Station (Foster, 1971). Extractable phosphorus, calcium and potassium were determined by ammonium lactate–acetic acid (AL) solution. This method, which is a modification of a method by Robinson and Semp (1968), was adopted for routine soil analysis in Kawanda Research Station. Organic matter was determined by the method of Walkley (1947).

Results and discussion

CHANGES IN SOIL CEC WITH THE ADDED ORGANIC MATERIALS

Changes in soil cation exchange capacity during incubation of soils A, B, and C with different levels of coffee husks, elephant-grass leaves, and cow manure are presented in *Figures 17.1, 17.2 and 17.3* and *Tables 17.1, 17.2 and 17.3*.

Generally, during incubation, the third levels (50 t/ha) of organic materials gave the highest CEC. The peaks occurred in the second month (i.e.

Table 17.1   CHANGES IN CEC DURING INCUBATION OF SOIL A WITH DIFFERENT LEVELS OF ELEPHANT-GRASS LEAVES (mEq/100g SOIL)

Incubation period (months)	Levels of elephant grass			
	1	*2*	*3*	*Mean*
0	8.50	8.40	6.75	7.88[b]
1	16.51	12.66	9.88	13.02[b]
2	25.03	40.15	24.10	29.76[a]
3	15.00	7.25	9.78	10.68[b]
4	13.75	8.00	8.50	10.08[b]
5	7.68	8.25	7.00	7.64[b]
6	13.40	8.43	7.25	9.73[b]
Mean	14.25	13.30	10.46	

SE = 2.390. Means with the same superscript are not significantly different at 1% level

Table 17.2   CHANGES IN CEC DURING INCUBATION OF SOIL B WITH
DIFFERENT LEVELS OF ELEPHANT-GRASS LEAVES (mEq/100g SOIL)

| Incubation period (months) | Levels of elephant grass | | | |
	1	*2*	*3*	*Mean*
0	16.25	13.70	11.75	13.90[c]
1	18.98	38.49	27.50	28.32[a]
2	29.98	20.90	25.30	25.39[ab]
3	26.00	17.25	25.85	23.03[abc]
4	20.25	15.00	19.75	18.33[bc]
5	16.50	15.75	16.50	22.36[abc]
6	16.75	15.00	18.00	16.58[bc]
Mean	20.67	19.44	20.66	

SE = 2.923. Means with the same superscript are not significantly different at 5% level

Table 17.3   CHANGES IN CEC DURING INCUBATION OF SOIL C WITH
DIFFERENT LEVELS OF ELEPHANT-GRASS LEAVES (mEq/100g SOIL)

| Incubation period (months) | Levels of elephant grass | | | |
	1	*2*	*3*	*Mean*
0	26.25	30.00	25.05	27.10[c]
1	50.05	38.26	43.86	44.06[b]
2	53.35	58.85	62.45	58.22[b]
3	47.25	45.00	36.50	42.92[b]
4	43.25	36.00	38.00	39.08[bc]
5	25.50	33.50	22.93	27.31[c]
6	28.50	41.00	34.50	34.67[bc]
Mean	39.16	40.37	37.61	

SE = 3.025. Means with the same superscript are not significantly different at 1% level

eighth week) of incubation. Elephant-grass leaves were the most effective in increasing soil CEC, but the effect was very short-lived. The very significant increases in CEC with elephant-grass leaves could have been due to the very high potassium (3.10 per cent) and calcium (0.35 per cent) coupled with a relatively narrow carbon/nitrogen ratio which probably enabled it to decompose more rapidly than the other two organic materials (*Figure 17.3*). The high potassium and calcium could have affected the CEC in two ways. Directly, these bases when released during decomposition were then adsorbed in the exchange complex, the sites of which have been enhanced by the additions of OH^-, $COOH^-$, and NH_2^- from the decomposing organic matter, thereby increasing soil CEC through pH-dependent charges (Broadbent and Bradford, 1952). Indirect effects were in accelerating the rate of decomposition. Russell (1976) reported that there was some evidence that the rate of decomposition of leaves is enhanced by their cation content. This effect could be due to the bases controlling the pH of the rotting material and, therefore, the composition of the microorganisms. This would agree with the findings of Kolenbrander (1974), that plant foliage, because of its low resistance to biological activity, decomposed very rapidly and its effects were very short-lived.

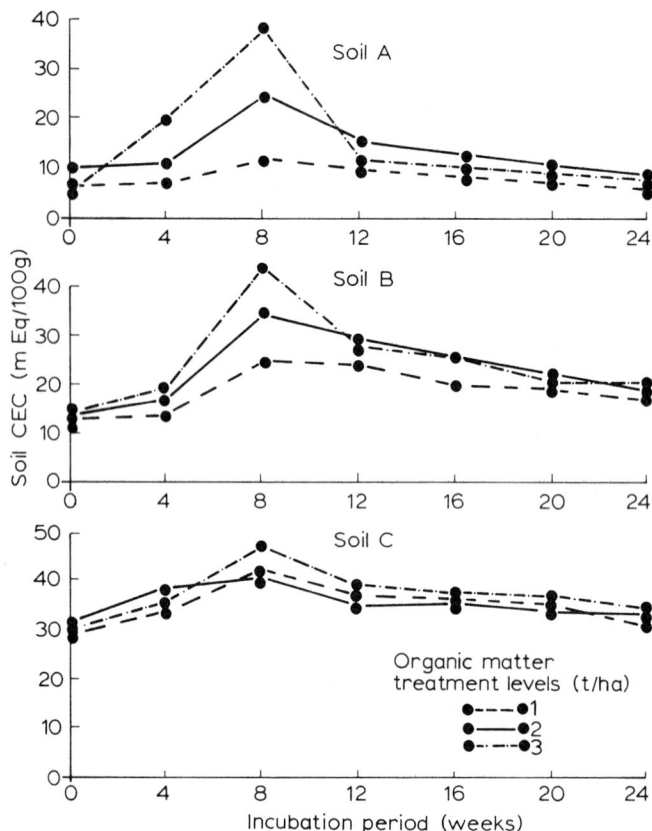

Figure 17.1   Changes in soil CEC during incubation of soils A, B and C, with different levels of coffee husks

Soil C had the highest CEC; soil B was medium; while soil A had the lowest CEC values throughout the experiment (*Figures 17.1–17.3*). However, the percentage increase due to added organic materials was highest for soil A (51 per cent); that for soil B was 23 per cent and for soil C was only 8 per cent. Analysis of variance showed that, on soils A and C, the three levels of organic materials did not have any significant differences in CEC. However, in soil B, coffee husks and cow manure increased the CEC significantly ($P < 0.05$).

These results showed that the amount of organic matter present in the soil influences CEC tremendously. Thus, soils with a very high organic matter content also gave a high CEC (soil C), and when organic carbon content was highest in the soil, that was also when the CEC was at its peak. However, the soil with an originally low CEC (i.e. also low organic matter content) benefited most from the addition of organic materials (soil A).

EFFECTS OF LENGTH OF INCUBATION ON SOIL CEC

Length of incubation had very marked effects on changes in soil CEC over the 6 months of the experiment. Generally, CECs of all the soils reached

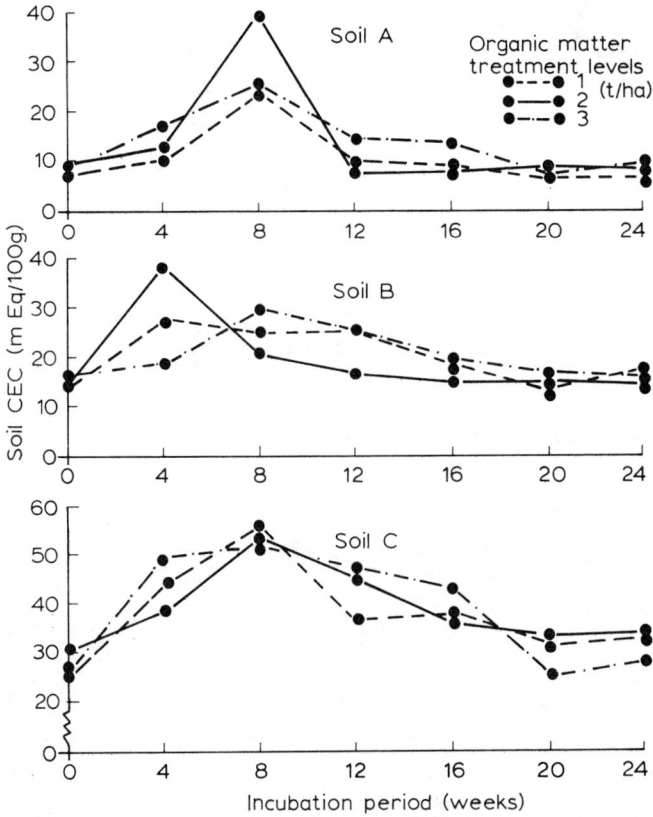

Figure 17.2 Changes in CEC during incubation of soils A, B and C with different levels of elephant-grass leaves

their highest peak during the second month of incubation (*Figures 17.1–17.3*). There was a sharp increase from the start of the incubation (almost linearly) to the second month. It then went down gradually to the last date of incubation. This decrease, however, did not reach the original level in most cases. The marked rise and fall in soil CEC could be attributable to the decomposition of added organic materials. The increases from the first month were probably brought about by the added organic materials which were lacking at the start. The drop after the second month indicates the possible exhaustion of the most readily decomposed portion of the added organic matter. It had earlier been noted that the decomposition of added organic materials was very rapid at the beginning, the rate decreasing rapidly also with time after the initial stages. Some results obtained by Jenkinson (1965) showed that there was a rapid loss of ^{14}C as carbon dioxide during the first 6 months of incubation, during which, two-thirds of the labelled carbon added as ryegrass to the soils had been lost. After these initial rapid stages of decomposition, the products of microbial activities appeared to be the principal substances undergoing decomposition.

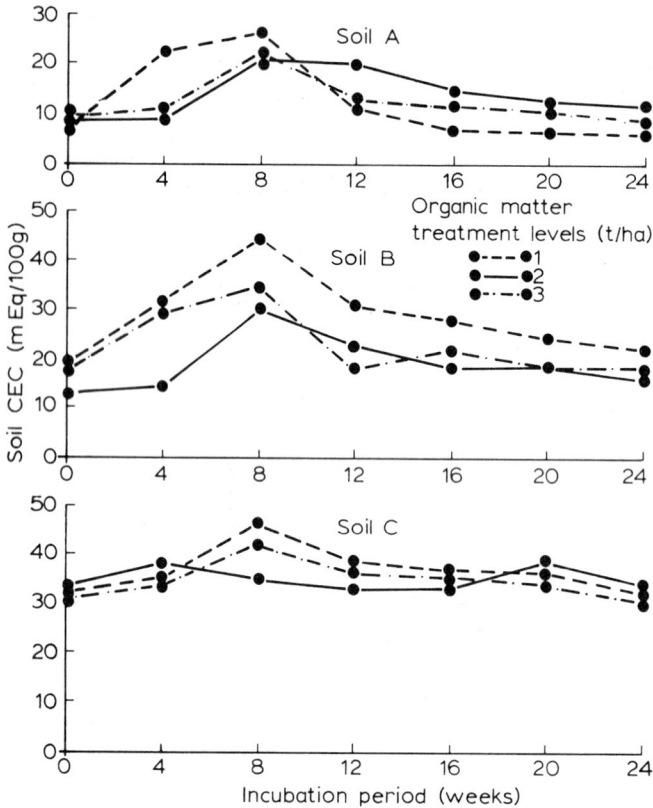

Figure 17.3 Changes in CEC during incubation of soils A, B and C with different levels of cow manure

EFFECTS OF SOIL pH ON CEC DURING INCUBATION

The original soil pH had some significant influence on the soil CEC derived from decomposition of organic materials. The effects of pH could have been indirect, by limiting the ability of soils to hold the released bases, because most of the charges which develop in organic matter are pH-dependent and thus increase with pH. Analysis of variance showed that, for soil A, only the second month's means were significantly different from the other means, whereas for soils B and C, generally the CEC at the beginning and at the end of the incubation were found to be significantly different from the other means. (Examples with elephant-grass leaves are indicated in *Tables 17.1–17.3*). The behaviour of soil A in this respect could be accounted for by the fact that it was the most acidic, consisting mainly of 1 : 1 clay/minerals and could not hold basic cations firmly long enough in exchangeable form. It could easily have been saturated (saturation capacity being 30 per cent), or allowed most of the divalent cations to leach out, its exchange sites being mostly pH-dependent (Coleman, Kamprath and Weed, 1958; Snyder, Molean and Franklin, 1969).

CONTRIBUTION OF ORGANIC MATTER TO SOIL FERTILITY THROUGH CEC

The results obtained in this experiment are proof enough that, in the tropics, the single most important contribution of organic matter to the soil is cation exchange capacity, with nutrient supplies second (Russell, 1963). Because of the very low CEC resulting from the silicate clays of the highly weathered and leached tropical soils, most of the charges which develop originate from soil organic matter. These charges are pH-dependent and increase as the pH increases. Tisdale and Nelson (1970) reported experiments which showed that certain organic anions, arising from the decomposition of organic matter, form complexes with iron and aluminium. Some humic substances form a coating over sesquioxide particles, which acts as a protective cover, thus reducing the fixing capacity of the soil. These two characteristics of organic matter enable the highly weathered and leached acidic tropical soils to retain, in exchangeable form, basic cations and phosphorus even at a low pH. Thus, as the nutrients are released by microbial breakdown of the added organic materials, they are retained by the charges that become released from clay particles and from the organic matter itself (OH^-, $COOH^-$, NH_2^-). This may explain why Stephens (1969) found an increase in the level of exchangeable potassium using farmyard manure, while potassium fertilizers did not raise the potassium level. This would indicate that the potassium added in the fertilizers leached out because of the low CEC.

Thus, in order to maintain nutrient balance in these soils, it is imperative to add some organic wastes to maintain the soil organic matter at a productive level (that is 3 per cent according to Foster (1969)). The CEC critical level is, however, dependent on several factors. For example, it could depend on soil types and soil pH. Thus, for the relatively acidic ferrallitic soils in Uganda, the critical CEC level is said to be 10 mEq/100 g (Kawanda Annual Report, 1971/72). However, such a value is rather theoretical. What is important is the percentage of the exchange sites that are really occupied by basic cations. This is especially important in acidic soils where aluminium and iron tend to dominate the exchange sites. Thus, the effects of added organic materials could be valued for the magnitude of change they bring about in percentage base saturation, in addition to enhanced CEC.

References

BROADBENT, F. E. and BRADFORD, G. R. (1952). Cation exchange groupings in the soil organic fractions. *Soil Sci.*, **74**, 447–457

COLEMAN, N. T., KAMPRATH, E. J. and WEED, S. B. (1958). Liming. *Adv. Agron.*, **10**, 475–522

FOSTER, H. L. (1969). The dominating influence of soil organic matter and soil reaction on the fertility of ferrallite soils in Uganda. *Proc. E.A. 11th Spec. Comm. Soil Fert. 1969*

FOSTER, H. L. (1971). Rapid routine soil and plant analysis without automatic equipment. I. Routine soil analysis. *E. Afr. agric. For. J.*, **37**, 160–170

JENKINSON, D. S. (1965). Studies on the decomposition of plant material in soil. I. Losses of carbon C^{14} labelled ryegrass incubated within the field. *J. Soil Sci.*, **16,** 104–115

KOLENBRANDER, G. J. (1974). Efficiency of organic manure in increasing soil organic matter content. *Trans. 10th. Inter Congr. Soil Sci. II (1974),* pp. 129–136

MILLS, W. R. (1952). *Nitrate Investigations: Inter-African Soils Conference B.C. (1952)*

MILLS, W. R. (1954). A review of recent trials with fertilizers in Uganda. *2nd Inter-African Soils Conference. Document 89, Sec. III B.C.,* pp. 1133–1142.

ROBINSON, J. B. D. and SEMP, G. (1968). Advisory soil or plant analysis and fertilizer use. I. Comparison of soil analysis methods. *E. Afr. agric. For. J.,* **34,** 117–139

RUSSELL, E. W. (1963). The role of organic matter in soil productivity (with special reference to tropical and arid regions). The use of isotopes in soil organic matter studies. *Rep. FAO/IAEA Tech. Meeting, Brunswick—Volkenrode,* (1963). Oxford; Pergamon Press

RUSSELL, E. W. (1976). *Soil Conditions and Plant Growth,* 10th Edition. London; Longmans Green and Co.

SNYDER, G. H., MOLEAN, E. O. and FRANKLIN, R.E. (1969). Interactions of pH-dependent and Permanent Charges Clays: II. Calcium and Rb Bonding to bentonite and illite suspension -Clay -Phase retention. *Proc. Soil Sci. Soc. Amer.,* **35,** 292–296

STEPHENS, D. (1969). The effects of fertilizers, manure and trace elements in continuous cropping, rotation in Southern and Western Uganda. *E. Afr. agric. For. J.,* **34,** 401–417

TISDALE, S. I. and NELSON, W. L. (1970). *Soil Fertility and Fertilizers,* 2nd Edition. London; Collier–Macmillan Ltd.

WALKLEY, A. (1947). A critical examination of a rapid method for determining organic carbon in soils—effect of variation in digestion conditions and inorganic soil constituents. *Soil Sci.,* **63,** 251–264

18

MINIMUM TILLAGE AND MULCH FARMING IN THE SEMIARID AREAS OF EAST AFRICA

R. K. JANA
University of Dar Es Salaam, Tanzania

Soil erosion by wind and water is a problem in the semiarid areas, especially where natural vegetation is removed. Conventional tillage (repeated ploughing and harrowing for clean cultivation) usually loosens and exposes the soil, thereby causing severe erosion.

Increased topsoil erosion degrades soil structure, increases the fluctuation of soil temperature, and decreases organic matter, nutrients and moisture-holding capacity. Eventually crop production falls drastically.

Tillage helps to remove weeds. Mulching also controls weeds and, in addition, helps to improve the soil moisture regimen, to regulate soil temperature and to reduce soil erosion and surface run-off. Intercropping practices also help to suppress weed growth and maintain soil fertility.

A series of experiments has been conducted at the Faculty of Agriculture, Morogoro (latitude 6 degrees South and altitude 525 metres) involving different tillage techniques (zero, minimum, and conventional), various mulches (grassy, woody and residual), cropping systems (intercropping, cereal–legumes, sole crop) and fertilizers (plus and minus). The results so far obtained indicate that minimum tillage–mulch farming techniques under intercropping systems are more scientific and productive on a sustained basis compared with conventional practices. This system thus appears to be an agriculture for the future in this area.

19

INDIAN AGRICULTURE—WAS ALBERT HOWARD RIGHT? A REASSESSMENT OF HIS PIONEERING CONTRIBUTIONS

I. D. CARRUTHERS and R. PALMER-JONES
Wye College (University of London), Ashford, Kent, UK

These authors gave a brief summary of the career of Albert Howard, noting his association with Wye College in the early years of this century and his appointment as the first economic botanist to the Government of India. The wide range of topics and the directly practical way in which he worked, in contrast to the specialized, isolated and often academic work of scientists today, was emphasized. Certain aspects of his work, in particular his basic scientific work on composting, were reviewed. There are many and varied reasons why this type and style of work is given little scientific prominence in low-income countries today, despite the evident importance of compost to countries increasingly dependent upon fertilizers derived from fossil fuel, and the emphasis given in government plans to the need for relevant and appropriate research.

20

BIOLOGICAL AGRICULTURE AND DEVELOPING COUNTRIES: PROBLEMS AND PROSPECTS

A. ABIDOGUN
National Cereals Research Institute, Ibadan, Nigeria

In recent years, global concern with environmental quality has led to the questioning of orthodox modern technology and a search for alternative production techniques. Biological agriculture represents a promising new line in this direction. The high productivity of orthodox modern agriculture depends on massive inputs of chemicals. On the other hand, biological agriculture relies on organic inputs and biological husbandry practices. Biological agriculture can thus be expected to be less productive than orthodox agriculture, at least with our present state of knowledge.

For industrialized countries, a reduction of the pace of productivity through the adoption of biological agriculture could well have salutary effects from the standpoint of the current problems associated with agricultural overproduction in these countries. But what of developing countries? In this paper the problems and prospects of biological agriculture in developing countries have been critically examined. On balance, the scope of biological agriculture in these countries would seem to be limited in the short run, on a number of grounds. The first reason is their current low level of agricultural productivity *vis-à-vis* a high rate of population growth. Secondly, virtually all developing countries are located within the tropics, an environment in which agricultural pest control is likely to be difficult in the absence of current methods of chemical control. Thirdly, alternative sources of food supply through importation from advanced countries may be blocked, or at least restricted, in the presence of the reduced productivity which may result when these countries themselves adopt biological agriculture.

IV

SYSTEMS OF AGRICULTURE

Introduction

Biological husbandry is not, in itself, a formalized system of production, although it includes clear principles on which most of its adherents would agree; as applied to agriculture, these are discussed in Dr R. D. Hodge's opening paper (Chapter 1 of this volume). The principles, which to many adherents represent little more than agricultural commonsense, are incorporated in the wisdom of several formal movements (for example bio-dynamics—*see below*), and to one degree or another appear wherever farming is practised with a conscious effort towards self-sufficiency. The five chapters in this section cover both formal and informal systems in which biological methods of agriculture play a part.

Biodynamics, discussed here by Dr H. H. Koepf, is a specialized form of biological husbandry based on the philosophy of Dr Rufolf Steiner (1861–1925). The movement has long been established as a comprehensive system in Europe, the United States and elsewhere, and has given rise to many successful farms on which it has been possible to carry out long-term research; Dr Koepf presents results of comparative studies that testify to the success of biodynamic methods in a wide range of farms, soils and climates. Professor C. R. W. Spedding, Dr J. M. Walsingham and Dr. M. A. Wagner of Reading University report on the merits of small-scale farming, for which biological methods are often the most appropriate. Dr W. Lockeretz discusses the results of a study of organically run farms in the US Cornbelt, conducted from Washington University, St Louis, Mo. between 1974 and 1979, and provides a useful summary of other recent investigations on biological husbandry and its economic and sociological implications in the US. Dr R. Elliott's paper introduces the work of APACE, an Australasian organization concerned with the development and application of low-impact, decentralized technology, and discusses APACE-sponsored agricultural development in the Western Solomon Islands. Dr K. S. Clough and Mrs J. Carrington show how biological husbandry principles are applied successfully in a low-energy food-production experiment on Prince Edward Island, Canada.

21

THE PRINCIPLES AND PRACTICE OF BIODYNAMIC AGRICULTURE

H. H. KOEPF
Emerson College, Forest Row, Sussex, UK

Origin and scope

Biodynamics arose from lectures which Rudolf Steiner, the founder of Anthroposophy, gave in 1924 at the request of farmers and gardeners, who felt that current methods of farming had resulted in a decline in quality of feed and seed. During the lecture course, those who were present organized themselves in an Experimental Circle, soon to be joined by others from various European countries. Thus the biodynamic movement became the first organized 'alternative group'.

From the outset the movement was concerned with far more than organic farming alone; it evolved as a complete system of farming—a point that must be stressed because it is not always seen correctly. Just as in any other serious 'alternative' endeavour, biodynamic farming encompasses all ecological, economic and social aspects of agriculture. Reports arising from the early days of this work include a wide range of topics, such as the handling of manures, composting, using biodynamic preparations, landscaping measures, crop rotations, making use of legumes, green manures, intercropping and companion planting, mulching, care of wildlife, farm-produced feedstuffs, herbs, and many other items of sound and sustainable plant and animal husbandry. In summary, we can say that biodynamic farming includes three major aspects:

1. Sound farming and gardening techniques, no matter whether old or new;
2. Such principles as diversification, recycling, excluding objectionable chemicals, decentralized production and distribution etc., ideas held in other biological movements. Since the 1920s, biodynamic farmers have developed the execution of such principles, and also reintroduced useful traditional techniques;
3. The specific biodynamic measures and concepts as they evolve from Steiner's spiritual teaching, which mould the method into a consistent whole.

The biodynamic movement was also the first in the field with another innovation—marketing foodstuffs for their nutritional quality, under a registered trade mark. Although there had for long been gradings and brand names for wines and cheeses, such basic foodstuffs as grains, vegetables and fruit, produced by definitive methods,were now for the first time marketed under the Demeter sign or an equivalent trade mark.

During two major periods of expansion, in the 1930s and from 1950 onwards, the biodynamic movement spread its activities through temperate and warm climates in both hemispheres. Compared with conventional farms their numbers are still few. Excluding all gardens, part-time growers, and farms which are owned by institutions etc., there remain several hundred biodynamic farmers in Germany whose produce sells under a Demeter arrangement, and similar situations are found in other countries. There are full-time advisors, independent research, training schemes and certification programmes, and the voice of the biodynamic movement is heard in many current discussions on resources and quality.

The need for complementary concepts

Steiner's guidance for farmers evolved from his spiritual science, which is fully discussed in his writings (1963, 1964, 1978). Here I limit myself to one of his more important underlying themes—the need for complementary concepts. The view of biodynamic philosophy is that, in terms of yields and labour requirements, conventional agriculture produces impressive results but pays too high a price in terms of pollution, waste of resources, impaired quality, and losses of human values. It is a fallacy to think that this price is the necessary consequence of high yields. Rather, it results from the theory that underlies conventional farming. To put it briefly and somewhat loosely, it follows from the analytical-quantitative method in scientific thinking. This is the method, familiar to scientists, of analyzing complex entities into simple components and functions, and it can be applied equally well to a plant, a farm or an ecosystem. Once the isolated components are understood they can be manipulated from outside—but in the process one tends to forget the very nature of the organism. Subfunctions of a living whole cannot be isolated without altering them in a way that is not precisely predictable. The observer interferes with the object under study. The analogy to quantum physics is obvious. Furthermore, functions cannot be manipulated in isolation without incurring unpredictable repercussions from the organism as a whole. The more complex the system (say an animal or human body, a higher plant, or a farm), the more likely are unknown side-effects to occur.

How then should one try to understand a living entity, and yet avoid these questionable effects? The answer must lie in an approach to living things which is complementary to the common reductionist train of thought. The wholeness of an organism is real, and must not be denied: it has its existence in the concept or type of the organism. A machine also is a functioning totality, but here the concept comes from without. Living organisms bear their reality within them. Hence the question arises—whether and how human thought can grasp that wholeness, which manifests itself in the form and functions of a living entity.

The answer is that one can draw closer to the reality or concept when detailed and comprehensive description and observation are accompanied by active thinking. One recognizes in the organic world the working of ideas which can be grasped by reason, although they evade causal explanation: this is exemplified by the Goetheanistic method of studying nature, which seems to contain an intuitive element. The objection that this approach can lead only to subjective notions is very close—but becomes invalid if one realizes that the observer can be fully aware of every step that he takes while observing and thinking. Thus I am postulating a complementary attitude, which supplements the reductionist thinking of our days. A. M. K. Müller, Professor of Theoretical Physics in Braunschweig, has recently condensed in one sentence what I am speaking of here. He said: 'To see takes us further than to explain, but explaining communicates more precisely than seeing' (Müller, 1977). In studying nature, both 'causal analysis' and 'understanding by seeing' have their proper functions. When one studies or works with living nature, then causal analysis must serve. Its limitations and restricted value must be established by the more comprehensive 'understanding by beholding'. By the very nature of his daily work and experience, the practising farmer is directed towards 'understanding by seeing'. He faces the full reality of his farm, which the scientist with his test tubes does not, unless he makes a continued and conscious effort. Furthermore, it is most unwise for the scientist to try to divert the farmer from developing his holistic approach.

The farm as an organism

To conceive the farm as an organism—Steiner says more precisely, 'a kind of individuality'—is one of the basic biodynamic concepts. The reality of this organism lies in the interaction between enterprises, natural conditions and man, who fashions the whole. To be aware always in one's work of the totality of the farm, will affect all facets of farm life, and also wider ecological and social aspects. This can perhaps best be demonstrated by comparing some of the objectives of the biodynamic system with those of conventional farming (*Table 21.1*).

From this abstract I now take up three points: (1) yields and economic viability; (2) the problem of lasting fertility, and (3) biodynamic preparations. In the space allowed I cannot describe how crop rotations, composting, feeding and housing of livestock, weed and pest control are handled according to biodynamic principles. It is important, however, that biodynamic farms manage without biocides; moreover such specialized crops as fruit and grapes for wine are also produced without the use of objectionable biocides. This is achieved by biodynamic soil and crop management, supplemented by a few biologically inoffensive controls.

Yields and returns

A full survey by Ronnenberg (1973) discussed yields and income from biodynamic farms, based on new and old data from several European countries. The conclusion was that '. . . biodynamic farms produce yields which are only slightly lower than in general agriculture (potatoes being an

Table 21.1   OBJECTIVES IN BIODYNAMIC AND CONVENTIONAL FARMING

'Biodynamic' objectives	*'Conventional' objectives*
A. Organization	
Ecological orientation, sound economy, efficient labour input	Economical orientation, mechanization, minimizing labour input
Diversification, balanced combination of enterprises.	Specialization, disproportionate development of enterprises
Best possible self-sufficiency regarding manures and feed	Self-sufficiency is no objective; importation of fertilizer and feed
Stability due to diversification	Programme dictated by market demands
B. Production	
Cycle of nutrients within the farm	Supplementing nutrients
Predominantly farm-produced manuring materials	Predominantly or exclusively bought-in fertilizers
Slowly soluble minerals if needed	Soluble fertilizers and lime
Weed control by crop rotation, cultivation, thermal	Weed control by herbicides (cropping, cultivation, thermal)
Pest control based on homoeostasis and inoffensive substances	Pest control mainly based on biocides
Mainly home-produced feed	Much or all feed bought in
Feeding and housing of livestock for production and health	Animal husbandry mainly orientated towards production
New seed as needed	Frequently new seed
C. Modes of influencing life processes	
Production is integrated into environment, building healthy landscapes; attention is given to rhythms	Emancipation of enterprises from their environment by chemical and technical manipulation
Stimulating and regulating complex life processes by biodynamic preparations for soils, plants, manures	No equivalent to biodynamic preparations; use of hormones, antibiotics, etc.
Balanced conditions for plants and animals, few deficiencies need to be corrected	Excessive fertilizing and feeding, correcting deficiencies
D. Social implication, human values	
National economy: optimum input : output ratio regarding materials and energy;	National economy: poor input : output ratio regarding materials and energy;
Private econ.: stable monetary results	Private econ.: high risks, gains at times
No pollution	Worldwide considerable pollution
Maximum conservation of soils, water quality, wild life	Using up soil fertility, often erosion, losses in water quality and wild life
Regionalized mixed production, more transparent consumer–producer relationship; nutritional quality	Local and regional specialization, more anonymous consumer–producer relation; interested in grading standards
Holistic approach, unity between world conception and motivation	Reductionist picture of nature, emancipated, mainly economic motivation

exception at times). Milk production was higher, in some cases up to 30 per cent, in the biodynamic farms from which data were available.' This summary covers family farms and larger ones. The findings are confirmed in a later survey of 9 pairs of biodynamic and conventional farms, by the Ministry of Agriculture, Baden-Württemberg (1977). Because of lower expenses and a premium on some of their products, the biodynamic farms produced higher net incomes. For grains alone (Alternativen, 1980) results from biodynamic in relation to conventional farms were: yields, 85 per cent; gross income, 110 per cent; variable production costs, 71 per cent; hence net income, 127 per cent. Had the same product prices applied to both, net income for biodynamic farms would have been only 93 per cent of that from conventional farms: the difference arises from the premium willingly paid by consumers for biodynamic produce.

Table 21.2 provides a further comparison, this time between conventional farms and a well-run farm, biodynamic since 1929, in Heidenheim, Württemberg. The data cover the year 1973.

Table 21.2 FARMING COSTS AND YIELDS: CONVENTIONAL AND BIODYNAMIC (BASED ON THE ANNUAL REPORT OF THE MINISTRY'S ACCOUNTING SERVICE (SATTLER, PERS. COM.))

Costs and yields	Talhof farm	Comparable conventional farms in the district
Expenses for fertilizers or materials for preparations, and straw (DM/(ha.y))	7.70	147
Yields: grains (kg/(ha.y))	3600	2900
milk (kg/(cow.y))	4399	3376
Bought-in concentrate (DM/(cow.y))	35.0	225.0
Hectares per worker	10.8	9.7
Income per hectare (DM)	1800	1111
Income per worker per year (DM)	18750.0	10760.0

Oelhaf (1978), surveying organic farms in the USA, found yields similar to the above, though labour requirements of the US organic farms appear to be 5–20 per cent higher than on conventionally managed farms.

It can be stated that commercial biodynamic farms and gardens the world over stand their ground in the present economic climate. Some have done so for over half a century, keeping abreast with the general increase of yields, and improving the fertility of their soils. However, yields per hectare has become a somewhat questionable criterion for assessing the merits of a system, if one considers the distorted situation regarding actual needs, yields and surpluses in Europe, in relation to such factors as material and energy input. On a global scale, true potential and needs must be evaluated on a regional basis.

Lasting fertility

To promote lasting fertility by harnessing the resources of the farm itself is a central concern in biodynamic work. This makes unnecessary the heavy applications of bought-in fertilizers that have become customary in conventional farming. High yields can be produced if the following factors are taken into account:

1. Land use and a cropping system which replenishes the organic and nitrogen content of the soil. Legumes and arable fodder plants are important. Of their arable land, biodynamic farmers usually plant about 50 per cent or more in grains.
2. Recycling of appropriate quantities of manures and composts, using biodynamic preparations. Manure handling is characterized by careful collection and storage, and aeration of slurry is increasingly practised.
3. Methods of soil cultivation which give due importance to soil biology.

Contrary to the frequently held view in conventional farming, animal manures seem to the biodynamic farmer to be indispensable for sustaining lasting soil fertility. This is the lesson of longstanding field experiments in many countries, and completely in accord with biodynamic experience. Bronner and Janick (1974), who studied the soils in the sugar-beet growing districts in Austria, found that only cattle manure could maintain the organic content of the soil; large amounts of vegetable matter and pig slurry alone could not.

Soil fertility, as understood in biodynamic farming, includes three components:

1. Short-term or actual fertility: the direct effects on yields of manures and composts.
2. Long-term fertility. This term signifies the contribution of the soil, according to soil type.
3. Medium-range fertility. This is the combined carry-over effects of cropping, manuring and cultivation. It accumulates and becomes exhausted (if erosion is not a problem) in periods of 10–15 years and more. It cannot really be measured in the laboratory, but farmers know it very well from experience.

To split up soil fertility in this way might seem artificial, but it encourages recognition of the third component—medium-range fertility—on which the success of biodynamic farming—as that of any biological method—crucially depends. Hence all measures of soil and crop management have to be looked at from two aspects: do they serve the primary end, to produce a good crop? What is their longer-term effect on soil fertility? It is essential to keep these two aspects in mind.

Table 21.3, drawn from a more complete report on a field plot experiment in the Biodynamic Institute in Järna, Sweden, illustrates the effects of manuring, for 18 years. The crop rotation included wheat, potatoes, vegetables and clover-grass. The figures show an improvement of biological factors under organic soil treatment. Although the treatment did not produce marked differences in the organic content and bulk density of the topsoil, it did in the subsoil: the plots which received manure, but no chemical fertilizer, grew 'physiologically deeper'.

Table 21.3 SOIL CHARACTERISTICS AS INFLUENCED BY DIFFERENT
SYSTEMS OF MANURING OVER 18 YEARS (PETTERSSON AND
WISTINGHAUSEN, 1979)

Treatment	Full bio-dynamic	Crude manure	¹/₂ manure: ¹/₂ NPK	Control	Fertilizer NPK	NPK
Fertilizer: kg N/(ha.y)	82	93	61	0	56	111
Yields, expressed as t/ha cereals	4.86	4.90	5.03	3.77	4.83	4.87
Bulk density of soil top layer	1.14	1.09	1.10	1.10	1.14	1.16
subsoil	1.33	1.29	1.42	1.50	1.51	1.48
Organic matter (total N) topsoil %N	0.24	0.24	0.25	0.25	0.26	0.26
subsoil %N	0.14	0.17	0.08	0.16	0.12	0.09
Earthworm burrows > 1.5 mm/m²	100	111	53	22	11	16
mg CO_2/100 g of soil	125	108	91	83	75	81
Dehydrogenase, TPF/10g soil	547	377	302	213	211	258

What of the effects of nitrogen on soil fertility? An adequate supply of nitrogen is crucial both for yields and for quality of crops. Can biodynamic soil management provide these without using chemical nitrogen? Some, but not all, biodynamic farms use commercial organic material at times. Practical experience shows that there can be enough nitrogen available for most crops. There is a possible reason for this. A holistic, rather than a reductionist mode of thinking takes phenomena for what they are. When chemical nitrogen is used as fertilizer, then the result will be a net loss of nitrogen from the soil–plant system. This simple fact has been known, but largely disregarded, for more than half a century. However, net gains to the system can be achieved under biodynamic or biological management: the total nitrogen content of the soil can be increased. 'Medium-range fertility', as I have called it, will build up. When fertilizer is used, the whole system moves in a somewhat different direction. Much of the recent work with labelled nitrogen (Allisson, 1955, 1966; Fleige and Capelle, 1974; Hannold *et al.*, 1975; Holstern, 1977, and many others) shows that, although theoretically the fertilizer should meet the needs of the crop, it does not do so. Instead the soil contributes from its own long-term reserves, which may total up to 3000–8000 kg of nitrogen per hectare. Year after year this is depleted as the slowly available fraction is used up. Nitrogen fertilizer seems to accelerate the trend rather than to slow it down. Increased chemical fertilizer results in increased mobilization of the soil reserve, higher denitrification and leaching, reduced biological fixation and nitrogenase activity. After a time the soil–plant system as a whole begins to work inefficiently.

If animal manures or composts are combined with appropriate cropping, then the whole system will work efficiently. Fertility, which to a large extent depends on the supply of nitrogen to crops, begins to build up year after year over periods of 10, 15 and more years. This is also the practical experience of biodynamic farms. To examine this process in more detail seems to be an important task for future research. I am quite sure that farmers should again begin to think in terms of longer time-spans, in spite of all the actual and alleged pressures that make them focus their attention on the next crop, but hardly at all beyond.

Preparations

There is no equivalent in other systems of agriculture to the so-called 'biodynamic preparations'. These include two sprays, 500 and 501, which are used for soils and plants. Preparation 500, based on cow manure, is applied at a rate of about 200–300 g/ha; preparation 501 (finely ground quartz) at 4 g/ha. Two to 4 ppm of yarrow, camomile, stinging nettle, oak bark, dandelion and valerian flowers (respectively preparations 502–507) are added to manures and composts. For details of how these substances are processed *see* Steiner (1974). Although it is known that minute quantities of a specific substance can influence the process of life, the notion that these preparations serve a useful purpose is not readily acceptable to a modern, conventionally trained mind. But to stop at this point would be too easy. There exists experimental evidence for the effects of these preparations, regardless of what established theoretical opinions might find acceptable. There is also a rational way to understand the underlying principles, which is fully discussed in Steiner's lectures and related writings (Corrin, 1960; Soper, 1976).

Watching the effects of preparations on their plants may provide practical farmers and gardeners with all the evidence they need. Their observations should not be discounted lightly: often they furnish a good point of departure for experimental work. Experimental data confirm that these substances exert several effects, significantly increasing the yields of a number of crops. They also influence the composition of products, that is, their quality. In addition, when added to manures or composts their effects show up during fermentation and in the finished material: the following examples are drawn from a wide series of reports.

Stearn (1976) has carried out growth experiments in the laboratory and in a growth chamber. He treated, with these preparations, maize and peas growing in Hoagland solution or soil. The strongest positive results on the lengths and weights of shoot and roots were achieved by 25–50 mg/ℓ of preparation 500 and 500–2500 mg/ℓ of preparation 501.

Table 21.4   EXPERIMENTAL TREATMENT OF WHEAT WITH PREPARATIONS 500 AND 501 (SPIESS, 1979)

Year	Relative yields 1973–76 Treatments			
	1	2	3	4
1973	100 (= 3.0 t/ha)	109	117+	121+
1974	100 (= 4.15 t/ha)	106	109	111+
1975	100 (= 4.1 t/ha)	105	102	102
1976	100 (= 4.2 t/ha)	105+	104	109++

1 = control; 2 = 3 × prep. 500; 3 = 3 × prep. 500 and 501; 4 = 6 × (1976 : 4 ×) prep. 500, 3 × prep. 501.
All plots received manure.

The figures in *Table 21.4* are extracted from four years of field trials with wheat, by Spiess (1979). The same author found also that preparations 500 and 501 increased yields of sugar beet by 8–14 per cent, stimulating growth of leaves by 8–26 per cent. These results confirm earlier work with the same plant by Abele (1973). In a number of cases, significant increases in yields of grains, root crops and vegetables have been reported by Klett (no date), Pettersson (1970), Thun and Heinze (1972) and Kotschi (1980). Regarding the quality of crops it has been shown by Klett (no date), Abele (1973), Pettersson (1977b); Wistinghausen (1979) and others that such characteristics as the percentage of true protein, nitrate and soluble amino acids, sugar and other carbohydrates, vitamins and the activity of various enzymes can be changed by these preparations. *Table 21.5* from Samaras (1977) concerns the keeping quality of carrots grown under the influence of preparations 500 and 501. Sprays applied during growth reduce losses during storage. They also reduce the production of carbon dioxide, some enzyme activities and the numbers of epiphytic bacteria.

Table 21.5   THE KEEPING QUALITY OF CARROTS, AS INFLUENCED BY PREPARATIONS 500 AND 501 (SAMARAS, 1977)

Keeping qualities	Control	Treatment	
		3 × prep. 500	*3 × 500; 4 × 501*
CO_2 in 94 h (mg/kg fresh matter)	2831	2755	2565
Catalase activity (μmol H_2O_2/(min. g dry matter))	397	375	346
Peroxidase (μmol GJ/(min. g dry matter))	15992	11051	7591
Saccharase activity m/U/g fresh matter	90	98	112
Bacteria (10^6/g fresh matter)	1.07	0.45	0.43
Fungi 10^3/g fresh matter	2.7	2.7	2.6
Loss of dry matter of mashed carrots (%)	56.1	46.6	29.2
Spoilage during 164 days (%)	28.2	23.0	20.4

Preparations 502–507 have been observed to smooth to some extent the fluctuations of the temperature in compost heaps. Heinze and Breda (1962), in three separate experiments on composting in cooperation with an agricultural school and a German experimental station, showed that preparations increased the exchange capacity of organic matter in the finished composts by an average of 25.4 per cent. Abele (1976), treating slurry from pigs and cattle, found that preparation 503 (camomile) reduced losses of nitrogen and established beneficial effects of other preparations on a number of crops in pot and field plot experiments. *Table 21.6* shows the differential effects of NPK fertilizer and biodynamically treated slurry on productivity of a ley. The results were remarkable, considering that 40 and 60 m^3 of slurry are rather heavy dressings. Koepf (1966) treated slurry from cattle with the preparations, and tested young wheat plants in growth experiments with the material. At least two effects of the preparations were

Table 21.6 PERCENTAGE OF CLOVER AND YIELDS OF A LEY, AS INFLUENCED BY MANURING AND BIODYNAMIC PREPARATIONS (ABELE, 1976)

Fertilizer:						
NPK : kg/ha N	*75*	*150*	*225*	*75*	*150*	*225*
Slurry: m³/ha	*20*	*40*	*60*	*20*	*40*	*60*
		% clover			*Yields* (t/ha dry matter)	
Fertilizer	9.9	4.2	2.9	12.8	15.1	16.4
Slurry aerated	13.5	12.5	9.2	14.2	15.8	17.5
Slurry aerated and with biodynamic compost preparations	18.6	14.8	15.9	14.8	17.2	18.0
					GD 5%: 0.46	1%: 0.81

detected. The lengths of the roots increased by 13.5 per cent, and their dry weight by 17.6 per cent, compared with the effect of untreated slurry.

These few examples must suffice to show the measurable positive effects on composts and crops of biodynamic preparations.

For the sake of completeness I should add a note on lunar rhythms, which interest many people—not only followers of the biodynamic approach. There is indeed an urgent need to put this problem into perspective. It is well established that a number of lunar biorhythms exist, and the question arises whether they are of any practical significance for growing crops. Lunar influences do not at present appear conspicuous, but neither were the effects of trace elements before more was known of them. Hence, in the biodynamic movement and elsewhere, experimental work is in progress that may help to steer a course between superstitious belief and narrow-minded rejection. The results so far available show that lunar rhythms are another environmental factor of growth, which deserves further critical attention.

The quality of feed and food

From the outset, the quality of feed and food has been a prime concern of biodynamic farmers and gardeners. It goes without saying that the term 'quality' also includes such items as appearance, grading, uniformity of produce and minimum residues. Treatment with biocides is not permitted on crops that are grown according to biodynamic or Demeter standards. Hence, on these products minimum residues can be expected. However, essentially the term 'quality' signifies the inherent biological or health value of a product.

That quality matters—or even exists at all—might be denied by a one-sided, analytical quantitative approach to nutrition which reduces foodstuffs merely to carriers of calories, protein and vitamins. Equally, it may happen that unripe apples are believed to be preferable, because their vitamin C content may be higher than that of ripe ones. (There is a true story behind this remark.) However, if one appreciates all nutritional values, including texture, flavour, taste, wholesomeness, also a number of other physiologically relevant factors, then each foodstuff can be appreciated as a specific combination of essential components, that is, as a wholeness. The objective therefore will be to produce the optimum quality of a given species and variety, rather than to go for a high content of one or another component.

Research along such lines supports the concept that soil management, aimed at creating optimum environmental conditions of plant growth, will result in superior quality of the species and variety in question. Because of varying natural site conditions this quality will not be the same everywhere, but it will be good in the circumstances. Individual characteristics of the product are valued as indicators of its overall quality.

I should mention that, almost without exception, biodynamic dairy farmers report improved health of their animals soon after they adopt the biodynamic method. Likewise, examples exist of improved performance of cows relative to the actual intake of nutrients from their homegrown ration. In part this is due to the fact that the quality of the vegetative parts of plants responds very much to the quantity and kind of manuring or fertilizing. It was an important addition to the methods of quality testing when Aehnelt and Hahn (1973) showed that high applications of chemical fertilizer to forage crops significantly contribute to reproductive disturbances in male and female cattle. This was demonstrated by surveys of cow herds and AI stations and also by feeding trials. In some of their experiments, feedstuffs from biodynamic field plot experiments were used with positive results (*see also* Schiller and Lengauer, 1967).

Table 21.7 YIELDS AND QUALITY OF POTATOES UNDER CONVENTIONAL AND BIODYNAMIC MANAGEMENT (PETTERSSON, 1977)

| *Results* | *Management* | |
	Conventional	*Biodynamic*
Yields, October (t/ha)	38.2	34.2
Losses by grading and storage (%)	30.2	12.5
Weight in April (t/ha)	26.6	30.0
Crude protein (% dry matter)	10.4	7.7
True protein (% of crude protein)	61.0	65.8
EAA (Oser)	58.9	62.8
Vitamin C (mg/100g fresh matter)	15.5	18.1
Darkening of extract E.10^3, 48h, 8 °C	462	354
Decomposition of extract (electrical conductivity)	30.9	22.0
Crystallization defects	5.2	4.2
Taste points (best = 4) December	3.0	3.1
April	2.3	2.7
Cooking defects December	4.1	1.8
April	9.2	2.1

Table 21.7, taken from a field experiment run by the Swedish Biodynamic Institute in cooperation with the University of Uppsala, shows the spectrum of quality tests applied to potatoes. (Details about soil treatment etc. may be found in the complete report.)

The higher yield of the conventionally treated plots is upset by losses during storage and grading. A high relative content of true protein, high

EAA index and vitamin C content, less darkening, better preservation of taste and fewer crystallization defects, are all indices of better quality.

Similar comparisons exist for most agricultural and horticultural crops. The spectrum of quality tests varies somewhat according to the product. They show the extent to which the potential quality of the cultivar has become manifest under environmental conditions of growth. One can describe in a general way the interplay between growth conditions on the one hand, and quantity and quality on the other (Koepf, Pettersson and Schaumann, 1976; Koepf, 1980). A plant completes its vegetative growth stage, and (if it has not been harvested) its reproductive phase as well, between two polar sets of environmental factors. It unfolds and moulds its form and substances in the interplay of the earthly and the cosmic poles. In every plant both poles are always active, and yield and quality depend on their ratio. The earthly pole causes the plant to produce more plant matter; it enhances the vegetative development, and may delay the completion of the growth stages. This tendency is augmented by water, humus, nitrogen, the clay content of the soil and some mineral elements. The cosmic pole, which is mediated by air, light, warmth, more dry conditions, also by the siliceous nature of the soil, works in a formative way in the realm of substance and shape. It enhances the 'ripening' of the plant and its constituents. Morphological and chemical characteristics of a plant can be associated with the balance between these poles, or the dominant influence of one of them.

Thus one has a key for putting the various environmental conditions and their effects in the plants into a rational system, relating them to the wholeness of the plant. For example, relatively high contents of nitrate, amides or free amino acids indicate that the plant could not transform the nitrogen compounds into protein. Often this condition coincides with high enzyme activity, poor keeping quality, poor taste and other factors. In this case, strong earthly growth conditions were not quite balanced by those which enhance 'ripening'. These and similar relationships have been established by field experiments, for example by growing crops in full sunlight, half shade, etc. in combination with various amounts and kinds of manuring and biodynamic treatment. Some of the factors may in part be substituted for one another. The effects on quality of soluble fertilizer resemble in some respects those of reduced light. When compared with crude manure, composts bring about more of the characteristics of a well-ripened product. So also does preparation 501, and to some extent 500—that is, they contribute also to better keeping quality, structure and taste. Thus the concept of wholeness is also valid for the problem of quality.

The biodynamic method of farming and gardening is now practised in many parts of the world. I hope that this outline may show that general agriculture is in need not only of further new technologies, but also of reflections on the principles on which the agriculturist works.

References

ABELE, U. (1973). *Comparative studies on conventional and bio-dynamic plant cultivation with particular reference to sowing time and type*. PhD Thesis, Giessen

ABELE, U. (1976). *Studies on the rotting of liquid manure and its effect on soil, plant yield and quality.* Report from Institut für Biologisch-Dynamische Forschung, Darmstadt

AEHNELT, E. and HAHN, I. (1973). Fertility of animals, a possibility for biological quality testing of feed and feedstuffs. *Tierärtztl. Umschau,* **4,** 1–16

ALLISSON, F. E. (1955). The enigma of soil nitrogen balance sheets. *Adv. Agr.,* **7,** 213–250

ALLISSON, F. E. (1966). The fate of nitrogen applied to soils. *Adv. Agron.,* **18,** 219–258

Alternatives to Contemporary Land Cultivation. (1980). Arb. DLG, Bd. 169, Frankfurt

BRONNER, H. and JANICK,V. (1974). Studies of soils in upper Austrian sugar beet areas which support and which do not support cattle. *Bodenkultur,* **25,** 223–251

CORRIN, G. (1960). *Handbook on Composting and the Bio-Dynamic Preparations.* 35 Park Road, London NW1; Bio-Dynamic Agricultural Association

FLEIGE, H. and CAPELLE, A. (1974). Field studies on the fate of labelled fertilizer nitrogen in the soil and in plants. *Mitt. Dt. Bodenk. Geo.,* **20,** 400–408

HANNOLD *et al.* (1975). Losses of labelled nitrogen from sandy loam. *Bodenkultur,* **26,** 221–232

HEINZE, H. and BREDA, E. (1962). Experiments on the composting of stable manure. *Leb. Erde,* **2,** 3–10

HOLSTERN, D. E. (1977). Nitrogen balance of a field plot cropped with perennial ryegrass. *Cal. Agric.,* **31,** 12–13

KLETT, M. (undated). *Studies of the quality of light and shadow in relation to the cultivation and testing of silicious preparation to increase quality.* Report to Stiftung Volkswagenwerk by Institut für Biologisch-Dynamische Forschung, Darmstadt

KOEPF, H. (1966). Experiment in treating liquid manure. *Bio-Dynamics,* **79,** 1–12

KOEPF, H. (1980). *Agriculture.* Stuttgart; Verlag Fries Geistesleben

KOEPF, H., PETTERSSON, D. and SCHAUMANN,W. (1976). *Bio-Dynamic Agriculture.* Spring Valley; Anthroposophic Press

KOTSCHI, I. (1980). Studies of the effect of spray preparation '500' and '501' used in the bio-dynamic method of agriculture, on agricultural plants. PhD Thesis, Giessen

MINISTRY OF FOOD, AGRICULTURE AND THE ENVIRONMENT. (1977) Baden-Württemberg, März Stuttgart: Evaluation of the three year investigation of nine bio-dynamic farms

MÜLLER, A. J. K. (1977). On the crisis of medical fundamentals. In G. Büttner, H. Hensel (eds), *Biologische Medizin,* p. 69–81. Heidelberg; Verlag für Medizin, Dr Ewald Fischer

OELHAF, R. C. (1978). *Organic Agriculture.* Montclair, New Jersey; Allenheld, Osmun

PETTERSSON, B. D. (1970). Effects of site conditions, manuring and bioactive substances on the quality of potatoes. *Leb. Erde,* **3/4,** 91

PETTERSSON, B. D. (1977a). Nutrients balance in Swedish agriculture. *Leb. Erde*, **3,** 85–86

PETTERSSON, B. D. (1977b). Comparative studies of conventional and bio-dynamic cultivation with particular reference to yields and quality. *Leb. Erde*, **5,** 175–180

PETTERSSON, B. D. and WISTINGHAUSEN, E. V. (1979). *Effects of organic and inorganic fertilizers on soils and crops.* Temple Maine; Woods End Agricultural Institute

RONNENBERG, A. (1973). *Economic Aspects of the Bio-Dynamic Method of Agriculture—Consequences for the Individual Production and for the Produce Market,* 2nd edn. Darmstadt; Fakultät Göttingen

SAMARAS, ILIAS. (1977). The behaviour after harvesting of differently manured vegetables with particular reference to the physiological and microbiological parameters. PhD Thesis, Giessen

SATTLER, F. (pers. comm. 1976). *see also*: 1977. From the development of the Valley Farm. *Leb. Erde*, **6** (1976) and (1977) **1,** 3–12

SCHILLER, H. and LENGAUER, E. (1967). Disturbance of fertility of cattle in connection with fertilizers, flora and the mineral content of the feed. Veröffentl. Landw. Chem. Versuchsanst. Linz Nr.7

SOPER, J. (1976). *Studying the Agriculture Course.* 35 Park Rd., London NW1; Bio-Dynamic Agricultural Association

SPIESS, H. (1979). On the effect of the bio-dynamic preparations Hornmist '500' and Hornkiesel '501' on the yield and quality of several cultivated plants. *Leb. Erde*, **4,** 126–131; *Leb.Erde*, **5,** 173–177

STEARN, W. C. (1976). Effectiveness of two biodynamic preparations on higher plants and possible mechanisms for the observed response. MS Thesis, Ohio State University, Dept of Agronomy

STEINER, R. (1963). *Occult Science—an Outline* (new translation). 35 Park Rd., London NW1; Rudolf Steiner Press

STEINER, R. (1964). *The Philosophy of Freedom,* 7th Edition. 35 Park Rd., London NW1; Rudolf Steiner Press

STEINER, R. (1974). *Agriculture: A Course of Eight Lectures,* 3rd Edition. 35 Park Rd., London NW1; Bio-Dynamic Agricultural Association (Distributed by Rudolf Steiner Press)

STEINER, R. (1978). *A Theory of Knowledge, Implicit in Goethe's World Conception,* 3rd Edition. Spring Valley, N.Y.; The Anthroposophic Press

THUN, M. (1977). Hornmist Preparation '500'. *Leb. Erde*, **6,** 215–219

THUN, M. and HEINZE, H. (1972). Experiments on connection between the position of the moon in the zodiac and cultivation of plants. *Leb. Erde, Herausg. von Forschungsring f. biol.-dyn.* Wirtschaftsweise, Darmstadt Bd I und II

WISTINGHAUSEN, E. V. (1979). *What is Quality?* Darmstadt; Verlag 'Lebendige Erde'

22

THE BIOLOGY AND TECHNOLOGY OF SMALLHOLDINGS

C. R. W. SPEDDING, J. M. WALSINGHAM and M. A. WAGNER
*Department of Agriculture and Horticulture, University of Reading,
Berks, UK*

Biology in agriculture

The theme of this conference is 'biological agriculture' and it may puzzle
those who find it difficult to imagine any other kind. Indeed, it would be a
strange definition of agriculture that avoided all reference to biology,
although it would be a very incomplete definition if it mentioned nothing
else.

Thus, most of us would consider that agriculture is essentially a human
activity, directed to the satisfaction of such human needs as food, fibre and,
increasingly, fuel, by the controlled use of plants and animals. The
biological essentials are the plants and animals of direct concern, those that
yield products of agricultural value.

In many agricultural systems, however (indeed, *most* in world terms) this
is not the only biological content of agriculture: perhaps the best example of
this other area is the soil. Some definitions of agriculture would, in fact,
include cultivation of the soil, and some people certainly regard food
production that does not involve the soil as an industrial rather than an
agricultural activity. Intensive poultry keeping is sometimes thought of in
this way, although much of the poultry feed will have been produced by
cultivation of the soil. More extreme in some ways is the soil-less cultivation
of plants, in water, nutrient films or bags of compost.

Whatever view we take of these procedures, they differ from 'traditional'
agriculture chiefly in the extent to which biological processes are involved
and the extent to which these are controlled.

It is worth considering this a little further, because the development of
agriculture could easily be described in terms of increasing control over
biological processes. Thus, the change from hunting and food-gathering
involved a substantial increase in the control of the plants and animals used
for food and materials for clothing, shelter, tools and weapons.

Control was exerted, not only over the ways in which plants and animals
grew, when and what they were fed, when they reproduced and which
individuals survived, but, eventually, on what kinds of plants and animals
existed. The modern breeds of livestock and the modern crop varieties are,
of course, quite unlike their ancestors and, to this extent, are the result of
considerable interference by human beings. In fact, agriculture *is* inter -
ference with the natural world, for the benefit of man; no-one would wish to

allow full and unfettered reign to all the pests, parasites and predators that would flourish in a completely natural system.

There is, however, considerable disagreement about the *ways* in which we should interfere and the *extent* of our interference, and it is useful to distinguish between these two.

The first concerns such issues as the use of 'artificial' fertilizers, drugs, hormones, pesticides and the like, partly on ecological grounds and partly to avoid contamination or alteration of 'natural' products. There is, of course, no sharp division between the two schools of thought, but rather a whole spectrum of views ranging from 'no use of anything artificial' to 'use of everything that proves to be economic'.

The second question is different, although, in practice, related. The development of agriculture has involved increasing interference, often resulting in substitution of nonbiological for biological processes (*see* Spedding, Walsingham and Hoxey, 1981). Castration of male animals was an early example of severe, and often extremely unpleasant, interference with a natural biological process (and, incidentally, not confined to agriculture, but still widely used in dogs, cats and horses). The methods employed are frequently less than humane, and 'traditional' in this context certainly does not represent 'better'. Not only are many modern methods far superior, especially from the points of view of the animal and the operator, but, as in the case of boars and bulls, modern husbandry methods often advocate avoidance of castration entirely.

However, in general, the trend in the development of modern agricultural systems has been towards a reduction in the biological content of agricultural systems. The battery hen represents the clearest example of this, although it is hard for many people to think about this objectively. The facts are that the battery hen no longer collects its own food, distributes its own faeces or incubates its own eggs. Nor does it have to cope with a whole range of pests, parasites and diseases that formerly accompanied most birds. It is almost the case that anything that can be done for the battery hen, *is* done; it does for itself only what cannot be done for it—and this is very little.

This particular issue is further confused by views about the kind of cages used, the number of birds per cage and the amount of space per bird.

However, the point being made here is simply to recognize that this process has taken place, involving the substitution of nonbiological processes for biological ones, in most 'developed' agricultural systems (*Table 22.1*). As a result, developed agricultural systems now use a great deal of support energy (*Table 22.2*). The reasons have been mainly concerned with increasing control—over the quantity and quality of feed, the incidence of disease, and many other factors that influence efficiency of production.

These developments were made possible by the availability of relatively cheap fossil fuels, a situation that has now of course greatly changed.

The greater economic efficiency of, for example, mechanization, depended upon cheap oil, both to make and operate equipment: however, it should also be recognized that these costs have to be related to those of human labour, and this may reflect the cost of oil indirectly. *Figure 22.1* shows how these costs have changed in the UK in recent years, but whether

Table 22.1   THE DISPLACEMENT OF (a) BIOLOGICAL PROCESSES AND
(b) OF MAN- AND ANIMAL-POWER, IN AGRICULTURE (COURTESY OF
SPEDDING, WALSINGHAM AND HOXEY, 1981)

(a)

Biological process displaced	Nonbiological process substituted
Natural fertilization in plant/seed dispersal	Plant breeding/harvesting of seeds
Fixation of atmospheric N by bacteria	Application of artificial nitrogenous fertilizers
Exploration of soil by roots for potash and phosphorus and water	Application of artificial fertilizers Irrigation
Natural control of pests and weeds	Use of pesticides and herbicides
Collection of feed by animals	Harvesting, processing and automated provision of compounded feed; forage conservation
Grazing	Zero-grazing (the cutting and carting of herbage)
Natural deposition of excreta on the land	Collection of excreta from housed animals and its disposal, treatment or distribution on land
Incubation of eggs by hen birds	Artificial incubation
Natural service by male animals	Artificial insemination
Natural hormonal processes	Control of light, daylength and temperature; use of synthetic hormones
Natural suckling (of calves and lambs)	Artificial rearing on milk substitutes
Natural immunity to disease in animals	Use of vaccines
Use of animal power	Use of machines and fossil fuel

(b)

Operation displaced (based on man- or animal-power)	Process or equipment substituted
Herding and guarding of animals	Fencing and housing
Feeding of animals	Machinery
Watering of animals	Automatic water supply
Hand milking	Milking machines
Carting and spreading of manure	Tractors and machinery
Hand, horse, oxen or buffalo cultivation sowing planting	Tractors and machinery
ploughing	Direct drilling and herbicides.
weeding	Herbicides applied by machinery
Harvesting of crops	Tractors and machinery
Hand application of fertilizer	Machinery and tractors
Drainage ditch digging	Machinery

Table 22.2 THE USE OF SUPPORT ENERGY IN
DEVELOPED AGRICULTURAL SYSTEMS TO THE
FARM GATE (SPEDDING AND WALSINGHAM, 1978)

Food product	Fossil fuel support energy input (GJ/ha)
Wheat	18.9
Potatoes	34.0
Milk	26.3
Beef	33.0
Tomatoes	40049.8

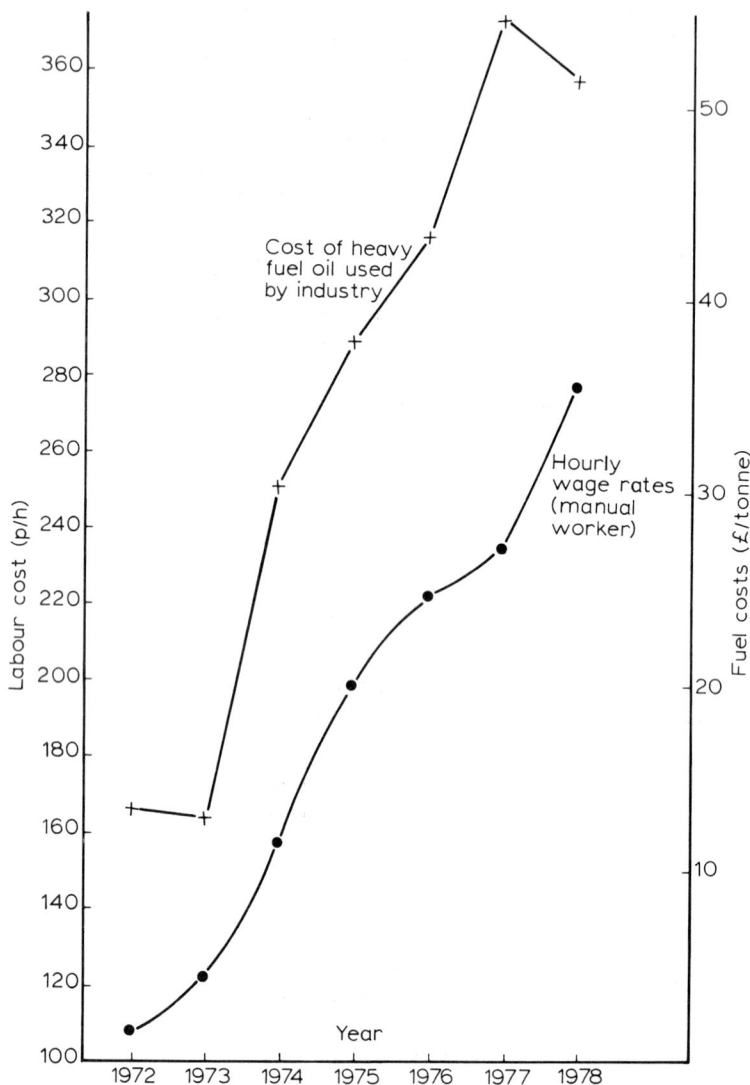

Figure 22.1 Recent changes in oil and labour costs

labour will become relatively cheaper in the future will be greatly influenced by whether we are talking about full-time or part-time labour.

Here is one important facet, therefore, that may be different for large-scale and small-scale farming.

Small-scale farming

Before discussing the characteristics of smallholdings, it may be useful to consider why we should be interested in small-scale farming. There are, in fact, several quite different reasons.

If we are thinking about *world* agriculture, then the answer is quite simple: the vast majority of the world's farmers are small-scale operators (*Table 22.3*). They may not be responsible for producing the majority of the

Table 22.3 THE SIZE OF THE WORLD'S FARMS: PROPORTION (%) OF FARMS UNDER 10 ha IN SELECTED COUNTRIES

United States of America[2]	10
United Kingdom[1]	31
Denmark[1]	33
Luxembourg[1]	36
Norway[2]	38
France[1]	41
Ireland[1]	41
Netherlands[1]	50
Brazil[2]	51
Costa Rica[2]	54
West Germany[1]	58
Belgium[1]	59
Sri Lanka[2]	66
Kuwait[2]	84
Italy[1]	86
Botswana[2]	88
Pakistan[2]	89
Malta[2]	98
Congo (Africa)[2]	100

Sources: [1] MAFF (1978); [2] FAO (1970)

food consumed—although this is hard to be sure about—and they certainly do not produce the major part of the food that enters into commercial trade. Many of them are subsistence farmers and, in bad years, may themselves and their families be among the world's hungry millions. For the most part, hungry people are also poor. Producing more food does not feed them, because they cannot buy it. Giving food away may help in an emergency but it cannot be a long-term solution and, indeed, may put local producers out of business.

One way to combat world hunger, then, is to improve the productivity of small-scale farmers—to grow food where it is needed and, as far as possible, by the people who need it.

Even a cursory examination will show that, in the past, most agricultural research effort has been concentrated on the needs of large-scale farming, and has been based on raising output by the increased use of inputs. Such

Table 22.4   NUMBERS OF PEOPLE PRODUCING
FOOD ON A SMALL SCALE IN THE UK

Number of people having agricultural holdings[1]	
1–5 ha	36 652
5–10 ha	34 034
Number of allotment holders[2]	479 271
Estimated number of people waiting for allotments[3]	121 000

Source: [1] MAFF (1978); [2] DOE (1980; personal communication); [3] National
Association of Leisure Gardeners (1980; personal communication)

research may be of little relevance to small-scale farmers, who cannot afford
the inputs required. Fortunately, there is growing recognition that the needs
of small-scale farmers are different, and research is increasingly being
directed to this end.

   This represents an extremely important reason for studying the problems
of small-scale farming, but it is not the only one. An important reason
relating to industrialized societies concerns the growth of interest in smallholdings, in part-time farming and, indeed, in food production from

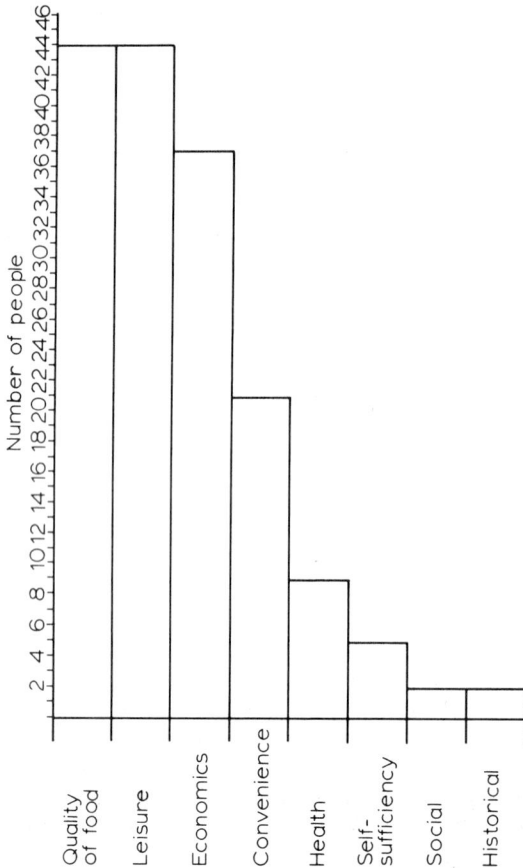

Figure 22.2   Garden and allotment survey: reasons given for growing own food (sample size
63 people)

gardens. The numbers of people involved are already substantial (*Table 22.4*) and seem likely to increase.

The reasons for increasing interest include:

1. A desire to produce some, at least, of one's own food (for economic reasons or a preference for fresh vegetables);
2. A desire to be active at or near home without much expenditure and, indeed, productively;
3. A desire to work the soil, to grow plants and keep animals (*Figure 22.2*).

It is possible that such interests will be combined, in the future, by increasing numbers of people who wish to reduce their dependence on one kind of job and thus to reduce their vulnerability to job difficulties (including strikes), to reduce their otherwise total dependence on purchased supplies of food and fuel, and their inability to control availability, quality or price.

In these circumstances, the economics of small-scale farming would look quite different and the arguments about economies of scale might differ, too.

Economies of scale

It is obvious that the economics of any operation *may* be influenced by the scale on which it is carried out—although this may not always, or necessarily, be so—but we should surely avoid the common assumption that this means 'the bigger the better'. This Dinosaur's Golden Rule *must* be nonsense: there must always be an upper limit and, when this is exceeded, it must then be better to become smaller.

Furthermore, it is inherently unlikely that the same scale of enterprise will be optimal for all resources. Thus, once a combine harvester of a given capacity is purchased it follows that the scale of cereal growing should be at a certain level, in order to make the best use of this machine. But this scale may not be best for disease control or in relation to the available labour or, indeed, other items of equipment.

Of course, it is necessary to recognize that scale of operation is a relevant factor to be taken into account, but there are many others.

It follows that it is very unlikely that we can usefully generalize about the relative efficiency of small- and large-scale farming. However, such generalizations already exist, so it is worth discussing them.

Relative efficiency of small- and large-scale farming

It is necessary to distinguish between the relative efficiencies (however measured) of farmers as they exist and the possibilities inherent in scale.

For example, most small-scale farmers are found in developing countries; they are poor and cannot afford inputs. They may also lack education and are often saddled with debts. Their land and climate may not be conducive to productivity and they may not be able to exercise much control over their

environment. Now, in these circumstances, it would hardly be surprising if
small farmers did not appear very productive or efficient. Even so, it
depends how productivity and efficiency are measured. For example, it is
well established that efficiency in the use of land and labour tends to be
higher in developed countries (Buringh and Van Heemst, 1977) but that
support energy is used with greater efficiency in less-developed agricultural
systems (*Table 22.5*). Indeed, it is extremely unlikely that all efficiency
ratios can be increased (and certainly not maximized) simultaneously
(Spedding, Walsingham and Hoxey, 1981).

Table 22.5   EFFICIENCY OF SUPPORT ENERGY USE IN DEVELOPED AND
LESS-DEVELOPED AGRICULTURAL SYSTEMS (LEACH, 1976)

Crop	Country	Comment	E^*
Corn (Maize)	USA	1945	2.02
		1970	2.58
	Mexico	By hand	30.6
		Using oxen	4.87
	Nigeria	By hand	10.5
Rice	USA	Intensive production	1.29
	Philippines		5.07

$$*E = \frac{\text{Gross energy in crop}}{\text{Support energy used to produce it}}$$

It is therefore necessary to decide on which resources are of most concern
(usually, that are scarce or costly) and, also, on the relative importance of
different outputs.

It is commonly assumed that this is the purpose of monetary expressions:
to allow us to combine, with accepted weightings, a variety of inputs and to
compare them with a combination of outputs. Thus measures of profitability have been used (Britton and Hill, 1975; Amies, 1978) to examine the
effect of scale on enterprises such as dairy farming. This may lead to
conclusions about the optimum size of a dairy farm but this usually has to be
expressed as quite a wide range and does not, of course, exclude the
possibility of exceptions to the average relationships. Sometimes it may
appear quite possible to achieve a similar level of profitability at widely
different enterprise sizes (Amies, personal communication; Kilkenny,
1980; *Table 22.6*).

It is worth considering one further, often-quoted example—a study (Best
and Ward, 1956) which compared the productivity of gardens in a residen-
tial area with that of the whole area (gardens + houses) if it had been
farmed. The results are shown in *Table 22.7*. The problem is that unless
output is in exactly the same form (e.g. wheat) it is difficult to compare.
Suppose the farmer grows wheat but the gardens produce spinach, peppers
and potatoes. Monetary values are an obvious, but not the only, basis for
comparison: they are the one most commonly used, however, and this is the
way in which this comparison was made. However, the outcome could be
altered simply by altering the value of the products or changing to different
products. Furthermore, this can be freely done, because costs do not enter

Table 22.6 VARIATION IN PRODUCTIVITY AND PROFITABILITY WITHIN FARM OR FLOCK SIZE CLASSES

Dairy farming† *Farm size* (ha)	*Milk yield/cow* (ℓ)	*Gross margin/ha* (£)*
< 45	3929–6828	297–1172
45– 64	3278–6441	179– 927
65– 84	4137–6320	255– 995
85–104	4283–6533	266– 819
> 105	4382–7157	304–1013
*constant valuation		

Sheep production‡ *Lowland flocks* *Flock size* (ewes)	*No. of lambs sold fat/ewe*	*Gross margin/ha* (£)
< 200	0.20–1.43	78–407
200–400	0.21–1.40	79–422
400–600	0.21–1.36	70–422
600–800	0.08–0.96	69–406
> 800	0.00–0.92[a]	68–402

[a] The zero value is caused by one of the recorded flocks selling all lambs as 'stores'
† Data from Amies (1978), and personal communication.
‡ Data from Kilkenny (1980)

Table 22.7 PRODUCTIVITY OF THE VEGETABLE-GROWING AREA OF GARDENS COMPARED WITH THAT OF THE WHOLE AREA USED FOR GARDENS AND HOUSES (AFTER BEST AND WARD, 1956)

	Output/cultivated acre (£)
Average gross agricultural output at farmgate prices for all types of farmland in England and Wales in 1953/54 was	36
Estimate for land most likely to be used for building	45
Specialized market-garden land	110
Domestic produce valued at farmgate prices	150
Retail value of domestically grown food	300
But if only 14% of a housing estate is cultivated for food crops then the value of production per house plot acre is approximately	42
If other urban land use is also taken into account, only 7% of the land is used for food production; then the output is approximately	21

into the calculation. It is hard to attach a meaning to comparisons that are based solely on cash output: nevertheless, this is a common basis for comparisons, including those concerned with such issues as relative self-sufficiency within a country. Furthermore, there is no entirely satisfactory single alternative.

Suppose we argue that productivity should be expressed in terms of the number of people who can be fed per unit area of land, and that we can get over such difficulties as agreeing on the diet (quantity and quality) that is to be supplied per person (taking into account weight, sex, age, activity, etc.)

and the fact that single crops (such as wheat) or even combinations of enterprises (as from a garden) may not provide balanced diets at all. Would this subsistence (or better) feeding capacity be a satisfactory basis for comparing gardens and cereal farms, which really have different purposes and objectives? In so far as it can be done, the values in *Table 22.7* have been expressed, in terms of how many people can be fed, in *Table 22.8*. The result looks rather different.

Table 22.8 COMPARISON OF THE FOOD PRODUCED, IN TERMS OF PROTEIN AND ENERGY OUTPUTS BETWEEN GARDEN PRODUCE AND WHEAT GROWN ON A FARM SCALE

Area and produce	Protein yield (kg)	No. of people this will supply with annual req. for protein	Energy yield (MJ)	No. of people this will supply with annual req. for energy
4 acres (1.6 ha) of vegetable-garden produce	521	22	55 549	12
4 acres (1.6 ha) of wheat (whole grain)	488	20	56 138	12
12 acres (4.9 ha) of wheat (whole grain)	1464	61	168 415	37

Using the example of Southampton (Best and Ward, 1956) 96 houses built at a density of 8/acre used 12 acres of land and resulted in 4 acres being used for growing vegetables and fruit.
Yields have been calculated for this area from the example of allotments at Wye.
The wheat yield is an average yield for that period.
The energy and protein outputs were calculated for the total edible crop raw i.e. whole potatoes, peas and beans less pods and so on, from *Composition of Foods*, USDA Handbook No. 8, 1963.
Protein requirement is taken as 65g/d or 24 kg/y and assumes adequate protein quality.
Energy requirement is taken as 12.6 MJ/d or 4955 MJ/y.

The fact is, that efficiency and productivity can be sensibly assessed only for stated purposes: we have to say why we wish to make the comparison and we will find that we have several different reasons, each leading to a different comparison.

The argument can easily become quite a sterile one. What we need to know is, what levels of productivity (for different products and resources) can be obtained from small-scale farming and what are the most appropriate methods to use. However, it is not only methods and tools (appropriate technology): there is also a need for appropriate *biology*.

Appropriate biology for small-scale farming

It may be the case that large- and small-scale farming could be identical in all other respects, provided that equipment could be scaled down and was available. In practice, there are usually many differences, including the outlet—the market to be supplied. This difference can be crucial and illustrates a number of important points.

Small-scale farms or smallholdings that exist in order to feed a household directly, require a sequence of harvests, either because fresh food is desired or because storage is not available or is wasteful or too costly.

Now, large farms tend towards large areas devoted to one crop (i.e. towards monoculture) and towards a single harvest by an expensive machine. The crop is grown to feed the machine rather than people and, for this reason, it is desirable for the product to be ready for harvest at one time. Where a household is being fed, however, food is required daily and there is a distinct advantage in a prolonged harvesting period. Furthermore, this is quite practicable because it involves very flexible human beings rather than inflexible machines.

The same general argument applies to many other operations, not just to harvesting, and to other cropping concepts, such as mixed and sequential cropping and intercropping.

Thus quite different crops and cultivars may be required for small-scale farms, and this has important implications for plant breeding.

Similar considerations apply to appropriate animals. Where small-scale farmers are relatively poor, large animals may be inappropriate because they represent too large a capital outlay, too great a risk and too much to eat or store when slaughtered. On the other hand, the use of animals for transport and draught purposes may be possible only if they are big enough to do the work, and ceremonial or celebratory occasions may be an effective means of consuming large animals.

Probably more important is the fact that animals often cannot be specially fed in small-scale agriculture: usually their function is to convert otherwise unusable byproducts. Sometimes they are a means of utilizing common grazing land, or the herbage available free on roadsides or waste land, and this may be an important means of importing fertility to a smallholding. Manure is retained on the holding although the feed is produced outside it.

To a very large extent, however, the role of the animal is to convert unusable byproducts and waste materials that cannot be consumed directly by man, into meat, milk or eggs that are directly consumable—although not necessarily without an expenditure on fuel for cooking.

The animals required, therefore, are mainly those able to live on the feeds available. The latter are, of course, unlikely to be ideal for feeding to animals and may not be well-balanced nutritionally. Thus, maximum performance per animal cannot usually be obtained and high genetic capacity to grow, give milk or produce eggs may be much less necessary than ability to thrive on the feeds available and to resist the prevalent diseases, pests and parasites.

In some respects, however, the reverse may be true. For example in sheep, reproductive rate is a key to productivity but, the higher the reproductive rate, the greater the need for high-quality feed and, usually, the greater the neonatal mortality. Yet it has been in the small flocks (<12) of Finland that the most prolific sheep have tended to develop, simply because far greater care and attention could be focused on the sheep at and around lambing time. This reflected the relative numbers of people and animals and is an example of cases where small-scale farming has advantages over large-scale farming, and can actually make use of more productive animals.

Small-scale animal keeping cannot, of course, justify high capital expenditure: one cannot afford an elaborate milking machine for one cow or one goat. Equally, the problem of milking by hand is quite different for very small numbers of animals.

Some inefficiencies of small scale can be overcome by cooperation. This is so for animal breeding, where only one male (bull, boar, ram, etc.) is required for a number of females. A farmer with only five sheep obviously should not keep a ram, which can cope with about 40 ewes.

Animals for smallholdings where self-sufficiency or subsistence are aimed at, will probably be of small size and low cost, unless used for work, and will generally live on herbage and crop byproducts rather than purchased feed.

Smallholdings that aim to compensate for their small size by operating sizeable businesses may, on the other hand, keep large animals (dairy cows and pigs, for example) or vast numbers (e.g. poultry) fed entirely on purchased feedstuffs and involving substantial capital investment. Part-time farming could resemble either of these patterns, depending on whether labour was employed or not.

Potential for small-scale farming in the UK

More information is needed in order to establish potentials for different soils and situations. Some data are available from smallholders who have meticulously recorded their results (Green, 1980) and an illustration of these possibilities is given in *Table 22.9*.

Table 22.9   PRODUCTION (FRESH WEIGHT PER UNIT OF LAND) BY SMALLHOLDERS AND GARDENERS

Crop	Production (t/ha)
Single crop	
Brussels sprouts 1979–80[1]	9.82
Mixed fruit and vegetable crops	
Small garden 1977–78[2]	16.29
Larger garden, South Wales 1977[3]	12.87

Sources: [1] Own data; [2] Yellowlees (1979); [3] Green (1980)

At Reading University we have established a research group to explore the problems of small-scale farming, partly because it is likely that the optimum use of a small area will require a high degree of integration of the enterprises undertaken. The complexity of these interrelationships can be considerable and the number of possible combinations and permutations of enterprises is very large indeed. Nevertheless, our aim is to try to elucidate the main principles on which small-scale farming should be based.

References

AMIES, S. J. (1978). *Herd Size and Business Efficiency*. Report No. 14. Reading, Berks; MMB LCP Information Unit

BEST, R. H. and WARD, J. T. (1956). *The Garden Controversy*. Wye College, Dept. Agric. Econ., Wye, Kent.

BRITTON, D. K. and HILL, N. (1975). *Size and Efficiency in Farming.* Farnborough; Saxon House

BURINGH, P. and VAN HEEMST, H. D. J. (1977). *An Estimation of World Food Production Based on Labour-oriented Agriculture.* Wageningen, The Netherlands; Centre for World Food Market Research

FAO (1970). *Report on the World Census of Agriculture.* Rome; FAO

GREEN, D. (1981). Allotments and gardens. In *Vegetable Productivity,* pp. 202–215. Ed. C. R. W. Spedding. Institute of Biology Symposium 1979. Macmillan

KILKENNY, J. B. (1980). *Meat and Livestock Commission.* Personal communication

LEACH, G. (1976). *Energy and Food Production.* London; IPC Science and Technology Press

MAFF (1978). *EEC Agricultural and Food Statistics* 1974–1977. London; HMSO

SPEDDING, C. R. W. and WALSINGHAM, J. M. (1978). Energy and the future of agriculture. *N.Z. agric. Sci.,* **12,** 76–80

SPEDDING, C. R. W., WALSINGHAM, J. M. and HOXEY, A. M. (1981). *Biological Efficiency in Agriculture.* London; Academic Press

YELLOWLEES, W. W. (1979). Ill fares the land. *J. R. Coll. Gen. Pract.,* **29,** 7–21

23

ORGANIC FIELD CROP PRODUCTION IN THE MIDWESTERN UNITED STATES

W. LOCKERETZ
Northeast Solar Energy Center, Boston, Massachusetts, USA

The position of organic farming in United States agriculture

Organic farming has an ambiguous status in the United States: while not very many people actually do it, a substantial number of people care a good deal about it. It is a topic that usually arouses strong feelings on the part of both supporters and critics. Until fairly recently, to the extent that these two groups talked to or about each other at all, it was often just to attack the views, and possibly also the character and integrity, of those in the opposite camp. This paper describes research undertaken to provide some empirical data on a topic that had largely been argued on the level of semantic quibbling or pointless diatribes. But because the relative merit of organic and conventional practices is more than a purely agronomic question, it seems desirable first to discuss some views of organic farming held by its advocates, its critics and the agricultural research community.

With regard to its importance in commercial agriculture—which is the sole focus of this chapter—organic farming has negligible significance. The number of organic farmers is not known precisely, but various rough estimates have generally indicated that only a tiny fraction of the nation's commercial farmers actually farm organically in the sense of not using any inorganic fertilizers (except for rock powders that have not been treated chemically) or synthetic pesticides. But far more numerous than those who practise organic agriculture on a commercial scale are two other groups of supporters: backyard gardeners or operators of small hobby farms, and those who are not food producers at all but who favour organic methods because they prefer to eat food which has been produced that way, or because they are disturbed by the environmental effects of the toxic chemicals widely used in conventional agriculture.

The supporters of organic farming tend to be outside the mainstream of agricultural research. Their attitudes towards agricultural research vary. Some—and their number may be growing—recognize the importance of learning as much as possible from whatever sources can provide worthwhile information, including the agricultural research establishment. They try also to reduce the separation between organic farming supporters and the research establishment by encouraging the latter to take an open-minded look at the performance of organic farms, and by providing opportunities for the different groups to understand each other's attitudes and assumptions. But some organic advocates seem happy with the polarization, and may even aggravate it by attacking the honesty of the majority of researchers. Thus, one sees allegations that researchers who criticize organic farming do so because their research is supported by fertilizer and pesticide manufacturers. In fact only a tiny portion of the agricultural research budget in the United States comes from these sources; most of it comes from public funds.

Among agricultural researchers and others with a professional involvement in agriculture, one finds very little clear, overt support for organic farming. Many probably do not have strong opinions one way or another and view it with curiosity, scepticism or indifference. In my view the most common opinion is that organic farming should not be recommended or promoted, but that one should at least take an open-minded look at any new data gathered on the topic. Moreover, most holders of this view would probably acknowledge that at least some aspects of organic practice should be more generally incorporated into our agricultural system.

However, some agricultural specialists take an extremely antagonistic and hostile position regarding organic farming. In part, this may reflect a sincere misunderstanding of what organic farming is all about, a misunderstanding that may be caused by, and also be the cause of, the lack of attention given to the subject by researchers. For example, a researcher might not distinguish a backyard garden from a commercial farm. Thus, because most people who use organic methods do so on a garden scale, and can therefore afford to use extremely labour-intensive methods like removing insects from plants by hand, it is sometimes assumed that organic farming necessarily involves such methods. If so, then it obviously would not work on a commercial scale, which has led some people to predict a several-fold increase in production costs under commercial organic agriculture. As this chapter will show, however, such an argument reflects an ignorance of the existence of a group of commercial organic farmers whose methods have far more in common with those of other commercial farmers than with organic gardeners.

Some agricultural specialists—I would like to believe only a very unrepresentative minority—make statements about organic farming that can hardly be explained as sincere misunderstandings. For example, because organic farmers do not use the pesticides and fertilizers that other farmers have adopted in the past 40 years, it is sometimes argued that yields with organic methods would automatically be the same as those that all farmers got 40 years ago. One would have to be very charitable to regard as a sincere misunderstanding the belief that chemicals are the only factor determining yields, as though no other advances of the past 40 years—improved crop varieties, mechanization, better tillage implements and a higher level of knowledge on the part of most farmers—had anything to do with it. Likewise it is hard to regard as a bona fide scientific undertaking some 'experiments' in which the 'organic' treatment is represented by doing nothing—that is, where a plot has not only received no pesticides and inorganic fertilizers, but has also not had any rotations, livestock manure applications, or mechanical weed control. Not surprisingly, 'organic' rarely fares very well in such casese. And there is even less excuse for a person with scientific training to attack supporters of organic farming with epithets like 'faddists', 'health-food nuts', or 'fanatics'.

Regardless of the relative prevalence of support, scepticism, or hostility towards organic farming among researchers, in any case until the mid-1970s organic farming was not regarded as a suitable topic for serious mainstream agricultural research. Several factors combined to change that situation, however, most notably the growing awareness of the environmental effects of massive use of agricultural chemicals and the sharp price increases and occasional shortages of the petroleum and natural gas used in their manufacture. Besides influencing researchers directly, these developments also strengthened the position of outside groups that had been promoting less resource-intensive and more environmentally benign technologies in agriculture, as in many other areas. The interest in organic farming among agricultural researchers—which, while still very modest in an absolute sense, represents a significant change relative to the past—may be one example of a greater responsiveness and a general opening-up of agricultural research to outside influences, including groups who formerly were excluded from determining research priorities.

Recent research on organic agriculture in the United States

Summarizing the status of organic farming research in the United States is complicated by the fact that different researchers use the term differently. It is sometimes used in a way that suggests that it is roughly interchangeable with 'alternative agriculture' or with small-scale production of specialty crops. Other researchers sometimes label as 'organic farming research', work on any of the specific components of an organic farming system, such as fertilization with livestock manures or integrated pest management. Indeed, the large amount of work on such practices is often used to answer the criticism that the agricultural research establishment has ignored organic farming, even though such research has usually examined such practices as a part of a farming system that would not qualify as 'organic' because it still involves major inputs of inorganic fertilizers or synthetic pesticides.

Confining ourselves to research on farming methods that would be considered 'organic' by organic farmers themselves, only a small amount of work has been done, all of it within the past few years. In addition to the work reported in this chapter, three other studies have looked at the energy use and economic performance of actual organic farms: those of Roberts (1977) on field crop production in the Corn Belt (the same farming system that will be discussed here), Berardi (1978) on wheat production in New York State and Pennsylvania, and Kraten (1979) on small grain farms in Washington State. Unfortunately, these studies were conducted with very limited resources (all as Master's Thesis topics, as it happened), so that the data covered a fairly small number of farms, or a short period, or both. Because of this, and because of the great differences in the methods used by the various investigators, and in some cases the wide range of farms within each study's sample, it is not possible to form a definitive overall conclusion concerning their findings. However, they suggest that organic farming is economically competitive, or at least is closer to being so than had been commonly thought. The reduced energy requirements of organic farming, for example, emerges from all of these studies.

Several researchers have examined the social and institutional relationships of organic farmers. Alexander (1977), in a companion study to that of Roberts (1977), looked at the entire industry of which organic farmers are one part, including input suppliers, sources of information and markets. Youngberg (1978) and Buesching (1979) discussed some of the philosophical bases of organic farming, its relation to a broader alternative agriculture movement, and the differences among its various groups of followers. Both of these papers were based mainly on semipopular literature, including that of the organic farming community itself, and on personal communications. A sociological study that collected primary data from farmers in Michigan (Harris, Powers and Buttel, 1979) investigated some of the same questions, and concluded that, in many of their personal characteristics and attitudes, organic farmers were more like conventional farmers than like certain stereotypes. Similarly, Vail and Rozyne (1980) found that, to a considerable degree, even small-scale organic farmers in a more marginal farming area (Maine) shared many of the economic and social values of the majority of US farmers.

There have been few studies of organic farming on a national scale. Oelhaf (1978) attempted to predict the economic consequences of a major national shift to organic production, something that he viewed fairly optimistically. However, he had very little solid empirical data to work with, and his extrapolations seem overambitious and premature. A very recent national examination of organic farming was carried out by the US Department of Agriculture (1980), which in the past has been criticized strongly by organic farming supporters for its neglect of this topic (indeed for its occasional outright hostility). This study commented favourably on the high level of management ability shown by many organic farmers. While its main findings in general are similar to those of the earlier work described above, it is likely that because of its prestigious sponsorship this report will be particularly influential in shaping future research activities and government programmes.

METHODS AND SCOPE OF THE WASHINGTON UNIVERSITY STUDY

The research described in this chapter was conducted at Washington University, St. Louis, Missouri, from 1974 to 1979. Because most of its methods and principal findings have already been reported in the references cited below, only the highlights will be presented here.

All the data in this research were collected on working farms in Iowa, Illinois and the adjacent parts of Minnesota, Nebraska and Missouri. This region lies in the Corn Belt, the extremely productive North Central region of the United States which is the nation's dominant producer of maize grain, soybeans and hogs. All the farms studied raised some or all of these products, together with the other field crops and animals that are important in the region, such as oats, hay and beef cattle. We did not examine production of vegetables, fruits or other products that are regionally of relatively minor significance.

The study had six main components, briefly described below.

Survey of the basic features of commercial organic farms

We attempted to identify as many organic farms in the region as possible by a 'snowball' technique in which each farmer gave us the names of others. A total of 363 farmers were found who operate at least 40 hectares organically. (In this research, we take 'organic' to mean crop production without any inorganic nitrogen fertilizer or urea, acidified phosphate, muriate of potash, or synthetic herbicides or insecticides.) Through a short mailed question-naire, we collected data on size, crop mix, materials used, and livestock inventories of these farms (Wernick and Lockeretz, 1977; Lockeretz and Wernick, 1980a).

Study of organic farmers' attitudes and motivations

A more detailed questionnaire was sent to 250 of these farmers, 174 of whom replied. This questionnaire covered the farmers' reasons for farming organically, their contacts with conventionally orientated and organically orientated institutions and information sources, their production and marketing practices and some personal characteristics (Wernick and Lockeretz, 1977; Lockeretz and Wernick, 1980b).

Analysis of economic performance and energy use of paired organic and conventional farms, 1974–76

This work involved 14 organic farms, each of which was matched with a nearby conventionally operated farm of approximately the same size and with similar soils. These organic farms were individually known to us beforehand, rather than having been systematically selected. In this, and the two studies described below, all organic farms had been managed organically for at least four years before being studied. Using a personal interview and follow-up mailed data forms, we collected data on planting practices, material applications, field operations and yields on each field for each of the three crop years (Klepper *et al.*, 1977; Lockeretz *et al.*, 1977; Lockeretz *et al.*, 1978a).

Analysis of economic performance and energy use of organic farms, 1977–78

The organic farms in this study were selected from the 363 respondents to the earlier survey. Twenty-three were studied in 1977, with 19 of these studied in 1978 as well. The comparison with conventional practice was done using county-average yield data and standardized crop production budgets. The data covered basically the same topics as the study just described (Shearer *et al.*, 1980).

Field studies of crop yields

Small plots of maize, soybeans and winter wheat were harvested on neighbouring organically and conventionally managed farms. The pairs of fields were matched by soil type, variety and planting data. A total of 26 pairs of maize fields were measured between 1975 and 1978, together with seven and four pairs respectively of soybean and wheat fields in 1977 or 1978 (Lockeretz *et al.*, 1978b, 1980).

Soil analysis

Samples of soil taken from 30 pairs of fields used in the yield measurements were analyzed for organic matter, nutrient content, pH and cation exchange capacity (Lockeretz *et al.*, 1980). Five additional pairs of fields were sampled for organic matter only.

RESULTS

Backgrounds and attitudes of organic farmers

Our study of 174 organic farmers revealed that they did not fit the stereotype of a back-to-the-land young beginning farmer who had taken up organic farming as part of a more general rejection of prevailing values (Lockeretz and Wernick, 1980b). In part, this was a consequence of our having chosen to study field crop producers who operated at least 40 hectares. Thus we automatically excluded small-scale production of special-ty crops, which might be more typical of young organic farmers just starting out. Still, while our selection method means we cannot say that all organic farmers are like those of our study (and without question, they are not), nevertheless we can say that not all organic farmers are like the stereotype. The farmers' median age (50 years) was the same as that of all farmers in the region. Moreover, over four-fifths of our sample had once farmed using conventional methods, something which was not required by our selection method.

They had various reasons for choosing to farm organically. The dominant reasons were pragmatic, such as concern over the health of their family and their livestock, and the cost and ineffectiveness of chemicals. More subjective or philosophical motivations, which would be the primary ones according to the stereotypes, were decidedly less prominent, although still significant.

These organic farmers were heavily involved in the larger economic system, both in their purchases of inputs and sales of their products through major marketing channels. Again, this contrasts with the stereotype of an organic farmer as striving for self-sufficiency and producing primarily for specialty markets and for direct sales. In general, our findings on organic farmers' backgrounds and attitudes were in fairly good agreement with the previously cited work of Vail and Rozyne (1980), Harris, Powers and Buttel (1979), and USDA (1980).

Practices

By definition, organic and conventional farmers differ in their use of inorganic fertilizers and synthetic pesticides, and it is often assumed that the two groups differ in many other areas. This was not generally true for the farmers we studied.

The organic farms were heavily mechanized and made full use of modern harvesting equipment and tillage implements. Consequently, they required only slightly more labour (Klepper *et al.*, 1977), in contrast to the stereotype of organic farming as being several times more labour-intensive. Their farms were about 20 per cent smaller than other farms in the region that raise similar crops, and that would also exceed our arbitrary minimum size of 40 hectares (Lockeretz and Wernick, 1980a).

To compensate for not using conventional fertilizers and pesticides, the organic farmers relied on crop rotations, mechanical weed control and applications of livestock manures from their own farms (Lockeretz and Wernick, 1980a). However, in this regard they differed only in degree from the majority of farmers in the Corn Belt, especially those with mixed crop/ livestock operations. Even though organic farming supporters sometimes contemptuously refer to conventional farms as 'chemical' farms, in fact such farms also use these same nonchemical methods, the difference being that they do not rely on them exclusively. In contrast to the image of organic farming as involving exotic or extremely labour-intensive techniques (an image which may reflect the failure to distinguish organic farming from organic gardening, as discussed earlier), there was very little use of any techniques that are not commonly found on conventional farms as well. The main exception was the frequent use of low nutrient-content commercial organic fertilizers, which made a negligible contribution to the overall nutrient balance but often entailed considerable expense.

Economic performance

The data on yields and crop production practices described earlier were analyzed to determine two quantities for all the cropland on each farm: market value of crops produced and operating costs per unit area (the latter

Table 23.1 ECONOMIC PERFORMANCE OF ORGANIC AND CONVENTIONAL FARMS ($/ha)[1]

Year	Value of production		Operating expenses		Net returns	
	Organic	Conventional	Organic	Conventional	Organic	Conventional
1974[2]	393	426	69	113	324	314
1975[2]	417	478	84	133	333	346
1976[2]	427	482	91	150	336	333
1977[3]	384	407	95	129	289	278
1978[3]	440	527	107	143	333	384

[1] All data are averaged over all cropland (including rotation hay and pasture, soil-improving crops and crop failures).
[2] Based on 14 organic and 14 conventional farms (Lockeretz *et al.*, 1978a).
[3] Based on 23 organic farms in 1977 and 19 in 1978, which were compared with county-average yield data and standard crop production budgets (Shearer *et al.*, 1980).

included only short-term expenses, such as labour, fuel and seeds, but did not include capital costs.) The difference between these two quantities, denoted here as net returns, is an approximate measure of the comparative profitability of various farms.

Consistent results were obtained for all three years of the first study and the first year of the second study (*Table 23.1*). The value of crops produced per hectare of cropland was slightly lower on the organic farms, but this difference was just about offset by their lower operating costs. (In this Table, 'value' refers to prevailing market values and does not reflect the occasional premiums obtained by organic farmers for the small portion of their output marketed through special channels.)

In 1978, however, there was a much sharper difference between the two groups' gross income, too great to be offset by the difference in expenses. This probably reflects the favourable growing conditions that year, in contrast to the four previous years when several of the farms were affected by drought. This explanation is corroborated by the yield measurements discussed later. Thus it appears that organic farming offers some protection against adverse conditions, while sacrificing some of the increase in income possible during favourable years.

Energy use

The data from the two economic studies were also used to compare fossil fuel consumption on both kinds of farms. This included energy consumed in operations performed on the farm, as well as in the production of short-term inputs such as fertilizers, but not in the manufacture of capital equipment. The results of this study, shown in *Table 23.2*, are quite clear: in all five

Table 23.2   ENERGY CONSUMPTION OF CROP PRODUCTION ON ORGANIC AND CONVENTIONAL FARMS[1]

| Year | Energy consumption per unit of market value (Mcal/$)[2] | | |
	Organic	Conventional	Ratio
1974	1.8	4.3	0.42
1975	1.7	3.8	0.45
1976	1.8	4.8	0.38
1977	1.9	5.1	0.37
1978	1.8	4.1	0.44
Average	1.8	4.4	0.41

[1] 1974–76 data from Lockeretz *et al.* (1978a); 1977–78 data from Shearer *et al* (1980).
[2] Average for all crops on each farm, with crop value figured at prevailing prices. Energy included fuel consumed on farm and in production of annual inputs (fertilizers, soil amendments and pesticides).

years, the organic farms required an average of only two-fifths as much fossil energy per unit value of production. This reflects their non-use of manufactured fertilizers and pesticides, and the lower proportion of cropland in maize, which on both kinds of farms requires the most energy.

Crop yields

The yield measurements reported in *Table 23.3* corroborate the two economic studies, which used farmers' self-reported yields. It is somewhat surprising that the two groups differed less in maize than in wheat, because their practices differ by much more in maize. (Conventionally grown maize routinely receives herbicide, a heavy application of complete fertilizer, and often insecticide as well: wheat, in contrast, receives a moderate fertilizer application and only occasionally herbicide or insecticide.)

Table 23.3 MEASURED CROP YIELDS ON ORGANIC AND CONVENTIONAL FARMS[1]

| Crop | Number of measurements[2] | Years | Yield (t/ha) | | Difference |
			Organic	Conventional	
Maize	26	1975–78	6.45	7.00	−0.55 (8%)
Soybeans	7	1977–78	2.44	2.57	−0.13 (5%)
Wheat	4	1977–78	1.88	3.28	−1.40 (43%)

[1] Maize data from Lockeretz *et al.* (1980); soybeans and wheat from Lockeretz *et al.* (1978b).
[2] Each measurement involved a pair of neighbouring fields planted to the same variety on or about the same day, with yields measured on plots of identical soil type.

The yield advantage of conventionally raised maize was greatest under the most favourable growing conditions; under very wet or dry conditions, in contrast, organic yields often equalled or even exceeded conventional ones (Lockeretz *et al.*, 1980). The organically managed maize fields had consistently less stalk lodging and *Diplodia* stalk rot.

Soil analysis

The clearest difference between the two kinds of fields was the slightly higher average organic carbon content on the 35 organic fields (organic carbon content of 2.35 per cent compared with 2.21 per cent on the matched conventional fields, $P > 95$ per cent). There were no other statistically significant differences, although there was a suggestion of lower available phosphorus content on the organic fields. An earlier nutrient budget indicated a gradual depletion of phosphorus and potassium on organic fields (Lockeretz *et al.*, 1977), but apparently this has not yet resulted in clearly detectable differences over the period of organic management (a median of 7 years in this study).

Possible implications for agricultural research

The project that has just been summarized was concerned with measuring specific characteristics and performance levels on a specific type of organic farm. Now that that work has been completed and its results reported, it may be appropriate to go beyond the limits of the actual data and to speculate on some additional lessons concerning agricultural research in general, and especially concerning research on systems that differ sharply from the main trends in agricultural development.

The conclusions reached in this work conflict strongly with a widely and often vehemently held view. At the risk of overgeneralizing, I offer the following as broader conclusions suggested by this conflict.

There are important differences between real farms and experimental plots, and each has its appropriate place in a well-balanced research programme.

The immense advantages in using experimental plots as a research site would hardly have to be noted if it were not that certain proponents of organic farming have sometimes viewed test plots very disdainfully. Because they lend themselves to carefully controlled, well-replicated experiments that can be analyzed by powerful statistical methods, test plots will always form the basis of any empirical research programme. However, researchers who work exclusively on test plots may tend to forget that many things change when one moves to a working farm. To a researcher investigating basic agronomic principles, such changes may be irrelevant, but they are critical when one is concerned with specific applications of these principles to practical farm management. A test plot in some sense is a model of a real farm, that is, a simplified and idealized version. Just as an engineer must be very careful in going from a bench-scale prototype to a full-size unit, or a medical researcher in going from clinical studies of a small group of subjects to broad public health programmes involving the general population, so too must agricultural researchers take explicit account of the somewhat artificial conditions of a test plot.

The work reported here was conducted entirely on working farms. Much information on crop response to inorganic fertilizers (often a basis for the common belief that a farm that did not use such fertilizers could not be very productive) was developed on test plots. Typically, variables other than the specific variable of the experiment are optimized on test plots: thus, in an experiment on fertilizer yield response, for example, a test plot would be tilled, planted, and harvested at the optimal times and would be kept well-weeded and free of insect pests. Under such conditions, the crop could be expected to show a strong response to added fertilizer. On a real farm, however, the farmer's limited time, equipment and management ability might lead to less than optimal practices, so that nutrient availability is not the factor limiting yields. It was not expected that the organic farms' yields in this study would be so close to those of conventional farms. The explanation may be, not that the organic farms' yields were unexpectedly high, but that the benefits the conventional farmers got from using fertilizer were unexpectedly low, at least compared to test plot results. (The particular conventional farmers we studied did at least as well as all farmers in the region, however; the previous remark refers to conventional farmers in general.)

While researchers have used both test plots and working farms as research sites, they have not always explicitly acknowledged that the two may be very different. Often the choice has been simply a matter of convenience. It seems highly desirable that a careful, systematic investigation be made of how studies of crop response to various practices are affected by the choice.

Ideas and innovations do not flow only from agricultural experts to working farmers; researchers may get interesting ideas concerning what to study from seeing what farmers are actually doing, even if such practices are not to be recommended to other farmers.

As noted earlier, organic farming is not a recommend management system. Whether this situation is justifiable or not is beside the point. But it is interesting that some agricultural experts were unaware that commercial organic farms even existed, and in some cases went as far as to state that it was *impossible* for them to exist, at least as economically competitive businesses. The main reason for undertaking the research described here was the striking fact that clearly successful farms indeed existed, despite this conventional wisdom. Even now, after 5 years of research, the project staff makes no particular recommendation one way or another concerning whether a farmer should adopt organic management. Despite this, we believe that valuable lessons were learned by studying such farms.

Apparently some agricultural researchers do not realize that one chooses a system for research according to what one can learn by investigating it, not just to be able to recommend it. Indeed, to some people the very fact of our having investigated organic farming meant that, whether we admitted it or not, we were advocates of organic farming. Agricultural researchers are being too limited if they restrict themselves to those systems which they think they eventually will be able to recommend. Nor should they assume that they themselves, and not working farmers, are the only source of ideas even worth studying, let alone recommending.

A diversity of sources of support for agricultural research can help assure that worthwhile research opportunities are not overlooked.

In the United States, virtually all agricultural research (except for development of specific products) has traditionally been conducted under the auspices of the so-called Land Grant College Complex, which includes the United States Department of Agriculture, one or more agricultural colleges in each state, and agricultural experimental stations operated by the states in cooperation with the Department of Agriculture. While the agricultural colleges are autonomous, they receive part of their research budgets from the Department of Agriculture, with whom they also frequently share personnel and facilities, and coordinate many of their programmes. Thus it is reasonable to regard the entire complex as in some ways constituting a single institution.

This complex has accumulated an impressive record of research accomplishments, although its overall approach has also received some sharp criticism (Hightower, 1972). It could be argued that it has sometimes shown excessive rigidity in excluding concepts that do not conform to its leaders' vision of what the future of United States agriculture should be. Organic farming would be one example of exclusion.

It is probably not a coincidence that the institution that conducted the first project on commercial organic agriculture, Washington University, is not part of the Land Grant College Complex, and that the research funds came from the National Science Foundation, which traditionally has not financed agricultural research. However, several of the later studies referred to above were done at the regular agricultural colleges, and in many cases (according to personal communication from the respective principal investigators) were motivated to a considerable extent by the Washington University work. This suggests the value of having at least a small portion of the total agricultural research effort located outside the main research establishment, where an investigator might be more likely—both for bureaucratic reasons and because of differences in training, background and outlook—to look at somewhat unconventional concepts. Depending on the early results of such work, these topics might then be pursued in a fuller way by the mainstream institutions which, in their superior technical resources, are better equipped to do so.

The social context of agricultural research

A common element in these three points is that they involve agricultural sciences' relationship to a broader set of conditions outside the subject matter of the research itself. This may not seem compatible with an idealized picture of research, according to which science develops by a purely internal logic. However, such a picture would omit the very important ways in which social, political and educational conditions affect researchers' ways of thinking.

To say that this applies to agricultural research is not to demean the field. Even theoretical physics, that most abstract of the natural sciences, has evolved in a way that reflects its social environment. Agricultural scientists would do well to acknowledge a similar situation. In the United States, agricultural researchers have not been given to introspection, and have tended to react with antagonism or defensiveness to critical outside judgment of the state of their field. It is noteworthy that, of all the disciplines concerned with agriculture, the one that seems to have shown the most interest in organic farming is rural sociology. Sociologists are more likely than natural scientists to study not only a particular set of phenomena, but also the people who are involved with those phenomena. Thus to rural sociologists, organic farming is interesting because of what it says not only about soil chemistry and pest control, but also about why farmers and agricultural experts behave as they do.

The striking achievements that the agricultural research profession likes to point to in answer to its critics—increased yields, reduction in labour requirements, and so forth—do not mean that its members are free from limitations in their outlook, that necessarily result from having received a particular kind of education and having acquired a particular kind of experience in a particular kind of organization. Such limitations may explain why agricultural researchers, until recently, refused to regard organic farming as a subject worth their effort, even though there was considerable interest in it in the broader population. The research reported here is not enough of a basis for recommending organic farming for widespread use. However, it certainly shows—if only by its contrast with the view of the subject held by many researchers, and by the subsequent research it helped to motivate—that the topic is one where interesting lessons can be learned.

As the energy situation deteriorates further, and as additional pressures are placed on agricultural resources and the environment, we will not want to miss other interesting research areas that may suggest the new approaches needed to meet these problems. At the very least, this will necessitate that vitriolic but unsupported denunciations of those who choose to farm in an unconventional way, should not be accepted in place of data. It will also require that more agricultural researchers accustom themselves occasionally to step outside their usual roles and to see whether they have been asking the right questions and using the right methods to answer them. If they are not willing to do so, there is no shortage of outsiders who will be very happy to do it for them.

References

ALEXANDER, C. (1977). *The Nature and Extent of the Organic Agriculture Industry.* MS Thesis, University of Missouri

BERARDI, G. M. (1978). Organic and conventional wheat production: examination of energy and economics. *Agro-Ecosystems,* **4,** 367–376

BUESCHING, D. (1979). Origins, development and current composition of the Alternative Agriculture Movement. Paper presented at the Annual Meeting of the Rural Sociological Society, Burlington, Vermont, August

HARRIS, C. K., POWERS, S. E. and BUTTEL, F. H. (1979). Myth and reality in organic farming: a profile of conventional and organic farmers in Michigan. Paper presented at the Annual Meeting of the Rural Sociological Society, Burlington, Vermont, August

HIGHTOWER, J. (1972). *Hard Tomatoes, Hard Times: The Failure of the Land Grant College Complex.* Washington, DC; Schenkman Publishing Co

KLEPPER, R., LOCKERETZ, W., COMMONER, B., GERTLER, M., FAST, S., O'LEARY, D. and BLOBAUM, R. (1977). Economic performance and energy intensiveness on organic and conventional farms in the Corn Belt: a preliminary comparison. *Am. J. agric. Econ.,* **59,** 1–12

KRATEN, S. L. (1979). *A preliminary examination of the economic performance and energy intensiveness of organic and conventional small grain farms in the Northwest.* Master's Thesis, Washington State University

LOCKERETZ, W., KLEPPER, R., COMMONER, B., GERTLER, M., FAST, S., O'LEARY, D. and BLOBAUM, R. (1977). Economic and energy comparison of crop production on organic and conventional Corn Belt farms. In *Agriculture and Energy,* pp. 85–101. Ed. W. Lockeretz. New York; Academic Press

LOCKERETZ, W., SHEARER, G., KLEPPER, R. and SWEENEY, S. (1978a). Field crop production on organic farms in the Midwest. *J. Soil Water Conservn.,* **33,** 130–134

LOCKERETZ, W., SHEARER, G., KUEPPER, G. and WANNER, D. (1978b). Yields of corn, soybeans, and winter wheat on 'Organic' and 'Conventional' farms in the Midwest. Presented at the Annual Meeting of the American Society of Agronomy, Chicago, December

LOCKERETZ, W., SHEARER, G., SWEENEY, S., KUEPPER, G., WANNER, D. and KOHL, D. H. (1980). Maize yields and soil nutrient levels with and without pesticides and standard commercial fertilizers. *Agron. J.,* **72,** 65–72

LOCKERETZ, W. and WERNICK, S. (1980a). Organic farming: a step towards closed nutrient cycles. *Compost Science/Land Utilization,* **21,** 40–46

LOCKERETZ, W. and WERNICK, S. (1980b). Commercial organic farming in the Corn Belt in comparison to conventional practices. *Rural Sociology*, (in press)

OELHAF, R. C. (1978). *Organic agriculture: economic and ecological comparisons with conventional methods*. Montclair, New Jersey; Allenheld, Osmun

ROBERTS, K. J. (1977). *The economics of organic crop production in the western corn belt*. Master's Thesis, University of Missouri

SHEARER, G., KOHL, D. H., WANNER, D., KUEPPER, G., SWEENEY, S. and LOCKERETZ, W. (1980). Crop and livestock production on beef and hog producing Midwestern organic farms: 1977 and 1978. (in press)

USDA (1980). *Report and recommendations on organic farming*. Study Team on Organic Farming, US Department of Agriculture. Washington, DC; US Government Printing Office

VAIL, D. and ROZYNE, M. (1980). The image and the reality of small organic farms: evidence from Maine. In *New Principles of Food and Agriculture*, Ed. D. Knorr. Westport, Connecticut; AVI Publishing Co (forthcoming)

WERNICK, S. and LOCKERETZ, W. (1977). Motivations and practices of organic farmers. *Compost Sci.*, **18,** 20–24

YOUNGBERG, G. (1978). The alternative agricultural movement. *Policy Studies J.*, **6,** 524–530

24

DEVELOPING AGRICULTURE OF VILLAGE COMMUNITY PROJECTS IN THE WESTERN SOLOMONS

R. ELLIOTT
*APACE, Box 81, Wentworth Building, University of Sydney
NSW 2006, Australia*

APACE is a voluntary charitable research institute concerned with the initiation, development and application of low-impact, decentralized technology and with the social consequences of such technology in both developing and developed countries. The work of APACE is organized by committees of honorary specialist consultants. The agriculture committee studies, as one field of interest, intensive tropical horticulture.

At Irriri village, Kolombangera Island of the Western Solomons, subsistence slash/burn garden practice on land that government advice suggested was agriculturally worthless, has changed to the establishment of more than six acres of permanent fruit and vegetable production. Above the narrow coastal plain, plantings ascend in diversity through tropical, subtropical and temperate zones.

The village people created priority groups for land selection, clearing, preparation, composting, propagation, planting, cultivating, irrigating, mulching, pest control, harvesting, seed collection, transportation and marketing. Jungle clearing is followed by levelling rough features, digging contour drains and establishing fruit and vegetables. On higher land, common contours feed vertical dams. In dry times, water is gravity-fed to crops. The cultured ecosystem occupies tilthed beds. The lower beds provide for disease-resistant crops and the higher beds accommodate disease-prone crops. Crop nutrition is provided by composted matter, seaweed, crushed coral rock and guano.

Today Irriri serves as a teaching model for the S.W. Pacific. The village accommodates apprentices who take the new skills to their home area and cultivate self-determination and awareness among their people.

25

AN ASSESSMENT OF THE ARK PROJECT AS A MODEL FOR ECOLOGICAL FOOD PRODUCTION

K. S. CLOUGH and J. CARRINGTON
The Ark Project, RR 4, Souris, Prince Edward Island, Canada

The Ark Project is a small group of people involved in research and demonstration of the application of biological principles to low-energy food production. Located on a coastal site in Maritime Canada, the project consists of outdoor gardens, a solar greenhouse and an aquaculture unit which, for the last 3 years, have been developing towards an integrated and sustainable food production system.

By examining elements of this system, such as nutrient cycling, energy use and diversity, an attempt has been made to assess the progress towards this goal. The interdependence of horticulture and aquaculture has been examined; the functioning of these systems has been described in reference to certain characteristics of natural ecosystems and compared with conventional vegetable and fish production in North America.

Individual techniques have already proved useful in practical application to commercial growers interested in the transition to a sustainable approach. However, the Ark Project as a whole can by no means be considered as a blueprint for others to follow: it is a demonstration of the application of biological principles from which individual techniques can be taken and applied elsewhere.

V
COMPARATIVE STUDIES

Introduction

Organic farming and horticulture movements during the first half of this century were supported by enthusiasts who saw in them better ways of living, better treatment of land and soils, and the production of better foods. To many supporters the results of these methods were self-evident and satisfactory, and needed no proof. Others saw the need both to demonstrate and to prove superiority, but their demonstrations failed to convince, and their experiments lacked authority. It became easy to write off the enthusiasts as cranks, unable to produce critical evidence for their claims in a field where science and scientific evidence were assuming increasing importance.

Biological husbandry today has its enthusiasts, many of them as convinced and unconcerned with proof as their forebears: others, however, recognize that the success of their methods should be measurable by the standards used to measure success in other kinds of husbandry. Is biological husbandry cost-effective? Are its yields adequate? Are its products superior to those of orthodox agriculture—do they taste better, store better, contain more nutrients, resist diseases more effectively . . .? These are the simple questions, and there are many more complex ones to be asked before scientists, farmers, economists and others concerned with evaluating methods of agriculture are satisfied.

The four chapters in this final section examine some of these points. Dr Hodges' paper reviews evidence from a wide field concerning supposed differences in quality between crops of organic and nonorganic origin. Dr P. N. Wilson and Mr T. D. A. Brigstocke analyze patterns of energy consumption in different systems of agriculture and apply their conclusions to biological husbandry. M S. Bellon and M J.-P Tranchant report on an analytical survey of four small mixed farms practising biological husbandry in France: their preliminary results suggest the need for more studies of this kind, using techniques of analysis already applied to orthodox farms. Finally, M D. Lairon and his colleagues report briefly on analyses of fruit and vegetables produced by organic and nonorganic methods: again, their work can at this stage do little more than point the need for many further studies, using the critical analytical techniques now available in well-equipped laboratories.

26

A QUALITATIVE COMPARISON BETWEEN CONVENTIONAL AND BIOLOGICAL AGRICULTURE

R. D. HODGES
Wye College (University of London), Ashford, Kent, UK

Introduction

It has long been accepted by those who practise biological husbandry that biological methods normally produce crops which are in some way qualitatively superior to crops produced by conventional agricultural methods. This acceptance has largely developed as the result of the wide experience of practising farmers and not on the basis of extensive research results. Because of the relative lack of factual evidence to support this claim, many agriculturalists and others have rejected the possibility of any qualitative advantage for biological produce (Allison, 1973; Leverton, 1973; Deutsche, 1974; Food and Nutrition Board, 1976). It is the intention of this chapter to try to draw together some of the evidence related to the qualitative effects of the two different husbandry systems and to see whether any conclusions can be drawn from this evidence.

Because the term quality can have such a wide range of meaning, it is essential to define precisely what is meant by quality in the present context. For many years one of the principal objectives of agricultural research has been to obtain maximum production, with only secondary attention being paid to developing crop quality (Davidescu, 1974; Beeson and Matrone, 1976; Steyn, 1977; Gabelman and Peters, 1979), even though it is well known that high yields usually do not result in high quality at the same time (Davidescu, 1974). Thus profitability tends to take precedence over quality (Bussler, 1974). Apart from increased yield, plant breeders have tended to concentrate their efforts on factors such as ease of mechanical handling, increased pest and disease resistance, improved transport or storage properties, and desirable marketing properties (Steyn, 1977; Gabelman and Peters, 1979). In comparison, little has been done to improve the nutritional quality. Most current quality standards are of a visual or 'cosmetic' type which provide a good appearance for marketing purposes. For example, the EEC standards of quality in agricultural produce are designed mainly to facilitate trade and marketing across borders (Schuphan, 1974b), and are largely based on visual characters such as size, colour and freedom from blemishes.

However, the concept of quality has much wider and much more important aspects than mere external appearances. Probably the most important aspect is that of 'inner' or nutritional quality, because people are primarily dependent for the continuing health and satisfactory functioning of their bodies on the quality of the food they consume, only secondarily on the quantity of the food, and not at all on the external appearance of the food, unless the latter is directly correlated with the nutritional quality. Davidescu (1974) has drawn attention to the relationship between the soil, food quality and health, as follows:

'The health of an individual, of a nation, depends largely on the variety of the products they live on, on their biological quality which, of course, is closely connected with the soil where they grow and with its state of fertility.'

Again, when evaluating the quality of an agricultural product intended for human food, Davidescu (1974) suggests that the idea of quality must include three types of characteristics, all of which are related to nutrition:

'a) the quality, from the point of view of the biological value, that is to say of the nutritional value, according to the content of the elements with high biological value;
b) the quality from the organoleptic point of view: taste, smell, colour, consistency, etc.;
c) the hygienic quality including both the absence of residues of toxic chemical products and pathogenic bacteria, and the absence of certain natural products with toxic character or inhibitors of certain physiological processes.'

And he adds: 'The quality of a plant product should never be judged only according to its exterior aspect and certain physical properties, but also according to its content of active biochemical elements, which assure the nutritional value of the product.'

Thus, from the point of view of human health and welfare, the most important criterion to be used when judging the quality of food produced seems to be a nutritional one—the ability of the food to provide a balanced intake of nutrients and not to provide or induce any nutritional imbalance or toxic substances. In the present context, therefore, the term food quality will be used primarily to refer to the ability of a crop to provide the consumer with food which enhances good health and development. Food of reduced or poor quality, on the other hand, is considered to reduce or detract from the good health and development of the consumer.

There are many factors that can act on a growing crop which may influence the nutritional quality of the final product. Among these are factors such as temperature, rainfall, insolation, soil type, altitude and aspect, and crop variety (Gray, 1959; Allaway, 1975). In any specific situation, however, most factors of this type will be common to whatever husbandry system is being used to grow the crop, and thus, in this discussion, only factors specific to either biological or conventional husbandry will be considered when assessing the potential effect of either system on the quality of food produced.

There are at least three ways in which one can consider the potential qualitative differences between the products of biological and conventional agriculture.

1. That both systems produce crops of variable but roughly equal quality and that there is no significant difference between them. If, however, one supports the organic view that this system does have a qualitative advantage over conventional agriculture then the following approaches seem to be the most likely alternatives.
2. Either that conventional farming produces crops which are of normal quality, and that biological farming produces crops which have something extra, qualitatively speaking.
3. Or that it is biological farming which produces crops of normal or optimal quality and that conventional agricultural techniques may result in produce of an inferior level of quality.

The hypothesis which is being considered in this paper is the third of these alternatives—that biological farming produces food of a good, but normal, quality and that conventional farming, on the other hand, produces food which is frequently in some way reduced or lacking in quality. However, in order to attempt a balanced account, at least some of the evidence in favour of all three of these approaches will be considered in subsequent sections.

The lack of any qualitative difference between biological and conventional systems

In recent years a number of studies have been conducted in which various aspects of nutritional quality were measured between plants grown by conventional and biological techniques, and which concluded that no significant quality differences were generated by either system. For example, Brandt and Beeson (1951) measured vitamin A and vitamin C levels in carrots, snap beans, potatoes and rye; Barker (1975) measured nitrogen levels in varieties of spinach; Michalik, Elkner and Fajkowska (1975) examined dry matter and vitamin C levels in tomatoes and sweet peppers; Svec, Thoroughgood and Hyo Chung (1976) examined six types of vegetables both by chemical analysis and organoleptic tests; and Nilsson (1979) measured a range of chemical values in carrots, cabbages and leeks. In the last case, Nilsson concluded that exclusive use of organic manures would not be likely to result in produce with a higher nutritive value, a more healthy chemical composition or better storage ability. In addition, several reviews (e.g. Allaway, 1975) have come to similar conclusions.

The qualitative superiority of a biological system

Over the period during which artificial fertilizers have gradually superseded organic manures as the main source of plant nutrients in agriculture and horticulture, there has been a continuing flow of reports supporting the qualitative superiority of organic produce. These reports have focused on a number of themes, such as better health and disease resistance, lower water content and thus increased nutrient density, better nutritional quality, better storage capacity, or better reproductive capacity in plants and/or animals produced by an organic system compared with the conventional system. Some of the more important examples of these reports are as follows.

McCarrison (1926) found that rats fed on a manured grain grew better and were more disease-resistant than those fed on grain produced with mineral fertilizers. In addition, the former grain was richer in vitamins than the latter.

Experiments carried out by Nath, Suryanarayana and McCarrison (1927) indicated that seed from a crop grown in manured soil was superior, particularly in its germination rate, to seed obtained from a similar crop grown on fertilized soil. These experiments also showed that the more composted the manure, the greater was the beneficial effect.

Rowlands and Wilkinson (1930), using groups of rats fed on a vitamin B-deficient diet, supplemented with seeds grown with either manure or fertilizers, found that the manured seeds promoted nearly twice the growth rate of the fertilized seeds.

Pfeiffer (1938) described a number of experimental tests comparing biodynamically grown produce with mineral fertilized produce; in all cases the former gave the best response. Among the tests used were: egg production, egg hatchability and egg spoilage in chickens; germination levels of wheat after heating to 100 °C for different times; food preferences of mice when offered a choice of types of grain; and increased viability and disease resistance in turkey poults.

In two books, Howard (1940, 1945) quotes many instances, largely based on observation and experience rather than experimentation, where organic methods gave an improved qualitative response in crops or livestock. Similar accounts have appeared in Balfour (1943, 1975).

It has been reported on several occasions that livestock require up to 15 per cent less of organically grown feeding stuffs, compared with conventional feeding stuffs, and that in spite of this they are frequently more productive (Hodges, 1977). For example, in the Haughley experiment (Balfour, 1975), the organic dairy herd consistently produced a higher milk yield on a lower ration of concentrates than did the conventional herd; similarly, the organic poultry flock needed less food per dozen eggs than did the conventional flock. One reason for this appears to be the higher dry-matter content frequently found in organic produce (*see* Hodges, 1977). A good example of this is found in an extensive, 12-year experiment comparing conventionally and biologically grown vegetables (Schuphan, 1974a). This demonstrated that, although productivity in the latter was, on average, 24 per cent down, the dry-matter content averaged 23 per cent more than in the conventionally produced vegetables.

Aehnelt and Hahn (1973) demonstrated the comparative effects of conventionally and organically fertilized fodder on the fertility of animals. In bulls given fodder from fertilized land, considerable reductions in semen quality were found, and these effects could be reversed or prevented by using fodder from manured pastures. Similar experiments in female rabbits, fed largely on vegetables produced by conventional or organic methods, showed detrimental changes in a number of fertility characteristics in those animals fed on the conventional produce.

The extensive work of Schuphan (1974a), mentioned above, also demonstrated differences in a wide range of 'nutritionally desirable' and 'undesirable' substances in conventional and organic vegetables, and the results generally favoured the latter.

In a qualitative comparison by Pettersson (1978) between conventionally and biodynamically grown potatoes, although the initial yield of bio-dynamic potatoes was lower than that of the conventional crop, the respective grading and storage losses were such that more biodynamic potatoes remained at the end of a 6-month storage period. Similarly, although the percentage crude protein was higher in the conventional potatoes, the more realistic quality measurements of true protein and Essential Amino Acid Index gave results more favourable to the bio-dynamic produce.

3. The qualitative inferiority of conventional agriculture

The evidence suggesting that products of conventional agriculture may be in some way qualitatively inferior to those of the biological system, is rather more extensive than the evidence supporting the previous section. There appear to be two main ways in which a quality reduction may arise in produce from conventional agriculture, and both of these are directly related to important differences between the two husbandry systems. First, there is the extensive use of soluble fertilizers, which may give rise to an imbalance in nutrients in the produce. Secondly, there is the use of pesticides and other toxic chemicals, which may leave residues in the crops. Because all pesticides are, by definition, toxic to various forms of plant and animal life and as many of them are also toxic to mammals, the use of such chemicals and the presence of their residues must always be at least a potential hazard to consumers of conventionally grown produce and thus must be considered a negative quality factor. Because of lack of space, only the first of these factors will be considered here.

In biological agriculture, nutrients lost from the soil in crops, etc. are generally replaced by the addition of manures and composts or by the use of relatively insoluble minerals such as rock phosphate. Thus, nutrients have to be cycled via the biological life of the soil before they become available to the crops to any extent, and even then they are not presented to the plants as considerable concentrations of soluble ions. Conventional agriculture, on the other hand, when making a fertilizer dressing, generally applies a moderate to large concentration of soluble nutrients to the soil in one application. It is generally accepted that there are 12 or 13 elements essential for normal plant growth and development (Mulder, 1953; Gilbert, 1957; Whitehead, 1966; McSheehy, 1971; Bussler, 1974). Both within the soil (Nielsen and Cunningham, 1962) and throughout the food chain from the soil to animals and man, these elements interact with each other, and any imbalance may profoundly affect the availability of essential elements to plants or animals, or the amount of these elements needed for normal growth or metabolism (Allaway, 1975). Although more than 100 years ago Liebig insisted that all nutrients removed from the soil by crops should be replaced (Bussler, 1974), modern conventional agriculture normally only replaces the four most important elements (nitrogen, phosphorus, potas-sium and calcium) and in doing so usually applies them in relatively heavy concentrations. This technique of fertilization may therefore give rise to soil imbalances in two ways:

1.   Interactions between large amounts of the major fertilizer elements and other major, minor or trace elements in the soil can result in an increase or decrease of the availability of some elements essential to crop growth (Schutte, 1964; Voisin, 1965). In the same way that soil pH changes can significantly alter the availability of a number of minor and trace elements, so an imbalance created by fertilizer application can make practically all the available quantity of a trace element 'disappear' (Voisin, 1965) even when there were originally adequate quantities of the element available in the soil. This reduction in availability can occur through a number of mechanisms, for example by chemical fixation or increased leaching. To quote Schutte (1964): 'Induced deficiencies are widespread and constitute a serious agricultural problem'.

2.   The increased productivity induced by extensive applications of nitrogen, phosphorus and potassium fertilizers may mean that, over a period of time, larger amounts of some trace elements may be removed in crops than can be replaced from the soil. Thus, over a number of seasons, deficiencies may develop which would not have occurred under a less intensive system, or under a system which replaces both major and minor elements.

Both these problems can be exacerbated by the trend in recent years for fertilizers to become more concentrated in their primary elements and thus to contain less impurities, and consequently less trace elements, than earlier formulations (Bussler, 1974). This is in contrast to organic manures, which contain a wide range of minor and trace elements as well as moderate amounts of the primary elements. There is thus a continuing tendency, under conventional agricultural systems using significant amounts of soluble fertilizers, for soil nutrient deficiencies to develop and increase. A good example of this tendency has been quoted by Takahashi, Kushizaki and Ogata (1977). In Japan, large numbers of dairy and beef cattle pastured on a relatively limited area of grassland (approximately 0.23 ha/head) have meant that heavy applications of fertilizers have been used to increase yields of herbage. The national survey carried out by Takahashi, Kushizaki and Ogata (1977) between 1971 and 1974, suggests that these intensive methods have resulted in extensive mineral deficiencies. Thus, in terms of plant mineral nutrition, 60 per cent of grasslands were deficient in potassium, 30 per cent in zinc, 50 per cent in copper, 20 per cent in boron, and 25 per cent were deficient in molybdenum. In addition, the percentages of grasslands where the nutrient content was below the recommended minimum for dairy cattle were: 50 per cent for phosphorus, 90 per cent for magnesium, 50 per cent for sodium, 50 per cent for iron, 85 per cent for zinc, 90 per cent for copper and 30 per cent for cobalt. Other examples of the increase of induced deficiencies have been quoted by Field (1956),Voisin (1963), Schutte (1964), Arnold (1967), McConaghy and McAllister (1967), Allaway (1968), Bussler (1974) and Van Nerum and Scheys (1976).

Deficiencies in plants, particularly of trace elements, may not show up as effects on yield but rather through changes in biochemical and physiological processes (Albrecht, 1958). Plants frequently show deficiency symptoms of most trace elements only when the concentration in the crop is well below normal (Wallace, 1961; Gauch, 1972). Thus, although a crop may be producing well and have no obvious deficiency symptoms, this does not mean that the plants contain enough of several elements to meet the dietary requirements of animals (Allaway, 1968). Bussler (1974) has categorized the process of development of a recognizable deficiency symptom, i.e. the external manifestation of a nutrient imbalance, into three phases. In summary, these are:

Phase A. Metabolic changes occur in the plant which can be detected only analytically. The pattern of mineral ions inside the plant cells is altered. As a result, changes in the pattern of organic constituents take place.

Phase B. Histological changes take place in the arrangement of organelles, cells and tissues. Long before the plant shows visible signs of damage, microscopic or submicroscopic changes have occurred.

Phase C. Symptoms of deficiency (chlorosis, necrosis, morphosis) become visible and reductions in productivity occur.

Phases A and B belong to the most common range of latent deficiencies which are seldom detected. The actual consequences of the consumption of this sort of product are known in only a few cases, mainly cases of stock nutrition, such as the desire to lick when a copper deficiency occurs, and tetany resulting from an imbalance of dietary magnesium (Bussler, 1974). Thus, for long periods before any symptom becomes visible, nonvisible disturbances may be acute, and the consumer, animal or man, will be receiving food which is nutritionally deficient. It is also possible that trace elements which are not essential for normal plant growth (Chapman, 1966), but are essential animal nutrients, such as cobalt, chromium and selenium, may be affected in this way. Allaway (1968) points out that '. . . it may be that a majority of the food and feed crops now produced contain too little chromium to meet the requirements of man and animals'.

A simplified scheme of the interactions, either stimulatory or inhibitory, that may occur between elements in the soil has been produced by Mulder (1953). Many specific examples of such interactions, between soil nutrients and applied fertilizer elements, mainly cases of inhibition, have been quoted by Voisin (1965) and Davidescu (1974). Thus increased nitrogen may cause a diminution of the assimilation of copper by plants; excess phosphorus can reduce the availability of zinc and copper; increased levels of potassium can inhibit the uptake of magnesium, calcium, sodium and boron; and additions of calcium may reduce the availability of manganese. One of the most widely investigated of these reactions is the inhibition of magnesium availability by applications of fertilizer potassium (*see* MAFF, 1967), and it is possible to trace the effects of this interaction in the soil through the forage produced (Kemp, 1971) to cattle feeding on affected pastures. Reduced serum magnesium levels in these animals can result in hypomagnesaemic tetany (*see* Voisin, 1963; Hodges, 1977). The opposite effect, stimulatory interactions between soil nutrients and applied fertilizers giving rise to nutritional problems, is exemplified by the relationship between phosphorus and molybdenum. Heavy applications of phosphate can significantly increase the molybdenum content of herbage and this can lead to copper deficiency in cattle grazing the pasture (Stout *et al.*, 1951).

Other aspects of quality reduction in relation to fertilizer application may be seen in a reduced disease resistance in plants and in an increased nitrogen content of crops. With regard to the reduction of disease resistance, Whitney (1976) considers it to be fairly generally accepted that high levels of fertilizer, particularly nitrogenous fertilizer, can increase the susceptibility of crops to some infections—for example, rusts, powdery mildews and blasts—but may have little effect on other diseases. Virus diseases may be made more serious by potassium and phosphate fertilizers. Chaboussou (1976) states that, in general, excess nitrogen increases the susceptibility of plants to fungus disease—for example, rust diseases, *Oidium* in wheat and vine, apple speckle, mildew and *Botrytis* on vine, and *Sclerotinia* on carrots. The basis for believing that plants provided with an abundant supply of nutrients, especially nitrogen, are more subject to aphid attack was provided by Markkula and Laurema (1967), who investigated the effect of amino acids and essential trace elements on the development of the pea aphid. Aphids use the mobile stream of soluble free amino acids and amides in the plant's vascular bundle as a protein source (Van Emden, 1972; 1973). Chaboussou (1978) supports the suggestion that susceptibility to pests and diseases is associated with increased concentrations of soluble substances in plants, and thus to the application of fertilizers. Chaboussou believes that organic manures give a much more balanced nutrient intake to the plant and result in fewer pests etc., while Schuphan (1974a) notes two factors which would favour a reduced incidence of aphid attack in organically fertilized crops: first, the decreased water content in the cellular tissue; secondly, the more solid collenchymatous thickening which increases the mechanical strength of the cell wall.

There is a wide range of research findings which indicate that resistance to disease organisms may be decreased in inorganically fertilized crops. For example, in cereals, Last (1953) found that wheat past its normal susceptible stage regained susceptibility to *Erysiphe graminis* when supplied with increasing amounts of nitrogenous fertilizer. Jenkyn (1976) has pointed out that the increased susceptibility of cereals to leaf pathogens as a result of increased nitrogen fertilizer necessitates the use of fungicides to obtain yields compatible with the nitrogen levels applied.

Van Nerum and Scheys (1976) drew the following conclusions from studies carried out in Belgium:

'We have been able to show that certain symptoms known to be due to microbial attack can also be caused by nutrient imbalance. Defective nutrition can render the plant more susceptible to microbial attack. Imbalanced nutrient solutions can weaken the plants . . . such plants are liable to be infected by parasites and perhaps saprophytes.'

This general conclusion is supported by Okamoto (1958), Reynolds (1974), Henis (1976), El-Fouly (1976), Ismunadji (1976), Kiraly (1976), Temiz (1976) and Trolldenier and Zehler (1976).

Specifically in the case of fruit production, defects such as cork spot and bitter pit in apples (Shear and Faust, 1971), as well as other defects and rots (Stoll, 1969) are directly related to a calcium deficiency attributable to excess of nitrogen. Magnesium deficiency has been ascribed to the application of fertilizer potassium (Boynton and Burrell, 1944), and Schultz (1976) relates this factor to an increase in susceptibility to rots.

In a detailed review of factors which influence the toxicity of food plants, Steyn (1977) considered that, of all the potential toxicants present naturally in human and animal foods, nitrate and nitrite constitute the greatest danger to health. In part, this is because their concentration in plants is markedly increased by various factors, particularly by the injudicious use of nitrogenous fertilizers applied on a large scale. With moderate to high applications of nitrogenous fertilizer, plants may accumulate considerable amounts of nitrate (Schuphan, 1974a; Lorenz, 1978; Maynard and Barker, 1979), sometimes up to 5 per cent of the dry weight (Allaway, 1975). In certain circumstances this accumulation of nitrogen may become a health hazard, particularly to human infants (Steyn, 1977) and to ruminants (Lorenz, 1978). The experiments of Schuphan (1974a) suggest that organically manured plants may have a lower nitrate content than mineral-fertilized plants. Increased nitrogen fertilization may not only reduce food quality by increasing the concentration of nitrate in plant tissues, but it may also detrimentally affect other quality aspects. Schuphan (1958) has shown with cabbage and spinach that, although crop yields and percentage crude protein may go on increasing up to quite high fertilizer nitrogen inputs, quality factors, such as the Essential Amino Acid Index and the percentage lysine in the crude protein, become reduced well before maximum productivity is attained. A similar situation with regard to fruit quality has been described by Stoll (1969).

Discussion and conclusions

It has been known for a long while that injudicious use of fertilizers can upset the mineral balance of growing plants, and numerous examples of the deleterious effects of specific major nutrients have been produced over the past five decades. Extensive reviews on the subject were produced 15–20 years ago by Voisin (1963, 1965) and Schutte (1964), and the present account has only reinforced their conclusions with some more up-to-date results. Voisin and Schutte, as well as other authors, have stressed Liebig's requirement that all nutrients removed in crops must be returned to the soil in order to maintain a balanced production. They have also stressed the need to supply extra elements in order to neutralize the effects of the major fertilizer elements added in excess. Nevertheless, conventional fertilizer practice generally only recommends the regular replacement of the four major elements, except where there is a specific deficiency. Although Voisin (1965) was a strong supporter of the need for mineral fertilizers, he was fully aware of their potential problems and was very outspoken against their careless and unbalanced use. From the results of investigations such as those of Takahashi, Kushizaki and Ogata (1977), and also from the continuing search by conventional agriculture for ever-greater yields, it seems likely that the situation has changed very little since Voisin reviewed the literature 20 years ago, and that induced deficiencies of minor and trace elements are no less widespread than when Schutte drew attention to them in 1964.

Superficially, it is difficult to explain the different results reported with regard to qualitative differences between biological and conventional systems: on the one hand there are clear-cut results which show little or no quality differences between crops grown organically or conventionally; on the other hand there are equally specific results which suggest a qualitative advantage of organic over conventional techniques. Close examination of the respective reports, however, suggests two possible explanations for these discrepancies. First, examination of the experiments which compared biological and conventional systems suggests that many of the so-called organic experiments were not set up correctly. For example, the addition of amounts of manure or compost to experimental plots or pots for one or two seasons does not necessarily result in true biological soil fertility. Alternatively, at least some of the most important results concerning the use of organic manures are from long-term experiments in which potential changes in soil fertility, either conventional or biological, had time to develop and stabilize. It is suggested that only from such stabilized conditions can valid comparative results be obtained, because the concept of biological soil fertility is totally different from the physicochemical approach of applying nutrients to plants via soluble fertilizers. For example, although the study by Nilsson (1979) showed that there were no consistent differences in dry-matter content between manured or fertilized vegetables, two very long-term experiments, the Haughley experiment which was begun in 1940 and Schuphan's 12-year experiments (1974a), did demonstrate a clear-cut increase in dry-matter content in organic produce. This concept, that valid results can be obtained only from 'mature' systems is supported by the work of Graf and Keller (1978, 1979) on biodynamic soils.

Secondly, whereas food quality is usually determined by chemical estimations of various nutrients, etc., in most of the investigations in the section on the qualitative superiority of a biological system, quality was assessed mainly by 'biological performance', usually by the growth and performance of livestock fed on the products of the different systems. Although simple growth rates are not necessarily a valid measure of nutritional superiority in diets, nevertheless measurements of aspects of biological performance are more likely to give an estimate of the true nutritional quality of a food rather than analytical measurements of a relatively few nutrients taken in isolation.

In conclusion, it is considered that there is much evidence to support the hypothesis that the use of NPK fertilizers can give rise to nutrient imbalances in the soil and crops, and that these imbalances are more likely to occur where large inputs of fertilizer are used to obtain maximum productivity. Such imbalances must clearly be considered as quality defects and these defects may manifest in a variety of ways. It is also considered that there is some evidence supporting the positive aspect of this hypothesis, that biological agriculture produces, or is more likely to produce, food of an optimum nutritional quality. However, this evidence is at present not strong enough to enable final conclusions to be drawn concerning the qualitative effects of biological agriculture. Further, detailed trials, based on similar principles to those of the Haughley experiment, must be performed before the full qualitative effects of the biological system can be ascertained.

References

AEHNELT, E. and HAHN, J. (1973). Fruchtbarkeit der Tiere—eine Möglich-keit zur biologischen Qualitätsprüfung von Futter- und Nahrungsmitteln? *Tierärztliche Umschau*, No. 4, 1–16

ALBRECHT, W. A. A. (1958). *Soil Fertility and Animal Health*. Iowa, USA; Fred Hahne Printing Co

ALLAWAY, W. H. (1968). Agronomic controls over the environmental cycling of trace elements. *Adv. Agron.*, **20**, 235–274

ALLAWAY, W. H. (1975). *The Effect of Soils and Fertilizers on Human and Animal Nutrition*. Agric. Inform. Bull. 378. Washington, DC; USDA

ALLISON, F. E. (1973). Soil organic matter and its role in crop production. In *Developments in Soil Science*, Vol. 3, p. 558. Amsterdam; Elsevier

ARNOLD, P. W. (1967). Magnesium and potassium supplying power of soils. In *MAFF Technical Bulletin No. 14*, pp. 39–48. London; HMSO

BALFOUR, E. B. (1943). *The Living Soil*. London; Faber & Faber

BALFOUR, E. B. (1975). *The Living Soil and the Haughley Experiment*. London; Faber and Faber

BARKER, A. V. (1975). Organic vs. inorganic nutrition and horticultural crop quality. *HortSci.*, **10**, 50–53

BEESON, K. and MATRONE, G. (1976). *The Soil Factor in Nutrition, Animal and Human*. New York; Marcel Dekker

BOYNTON, E. and BURRELL, A. B. (1944). Potassium-induced magnesium deficiency in the McIntosh apple. *Soil Sci.*, **58**, 441–454

BRANDT, C. S. and BEESON, K. C. (1951). Influence of organic fertilization on certain nutritive constituents of crops. *Soil Sci.*, **71**, 449–454

BUSSLER, W. (1974). Experimental contributions to the problem of improving the nutritional quality of food plants. In *Fertilizers, Crop Quality and Economy*, p. 503. Ed. V. H. Fernandez. Amsterdam; Elsevier

CHABOUSSOU, F. (1976). Cultural factors and the resistance of citrus plants to scale insects and mites. In *Proc. 12th Colloquium of the International Potash Institute*, pp. 259–280. Worblaufen—Bern, Switzerland; Int. Potash Inst

CHABOUSSOU, F. (1978). La resistance de la plante vis-à-vis de ses parasites. In *IFOAM International Conference, Sisaach, 1977*, pp. 56–59. Eds J-M Besson and H. Vogtmann. Aarau; Verlag Wirz

CHAPMAN, H. D. (1966). *Diagnostic Criteria for Plants and Soils*. Riverside; University of California

DAVIDESCU, D. (1974). Chemical fertilizers and crop quality. In *Fertilizers, Crop Quality and Economy*, pp. 1073–1103. Ed. V. H. Fernandez. Amsterdam; Elsevier

DEUTSCHE, R. (1974). Where you should be shopping for your family. *Nutrition Rev.*, **32**, 48–56

EL-FOULY, M. M. (1976). The effect of nitrogen fertilizers on growth of cereals and the impact on diseases. In *Proc. 12th Colloquium of the International Potash Institute*, pp. 69–76. Worblaufen—Bern, Switzerland; Int. Potash Inst

FIELD, H. I. (1956). Soil, herbage and animal. *Agric. Rev., Lond.*, **1**, 31–35

FOOD AND NUTRITION BOARD (1976). Soil fertility and the nutritive value of crops. *Nutrition Rev.*, **34**, 316–317

GABELMAN, W. H. and PETERS, S. (1979). Genetical and plant breeding possibilities for improving the quality of vegetables. *Acta Hort.*, **93**, 243–270

GAUCH, H. G. (1972). *Inorganic Plant Nutrition.* Strondsberg, Pa; Dowden, Hutchinson and Ross

GILBERT, F. A. (1957). *Mineral Nutrition and the Balance of Life.* Norman; University of Oklahoma Press

GRAF, U. and KELLER, E. R. (1978). Zusammenhänge zwischen kosmischen Konstellationen und dem Ertrag landwirtschalftlicher Kulturpflanzen auf konventionell und biologisch-biodynamisch bewirtschafteten Böden. 1. Feldversuche mit Kartoffeln. *Z. Acker-u PflBau*, **147**, 40–59

GRAF, U. and KELLER, E. R. (1979). Zusammenhänge zwischen kosmichen Konstellationen und dem Ertrag landwirtschaftlicher Kulturpflanzen auf konventionell und biologisch-biodynamisch bewirtschafteten Böden. II. Klimakammerversuche mit Radies. *Schweiz. Landw. Mhef.*, **57**, 325–336

GRAY, L. (1959). Factors affecting the nutritive value of foods of plant origin. In *Food. The Yearbook of Agriculture*, pp. 389–395. Washington DC; USDA

HENIS, Y. (1976). Effect of mineral nutrients on soil-borne pathogens and heat resistance. In *Proc. 12th Colloquium of the International Potash Institute*, pp. 101–112. Worblaufen—Bern, Switzerland; Int. Potash Inst

HODGES, R. D. (1977). Who needs granular inorganic fertilizers anyway? The case for biological agriculture. *Proc. Internat. Conf. on Granular Inorganic Fertilizers and their Production, November 1977*, Paper 18, pp. 224–244. London; British Sulphur Corpn

HOWARD, SIR A. (1940). *An Agricultural Testament.* London; Oxford University Press

HOWARD, SIR A. (1945). *Farming and Gardening for Health or Disease.* London; Faber and Faber

ISMUNADJI, M. (1976). Rice diseases and physiological disorders related to potassium deficiency. In *Proc. 12th Colloquium of the International Potash Institute*, p. 33. Worblaufen—Bern, Switzerland; Int. Potash Inst

JENKYN, J. F. (1976). Nitrogen and leaf diseases of spring barley. In *Proc. 12th Colloquium of the International Potash Institute*, pp. 119–128. Worblaufen—Bern, Switzerland; Int. Potash Inst

KEMP, A. (1971). The effects of potassium and nitrogen dressings on the mineral supply of grazing animals. In *Potassium and Systems of Grassland Farming. Colloquium No. 1*, pp. 79–92. Henley-on-Thames; The Potassium Institute Ltd

KIRALY, Z. (1976). Plant disease resistance as influenced by biochemical effects of nutrients in fertilizers. In *Proc. 12th Colloquium of the International Potash Institute*, pp. 46–57. Worblaufen—Bern, Switzerland; Int. Potash Inst

LAST, F. T. (1953). Some effects of temperature and nitrogen supply on wheat powdery mildew. *Ann. appl. Biol.*, **40**, 313–322

LEVERTON, R. M. (1973). Nutritive value of organically grown foods. *Comm. J. Amer. Dietet. Assoc.*, **62**, 501

LORENZ, O. A. (1978). Potential nitrate levels in edible plant parts. In *Nitrogen in the Environment*, Vol. 2, pp. 201–219. Eds D. R. Nielsen and J. G. MacDonald. New York; Academic Press

McCARRISON, R. (1926). The effect of manurial conditions on the nutritive and vitamin values of millet and wheat. *J. Indian Med. Res.*, **14**, 351

McCONAGHY, S. and McALLISTER, J. S. V. (1967). The determination in soils of potassium and magnesium and their uptake by crops. In *MAFF Technical Bulletin No. 14*, pp. 63–77. London; HMSO

McSHEEHY, T. W. (1971). Is there any correlation between health of a population and the method of husbandry used to produce its food? In *Just Consequences*, p. 151. Ed. R. Waller. London; Charles Knight

MAFF (1967). *Soil Potassium and Magnesium*. Technical Bulletin No. 14. London; HMSO

MARKKULA, M. and LAUREMA, S. (1967). The effect of amino acids, vitamins and trace elements on the development of *Acyrthosiphon pisum* Harris (Hom., Aphididae). *Ann. Agric. Fenn.*, **6**, 77–80

MAYNARD, D. N. and BARKER, A. V. (1979). Regulation of nitrate accumulation in vegetables. *Acta Hort.*, **93**, 153–162

MICHALIK, H., ELKNER, K. and FAJKOWSKA, H. (1975). Quoted by NILSSON (1979)

MULDER, D. (1953). Les elements mineur en culture fruitière. *1 convegno nazionale di Frutticoltura, Montana di St. Vincent*, pp. 188–198. Quoted by McSHEEHY (1971)

NATH, B.V., SURYANARAYANA, M. and McCARRISON, R. (1927). The effect of manuring a crop on the vegetative and reproductive capacity of the seed. *Mem. Dept. Agric. India*, **9**, 85–124

NIELSEN, K. F. and CUNNINGHAM, R. K. (1962). *Rothamsted Experimental Station Annual Report*, p. 65

NILSSON, T. (1979). Yield, storage ability, quality and chemical composition of carrot, cabbage and leek at conventional and organic fertilizing. *Acta Hort.*, **93**, 209–223

OKAMOTO, H. (1958). Relation between rice blast (*Pyricularia oryzae*) and potassium. In *Second Japanese Potassium Symposium, Tokyo*, pp. 76–89. Worblaufen—Bern, Switzerland; International Potash Institute

PETTERSSON, B. D. (1978). A comparison between the conventional and biodynamic farming systems as indicated by yields and quality. In *IFOAM International Conference, Sissach, Switzerland*, p. 105. Aarau; Verlag Wirz

PFEIFFER, E. (1938). *Bio-Dynamic Farming and Gardening*. New York; Anthroposophic Press

REYNOLDS, S. G. (1974). *The Effect of Compost plus Fertilizer on Vegetable Yields and Soil Fertility*. Unpublished work

ROWLANDS, M. J. and WILKINSON, B. (1930). Vitamin B content of grass seeds in relation to manures. *Biochem. J.*, **24**, 199–204

SCHULTZ, F. H. (1976). Effect of potassium on the catabolism of rot-infected apple fruit callus. In *Proc. 12th Colloquium of the International Potash Institute*, pp. 61–68. Worblaufen—Bern, Switzerland; Int. Potash Inst

SCHUPHAN, W. (1958). Quoted by VOISIN (1965)

SCHUPHAN, W. (1974a). Nutritional value of crops as influenced by organic and inorganic fertilizer treatments. *Qual. Plant. -Pl. Fds. hum. Nutr.*, **23**, 333

SCHUPHAN, W. (1974b). Experimental contributions to the problem of improving the nutritional quality of food plants. *Qual. Plant. -Pl. Fds. hum. Nutr.*, **24**, 1–18

SCHUTTE, K. H. (1964). *The Biology of the Trace Elements. (Their Role in Nutrition)*. Philadelphia; Lippincott

SHEAR, C. B. and FAUST, M. (1971). Value of various tissue analyses in determining the calcium status of the apple tree and fruit. In *Recent Advances in Plant Nutrition*, vol. 1, pp. 75–95. Ed. R. M. Samish. New York; Gordon and Breach

STEYN, D. G. (1977). *Modern Trends in Methods of Food Production, Food Processing and Food Preparation which constitute a Potential Hazard to Human and Animal Health*. Technical Communication No. 136. Republic of South Africa; Department of Agricultural Technical Services

STOLL, K. (1969). Höchsterträge und Qualitätserzeugung bei Obst und Gemüse als Düngungsproblem. *Qual. Plant. Mater. Veg.*, **18**, 206–226

STOUT, P. R., MEAGHER, W. R., PEARSON,G. A. and JOHNSON, C. M. (1951). Molybdenum nutrition of crop plants. I. The influence of phosphate and sulphate on the absorption of molybdenum from soils and solution cultures. *Pl. Soil*, **3**, 51–87

SVEC, L. V., THOROUGHGOOD, C. A. and HYO CHUNG, S. M. (1976). Chemical evaluation of vegetables grown with conventional or organic soil amendments. *Comm. Soil Sci. Plant Anal.*, **7**, 213–228

TAKAHASHI, T., KUSHIZAKI, M and OGATA, T. (1977). Mineral composition of Japanese grassland under heavy use of fertilizers—A review of two recent cooperative works. In *Proc. Internat. Symposium on Soil Environment and Fertility Management*, pp. 118–125. Japan; Soc. Science of Soil and Manure

TEMIZ, K. (1976). Interaction of fertilizers with *Septoria* leaf blotch of wheat. In *Proc. 12th Colloquium of the International Potash Institute*, pp. 129–132. Worblaufen—Bern, Switzerland; Int. Potash Inst

TROLLDENIER, G. and ZEHLER, E. (1976). Relationships between plant nutrition and rice disease. In *Proc. 12th Colloquium of the International Potash Institute*, pp. 83–93. Worblaufen—Bern, Switzerland; Int. Potash Inst

VAN EMDEN, H. F. (1972). Aphids as phytochemists. *Ann. Proc. Phytochem. Soc.*, No. 8, 25–43

VAN EMDEN, H. F. (1973). Aphid host plant relationships. Some recent studies. In *Perspectives in Aphid Biology*, pp. 54–64. Ed. A. D. Lowe. Bulletin No. 2., The Entomological Society of New Zealand

VAN NERUM, K. and SCHEYS, G. (1976). The nutrient status of the soil and the appearance of symptoms of microbial disease. In *Proc. 12th Colloquium of the Int. Potash Institute*, pp. 133–140. Worblaufen—Bern, Switzerland; Int. Potash Inst

VOISIN, A. (1963). *Grass Tetany*. London; Crosby Lockwood

VOISIN, A. (1965). *Fertilizer Application: Soil, Grass and Animal*. London; Crosby Lockwood

WALLACE, I. (1961). *The Diagnosis of Mineral Deficiencies in Plants by Visual Symptoms: A Colour Atlas and Guide*. London; HMSO

WHITEHEAD, D. C. (1966). *Nutrient Minerals in Grassland Herbage*. Farnham Commonwealth Agricultural Bureaux of Pastures and Field Crops

WHITNEY, P. J. (1976). *Microbial Plant Pathology*. London; Hutchinson

27

ENERGY UTILIZATION IN ORTHODOX AND BIOLOGICAL AGRICULTURE: A COMPARISON

P. N. WILSON and T. D. A. BRIGSTOCKE
BOCM SILCOCK Ltd., Basing View, Basingstoke, Hants, UK

Introduction

During the post-war period there have been vast changes in the pattern and productivity of agriculture in developing countries, achieved mainly through the replacement of manual labour by mechanization. Coupled with the increasing use of artificial fertilizers and herbicides, this has resulted in an increased dependence on fossil fuels, the chief among them being oil, which is used both directly for self-propelled machinery and also indirectly in the manufacture of machinery, fertilizers and plant protection chemicals. In consequence, food production systems are now very vulnerable to changes in the price and availability of oil. It is therefore essential that agriculture, like every other industry, should reconsider its dependence on energy consumption. This will become increasingly difficult as more complex industrial technology is involved (Wilson and Brigstocke, 1980). Kenwood (1976) believed that it was important to create a wide portfolio of energy research and development in order to retain maximum flexibility. At present, however, a high energy use still correlates closely with a high gross national product (Cook, 1971).

Oil has undoubtedly increased agricultural output per unit of land, which in turn has been accompanied by a reduction in the use of biological processes. This trend has led to higher levels of economic if not energetic efficiency and has, indeed, been designed to give greater control over the agricultural production system, but it does depend on large inputs of energy, notably oil. If we have to manage with less oil or other similar forms of easily transportable energy, it seems likely that agriculture must be made to depend more on biological processes, which are directly or indirectly solar-powered. This will be particularly so in small-scale farming, on which so much of the world population depends, and where a direct contribution to the alleviation of hunger is possible. What is required in these situations is not oil or an equivalent fuel but appropriate biotechnology—plants and animals that fit the needs of small farmers and small plots (Spedding, 1978).

Energy methodology

Energy use has recently become a closely studied factor in agricultural production systems. All manufacturing processes consume energy directly as fuel and also use energy indirectly in the provision of basic materials. The technique of energy analysis assesses all the energy inputs involved in the manufacturing process, including the energy used to make the manufacturing plant (Lewis and Tatchell, 1979). The sum of the energy requirements for the whole hierarchy of activities supporting a productive process such as agriculture has been termed the Gross Energy Requirement (GER) (IFIAS, 1974). Gifford (1976) believes that the details of the interactions between fossil-fuel burning and incoming solar energy capture in agriculture are known only vaguely, and satisfactory comparisons are not yet possible. Consequently a large number of studies of different agricultural systems have examined the energy ratio (ER) between the energy content of the food output from the system and the gross energy requirement for its production (Leach, 1976a). Many energy ratios have now been published, but Leach (1976b) has warned that because of differences in methodology, the use of variable assumptions and a changing range of inputs covered, the comparisons made by different authors can be misleading. In spite of this limitation, the differences in E ratio between temperate and tropical agricultural systems often reach several orders of magnitude, and so the comparisons of energetic efficiency are likely to be statistically valid.

Table 27.1   ESTIMATES OF AGRICULTURAL USE OF SUPPORT ENERGY (FROM WHITE, 1975)

Commodity or product	Energy input or support energy (GJ/(ha.y))	Energy output or ME (GJ/(ha.y))	ER $\left(\dfrac{Col.\ 3}{Col.\ 2}\right)$	Prtein output (kg/(ha.y))	Energy input to produce protein (MJ/kg)
Wheat	19.6	61.0	3.11	435	45
Barley	18.1	60.6	3.36	310	58
Potatoes	52.0	69.3	1.33	460	113
Carrots	25.1	32.5	1.30	234	107
Brussels sprouts	32.4	10.9	0.34	296	109
Tomatoes (glasshouse)	1300.0	62.0	0.05	945	1360
Milk	17.0	12.0	0.70	145	118
Beef (from beef herd)	10.6	2.4	0.23	31	348
Pigs (pork and bacon)	18.0	11.4	0.63	76	238
Sheep (lamb and mutton)	10.1	2.5	0.25	22	465
Poultry (eggs)	22.5	6.0	0.26	113	200
Poultry (broilers)	29.4	4.3	0.15	145	203

Table 27.1 presents data for the energetic efficiencies of the production of some temperate agricultural products up to the farm gate. *Table 27.2* reveals a very different picture when the efficiency of support energy used in less-developed tropical agricultural systems is quantified. The very favourable E ratios achieved with cassava, maize and bananas should be particularly noted. The comparison indicates that extensive, more primitive

Table 27.2   EFFICIENCY OF SUPPORT ENERGY
USE IN LESS-DEVELOPED AGRICULTURAL
SYSTEMS (SPEDDING AND WALSINGHAM, 1975)

Product	Location	E ratio
Upland rice	Gambia	3
Sweet potatoes	Zambia	4–9
Cowpeas	Zambia	9
Sugar-cane	Mauritius	11
Sweet potatoes	African rain forest	16
Rice	Fiji	20
Dryland rice	Sarawak	34
Maize	Zambia	40
Cassava	Fiji	71
Banana	Fiji	130

$$E \text{ ratio} = \frac{\text{Gross energy in food end-product}}{\text{Support energy input}}$$

systems of agricultural production are very efficient energetically, and can produce human food at a fraction of the energy input cost of more sophisticated systems. The other important feature to note from these two Tables is the very large variation in energy conversion efficiency between crop and livestock systems. This is a point which will be referred to later in this chapter. In hunter–gatherer and subsistence farming systems, the energy content of the feed output exceeds the muscular energy input from the workers by a factor of between 10 and 30. In modern agricultural systems, when Metabolizable Energy (ME) input of the feed is supplemented with fossil-fuel energy, the E ratio may be considerably lower and may even reach unity. However, Gifford (1976) has questioned the practical usefulness of such a ratio for four reasons:

1.   It may not be applicable to add together different forms of energy, such as the heat contents of fossil fuel, muscle power and the ME of feed or food.
2.   It may not be reasonable to relate the GER to the energy content of the output. A more appropriate way to express energetic efficiency might be to compare the minimum theoretical energy requirement to produce the product (within specific boundary conditions) with the actual energy being used.
3.   A comparison of different products and different systems on an E ratio basis may not be realistic. If support energy inputs are relatively cheap, then an E ratio of 1 is not necessarily technically wrong or uneconomic.
4.   De Wit (1975) believed that it is not meaningful to study energy inputs except in conjunction with labour inputs since the two are substitutable.

Historical perspectives

Othmer (1970) has estimated that in 1860 only 5 per cent of the energy used in the world came from fossil fuels; the remaining 95 per cent came from the muscular power of man and draught animals. However the 1950s and 1960s were the decades of cheap and assured fuel supplies, when oil cost about 1.5 $ US per barrel (198 litres), equivalent to having one human 'energy slave' working for 4000 hours for a US dollar (4000 MJ/$) (Leach, 1976a). Thus it was understandable that agricultural strategies in developed countries increased yield and reduced manpower by increasing fertilizer usage and by more intensive use of machinery (Wilson and Brigstocke, 1977). Indeed, during the period 1956–70, UK fertilizer inputs increased dramatically. The use of nitrogen increased 2.5 times, phosphate by one-third and potash by one-third (Cooke, 1971). During a similar period, White (1976) noted that increased fertilizer usage produced increasing yields and enhanced outputs of ME.

Up to 1950 the number of horses in the UK exceeded the number of tractors, and it was not until 1962 that the latter began to exceed the number of full-time agricultural workers (MAFF, 1968). Some workers have suggested that horses may again begin to replace tractors, as fuel costs rise. However, Stansfield *et al.* (1974) dismissed the idea of reintroducing draught animals, noting that when horses provided motive power in British agriculture, one-tenth of the total farm acreage was used to feed them.

Table 27.3 A HISTORICAL PERSPECTIVE OF THE ENERGETICS OF CORN PRODUCTION (OSBOURN, 1976)

Production figures	*System of corn production*		
	One man and his hoe	*One man and two horses*	*One man and his tractor in USA 1970[a]*
Area cultivated (ha)	0.5	10	40
Annual grain yield (t/ha)	2	2	5
Total annual output of grain			
tonnes per man	1.0	20	203
MJ × 10³	16.6	332	3378
Total annual input (MJ × 10³)			
food for man and animals[b]	4.5	134	4.5
fossil fuels	—	—	1198
Surplus food MJ × 10³	12.1	198	3373
man years	2.7	44	749
Ratio MJ output/MJ input	3.68	2.48	2.82
No. men sustained with food/ha	7.4	4.5	18.7

[a] Data from Pimentel *et al.* (1973)
[b] Data from Brody (1945)

Table 27.3 sets out in historical perspective the energetics of corn production (Osbourn, 1976). It is instructive to compare an area of corn cultivated with the use of machines, with a similar area cultivated by hand, in terms of the number of men theoretically sustained per cereal hectare. Over twice as many people can be fed when corn is produced mechanically, and it is for these reasons that the trend towards more mechanization is continuing, despite the increase in energy costs. Thus in the USA, 75 per cent of all inputs were purchased off the farm in the form of 'support energy' in 1970, compared with 25 per cent in 1910 (USDA, 1975). Shannon (1977) reported that the proportion of tractors over 100 hp (74 600 W) sold in the US had increased from 2 per cent in 1965 to 46 per cent in 1976.

It follows that, while agriculture has become more efficient techno-logically, it has become less efficient in its use of energy. Indeed, Black (1971) has pointed out that modern methods of agriculture, which may appear much more productive than primitive methods, are probably very similar in terms of the efficiency with which the total energy resources are used.

Stanhill (1977) has quantified the example of nineteenth-century Paris as a very successful closed urban agroecosystem. One-sixth of the total Parisian city area was used to produce more than 1 Mt of high-value, out-of-season, salad crops annually. This cropping system was sustained by the use of approximately 1 Mt of stable manure produced each year by the horses which powered the city's transport system. This 'marais system' not only provided a profitable solution to the problem of waste disposal, but also yielded high-quality crops, albeit with a very low E ratio of 0.25. In fact this does not compare very favourably with present-day UK allotment gardens (Leach, 1976a). However, with UK winter lettuce crops grown in heated glasshouses, the energy input for a single crop was 20–30 times more than the yearly requirement of the marais, and the food energy produced was only one-fifth. The real energetic significance of the marais system was that the inputs were largely of a biological origin, representing a renewable resource, in marked contrast to modern industrialized food production systems where energy inputs are dependent on nonrenewable fossil fuel sources (Stanhill, 1977).

Modern industrial agricultural systems

Food production can be a very energy-intensive process in which certain input substitutions have been made, such as mechanization to reduce manual labour, and fertilizer application to reduce the need for organic manure. Numerous authors (e.g. Pimentel *et al.*, 1973; Hirst, 1974; Steinhart and Steinhart, 1974) have queried the wisdom of the energy-intensive approach adopted by most developed nations. On the other hand, Ruttan (1975) believed that the data of Pimentel *et al.* (1973) indicated that, if anything, US corn producers were using less than the optimum amount of energy input per unit of corn produced. However this is not a point shared by all authors and Leach (1976b) is confident that 'advanced' farming systems, measuring only to the farm gate, are already running into difficulties because of the law of diminishing returns. To quote an example of a very energy-intensive food production system, it takes over 12 tonnes of oil-equivalent to land an edible tonne of 'luxury' shrimps (Leach, 1976a).

It is important to place the total energy consumption of the agricultural sector in perspective. UK agriculture uses less than 4 per cent of national power energy consumption (White, 1977). As the agricultural industry provides over 55 per cent of the raw food for the UK, it has a good claim on valuable energy resources even if these should become yet scarcer. However, when the food processing part of this cycle is taken into account, the situation appears very different. Several analyses (e.g. Gifford and Millington, 1973; Hirst, 1974) have been made of the total consumption of energy involved in the overall food production and processing system.

Table 27.4 compares energy consumption in the production and preparation of food as a percentage of total national energy consumption in three countries. There is a surprising similarity between the UK, Australia and the USA, but it will be noticed that agriculture requires less support energy input up to the farm gate than that required in the processing, distribution and preparation of food. Steinhart and Steinhart (1974) showed that, in the USA, the overall energy input was 2.2×10^{18} J in the farm sector compared with 3.5×10^{18} J in both the food-processing industries and the home.

Table 27.4 PRIMARY ENERGY CONSUMPTION IN THE PRODUCTION AND PREPARATION OF FOOD AS PERCENTAGES OF TOTAL NATIONAL ENERGY CONSUMPTION

Energy consumption in	Percentage of total national consumption		
	UK (White, 1977)	Australia (Gifford, 1973)	USA (ASAE, 1974)
Agriculture and fishing* (to farm gate)	3.9	2.1	2.2– 2.7
Food processing and distribution	7.0	5.8	6.2– 7.8
Food preparation (domestic)	4.9	7.1	3.6– 4.5
Total	15.8	15.0	12.0–15.0

*White (1977) does not include any allowance for fishing

These authors calculated the energy subsidy (primary energy input : food energy output) over several decades in the USA food production system and gave a figure of 8.8 for 1970. Blaxter (1977) calculated that the energy subsidy for the UK food production system was 9.6, using as his basis the gross (rather than metabolizable) food energy requirement of the population. Leach and Slesser (1973) quantified the energy use of the total US food system and their data are given in *Table 27.5*. The figures presented represent 12.8 per cent of all US energy used. The most striking feature is that agricultural energy inputs are small compared with the total food system use.

Many studies, notably those of White (1975), Heichel (1976) and Leach (1976a), have enabled comparisons to be made between a wide range of

Table 27.5 ENERGY USE WITHIN THE US FOOD SYSTEM (LEACH AND SLESSER, 1973)

Food production and energy use	Year			
	1940	1950	1960	1970
Agricultural production (% of total use)	18.2	26.7	25.9	24.2
Food processing and distribution (% of total use)	60.8	55.4	54.8	53.7
Domestic refrigeration and cooking (% of total use)	21.0	17.8	19.2	22.1
Total use (MJ $\times 10^{12}$)	2.87	4.75	6.03	9.10
Energy subsidies (MJ external input to produce 1 food MJ)	4.5	6.3	6.8	8.8

agricultural products. Other workers have studied in depth one individual production process, such as bread making (Leach, 1976a; Beech, 1980), meat production (Jacques and Blaxter, 1978), potato growing (Olabode, Standing and Chapman, 1977) and canned corn production (Brown and Batty, 1976). Beech (1980) found that in comparison with mashed potatoes, roast beef and reheated canned corn, standard white bread showed the lowest energy subsidy by a factor of at least 5. This finding suggests that bread is one of the most energy-efficient staple food products of an industrialized food-production system (*Table 27.6*).

Table 27.6 TOTAL PRIMARY ENERGY USED TO PRODUCE FOOD ITEMS IN THE FINISHED STATE, READY FOR CONSUMPTION (BEECH, 1980)

Process and energy	*Primary energy consumed* (MJ/kg *finished food*)			
	Standard bread (UK)	*Mashed potato (USA)*[a]	*Kernel corn (USA)*[b]	*Roast beef (UK)*
Agriculture	4.02[c]	6.51	4.15	79.95[e]
Processing, distribution and marketing	10.78[d]	5.80	27.53	23.45[f]
Domestic preparation	0	12.28	3.96	9.21[g]
Total primary energy input	14.80	24.59	35.64	112.61
Metabolizable food energy output (MJ/kg finished food)	9.91	3.11	2.49	10.35[h]
Primary energy input: food energy output	1.49	7.91	14.31	10.88

[a] Prepared from fresh potatoes; from Olabode *et al* (1977)
[b] A commercially canned product; from Brown and Batty (1976)
[c] Data from Leach (1976a)
[d] A combination of data from Beech (1980) and from Leach (1976a).
[e] Calculated from the figure given by White (1975) of 348 MJ/kg protein as beef from a beef herd, and assuming a protein content of 19.3% (w/w) for raw, lean beef (Long, 1961). Adjusted for cooking losses.
[f] Based on the figure given by Jacques and Blaxter (1978) of 19.7 MJ/kg edible meat slaughtered at an abbatoir and processed for sale in a retail butcher's shop. Adjusted for cooking losses.
[g] Supplied by British Gas Corporation, Research and Development Division (Barnard, J., private communication) and relates to the roasting of a beef topside joint (1.82 kg, raw) in a domestic gas oven. Sixteen per cent weight loss was incurred in cooking.
[h] Data from Long (1961).

Biological agriculture

The case for biological (organic) agriculture as a realistic and viable alternative to conventional agriculture has been considered (Hodges, 1978), and certainly in terms of support energy expenditure it is relatively attractive. Indeed Lockeretz *et al.* (1976) have shown that conventional agriculture can be about 2.3 times as energy-intensive as organic agriculture, while Crouau (1977) calculated the figure as 2.5 times.

Berardi (1978) compared organic and conventional wheat production and showed that the latter system averaged 48 per cent higher energy inputs but only 29 per cent higher yields of wheat. Furthermore, Berardi considered that the conventional agricultural economic-cost figures are inadequate when evaluating fossil fuel energy use in food production systems and the environmental impact of such use.

However, it is unwise to suggest a reversion of the trend from manpower to mechanization, and we believe that such a substitution would result in a reduction of farm productivity. Nevertheless, a number of workers disagree and Newcombe (1976) has compared the energy costs of vegetable production (*Brassica* spp) in Hong Kong in the late 1950s and 1970s (*Table 27.7*). It will be seen that the gross output of energy foods has hardly altered in the 15-year period. However, by comparing the energy input–output ratio of 1:0.23 for 1970s production with that for the 1950s, a fivefold

Table 27.7 THE ENERGY COSTS OF VEGETABLE PRODUCTION IN HONG KONG IN THE LATE 1950s AND THE 1970s (NEWCOMBE, 1976)

Energy balance	Inputs/(ha.y)	
	in 1950s	*in 1970s*
Input of labour (full-time labourer)	15	3
Fertilizers and insecticides (MJ × 10³)	17.0	141.1
Total support energy (MJ × 10³)	35.4	417.3
Gross output (MJ × 10³)	89.7	95.6
Ratio of energy input to output	1:1.23	1:0.23

reduction in overall energetic efficiency (pre-farm gate) is revealed. The argument that the substitution could be reversed is persuasive in such situations, where manpower is not limiting. However, the argument that this reversion is applicable to other modern agricultural systems is dubious. The advocates of organic farming are fully aware of this (e.g. Lockeretz *et al.*, 1978).

Nevertheless, the likelihood of further energy problems, and stricter environmental regulations for agriculture in the future, make it desirable to have alternative farming systems available which are not based on large inputs of artificial fertilizers, herbicides and pesticides. The USDA (1980) have produced a useful report on the potential for organic farming and have recommended a wide programme of research, noting that organic agriculture covers a wide spectrum of practices and philosophies, from the extreme where chemicals are not used under any circumstances, to more flexible systems where fertilizers, herbicides etc. are used selectively and sparingly.

However, a straight comparison between organic farmers and conventional management systems is not wholly satisfactory. Wernick and Lockeretz (1977) have noted that environmental factors may cause conventional farmers to make greater use of techniques that organic farmers are already using, such as making increased use of legumes and full use of livestock manure.

The extreme view is propounded by Chapman (1973) who argues for the abandonment of 'chemical farming' in the USA. This would result in such a reduced yield that much of the soil bank in the US would have to be brought back into production. However, Chapman calculated that output would fall by only 5 per cent and prices of farm products would increase 16 per cent. He claimed that incomes would rise by 25 per cent and so nearly all the farm subsidy programmes would disappear.

A large number of practical studies have been published, both in the UK and the USA, on intensive organic vegetable gardening. Jeavons (1979) has developed a method called the 'Biodynamic/French Intensive method' which, it is claimed, is considerably more productive than equivalent orthodox systems in California and has a very low support energy usage. However it is a labour-intensive gardening technique and it has yet to be demonstrated that this method could remain viable on a normal farm scale.

Less-developed agricultural systems

Leach (1974; 1976a,b) argued that less-intensive agricultural systems provide better food output to fossil-energy input ratios. *Table 27.8* provides a comparison between energy and protein outputs, energy ratios and energy inputs per unit protein output for selected agricultural systems. In primitive agricultural systems, where the only significant input is manual, an ER of

Table 27.8   ENERGY AND PROTEIN OUTPUTS, ENERGY RATIO AND ENERGY INPUTS PER UNIT PROTEIN OUTPUT FOR SELECTED AGRICULTURAL SYSTEMS (LEACH, 1976b)

Type of input and product		Annual energy output (MJ/ha)	Energy ratio (ER)	Annual protein output (kg/ha)	Energy inputs per unit protein output (MJ/kg)	Source
Low input						
1. Swamp rice, Dayak	S	28 100	20	62[a]	22.7[a]	Geddes, 1954
2. Rice, Philippines	S	21 000	17.3	46[a]	26.6[a]	Heichel, 1973
3. Taro-yam gardens, Tsembaga	S	23 300	16.4	93[a]	15.3[a]	Rappaport, 1968
4. Pigs in gardens, Tsembaga	S	3 100	2.1	62[a]	41[a]	Rappaport, 1968
Intermediate input						
Corn silage, Iowa, 1915	FG	66 100	5.9	—	—	Heichel, 1973
6. Corn grain, Iowa, 1915	FG	25 900	4.8	167[a]	32.6[a]	Heichel, 1973
7. Corn grain, USA, 1945	FG	35 400	3.7	222[a]	43.1[a]	Pimentel *et al.*, 1973
High input: crops						
8. Corn grain, USA, 1954	FG	42 700	2.67	268[a]	59.7[a]	Pimentel *et al.*, 1973
9. Corn grain, USA, 1970	FG	84 400	2.82	529[a]	56.6[a]	Pimentel *et al.*, 1973
10. Wheat, UK, 1970	FG	46 900	2.3	365	55	Leach, 1976b
11. Bread, UK, 1970	S	44 500	1.4	346	92	Leach, 1976b
12. Sugar beet + tops, UK, 1970	FG	115 100	4.2	n.a.	n.a.	Leach, 1976a
13. Sugar from beet, UK, 1970	(S)	82 900	0.67	n.a.	n.a.	Leach, 1976a
14. Maincrop potatoes, UK, 1970	FG	56 900	1.57	376	96	Leach, 1976a
High input: animal products						
15. Battery eggs, UK, 1970	FG	7 030	0.14	137	353	Leach, 1976b
16. Broiler poultry, UK, 1970	FG	5 870	0.10	203	290	Leach, 1976a
17. Milk, average, UK, 1970	FG	10 000	0.37	129	208	Leach, 1976a
18. Fish, all commercial vessels, UK, 1969	(S)	n.a.	0.05	n.a.	489	Leach, 1976a

[a] Not given in quoted source calculated from standard crop data
FG: to farmgate; S: to consumer or shop; (S): to factory gate (13) or dockside (18); n.a.: not applicable

up to 30 : 1 may be achieved. Lewis and Tatchell (1979) calculated that if primitive agricultural systems used UK labour, the ER would be reduced to 1 : 1 (compared with over 2 : 1 for cereals in intensive UK agriculture). Therefore, these authors concluded that primitive agricultural systems will not save energy unless the other components of living also become more primitive.

Norman (1978) reviewed energy inputs and outputs in hoe and draught animal subsistence farming systems in the tropics, following the first comparative survey of energy relationships for nonmechanized tropical crop production by Black (1971). This latter author found with several primitive systems that the ER falls between 3 and 34, with most crops in the 15–20 range. *Table 27.9* from Clark and Haswell (1970) shows a wide range

Table 27.9   ENERGY INPUTS AND OUTPUT/INPUT RATIOS FOR RAINFED HOE CULTIVATION CROPPING ENTERPRISES, AFRICA (ADAPTED FROM CLARK AND HASWELL, 1970)

Country	Crop	Net human energy input (MJ/ha)	Output/input ratio
(a) Cereals			
Ghana	Corn	488	16
Upper Volta	Corn	312	23
Nigeria	Corn	229	24
Gambia	Millet	408	7
Upper Volta	Millet	292	13
Uganda	Millet	615	8
Cameroon	Sorghum	525	22
East Africa	Sorghum	341	39
Gambia	Sorghum	124	16
Upper Volta	Sorghum	292	17
Mean		363	18.5 (unweighted)
(b) Non cereals			
Upper Volta	Sweet potato	439	21
East Africa	Cassava	772	20
Upper Volta	Cassava	195	54
Nigeria	Cassava	266	27
Ghana	Yams	1249	26
Upper Volta	Yams	443	9
Nigeria	Yams	814	44
Uganda	Banana	870	35
Mean		631	29.5 (unweighted)

of values, pointing to the difficulty in quantifying human energy inputs and crop production 'tasks'. Other ERs have been provided by the anthropologist Rappaport (1971) in his study of the Tsembega tribe in New Guinea, and may be computed from other sources (e.g. Ruthenberg, 1971). Chandra, Evenson and De Boer (1976), with data from Fiji, have given ERs from 71 to 130 for cassava, sweet potato, yam and banana cultivation.

Norman (1978) concluded that, in cropping systems where the major input was human energy, the production of noncereal carbohydrate crops was energetically more efficient than that of cereals. In these systems few, if

Table 27.10 ENERGY INPUTS IN CORN
PRODUCTION IN MEXICO USING ONLY MANPOWER
(PIMENTEL, 1977)

	Energy	
	Quantity/ha	kcal/ha
Input		
Labour	1144 h	—
Axe and hoe	—	16 500
Seeds	10.4 kg	36 608
Total		53 108
Corn yield	1944 kg	6 842 880
kcal return/kcal input		128.8
Protein yield	175 kg	

any, energy costs are involved in the processing, transporting and distribution parts of the food chain. Furthermore, human energy inputs,
particularly for crop irrigation, are difficult to quantify, as also are energy
inputs when draught animals are employed.

Pimentel (1977) has investigated slash and burn technology in Mexico.
His results are shown in *Table 27.10*. A total of 1144 h of labour was
required to raise a hectare of corn. The yield of 1944 kg/ha provided
sufficient food for about four people eating primarily as vegetarians. The
return of some 129 kcal of food energy per input kcal indicates just how
productive labour-intensive forms of agriculture can be.

Table 27.11 OVERALL ENERGY BUDGET FOR UK AGRICULTURE UP TO THE
FARM GATE (PJ PER ANNUM) (WHITE, 1976)

Input or energy subsidy		*Output or energy available to man*			
Solid fuels	4.1	Cereals	56.8	Arable crops	84.4
Petroleum	85.0	Potatoes	13.0		
Electricity	33.1	Sugar beet	14.6		
Fertilizers	83.5				
Machinery	52.0	Vegetables	2.4	Horticulture	3.2
Feedstuff processing		Fruit	0.8		
(off-farm)	51.3	Milk	38.0		
Chemicals	8.5	Beef	13.8		
Buildings	22.8	Pigs (pork and bacon)	17.3		
Transport distribution		Sheep (lamb and			
and services	16.3	mutton)	4.5	Livestock	80.6
Miscellaneous	4.3	Poultry (eggs)	4.5		
Imported feedstuffs	53.2	Poultry (chicken)	1.9		
		Poultry (turkey)	0.6		
Total	414.1		168.2		168.2

$$ER = \frac{\text{Output energy}}{\text{Input energy}} = \frac{168.2}{414.1} = 0.4$$

National agricultural systems

Energy flows in national agricultural systems can also be criticized as providing an incomplete picture, particularly as different authors use slightly different approaches. Nevertheless, work by Slesser (1973), Leach (1974; 1976a, b) and White (1975) should be studied for details of the evaluation of energy-accounting methodology. *Table 27.11* shows that the overall ER for British agriculture is 0.4 (White, 1976). Other authors, such as Leach (1974) and Blaxter (1974; 1975) have suggested similar figures. More recently, Noble (1980) has given an ER for UK agriculture of 0.87. In order to obtain the energy consumed in the 50 per cent of the unprocessed food that is produced in the UK, an energy input 2.5 times as great is required. This input is in the form of fossil fuels and imported feedstuffs (White, 1976).

Gifford (1976) has compared the overall energetics of five national agricultural systems for the mid to late 1960s. The difference between Australia and the other four countries is striking (*Table 27.12*). Indeed, Gifford concluded that 'Because Australia has not adopted a nitrogen and

Table 27.12   ENERGETICS OF FIVE NATIONAL AGRICULTURAL FOOD PRODUCTION SYSTEMS (GIFFORD, 1976)

Energy resources, requirements and output	USA	Australia	Country[1] UK	Holland	Israel
Areas ($\times 10^6$ ha):					
Cultivated and pastoral	530	490	19	2.3	1.2
Crops and fallow (incl. fruit and veg.)	184	20	4.5	1.0	0.35
High-grade pasture (incl. temporary)	?	21	8	1.3	0.06
Nutritional energy output ($\times 10^{15}$ J/y):					
Total	1750	270	135	90	9.9
as crop product	1260	230	70	60	8.4
as animal product	490	40	65	30	1.5
Gross energy requirements ($\times 10^{15}$ J/y) for:					
Direct farm fuel and electricity	1186	55	108	98	3.2
Fertilizer nitrogen	434	5.3	63	25	2.0
phosphorus	36	13	6.7	2.3	0.18
potassium	29	0.53	4.1	0.6	0.08
lime	?	?	8.4	?	?
Machinery and metal products	410	18	32		0.43
Chemicals	150	4.4	8.5	14	1.2
Irrigation	146	?	—		12.2
Other miscellaneous costs	—	1	68		0.2
Sum of inputs listed	2391	97	299	140	19.5
Energy ratio (Output/Input)	0.7	2.8	0.5	0.6	0.5

[1] Main sources for the table were: for USA, Steinhart and Steinhart (1974); for Australia, Gifford *et al.* (1975), Gifford and Millington (1975); for UK, Leach (1976a); for Holland, Dekkers *et al.* (1974); for Israel (pre-1967 borders), Stanhill (1974). All data apply to some period in the mid-to-late 1960s.

irrigation-intensive strategy and has not raised yarded cattle, it is in the fortunate position of being able to withstand increased fossil-fuel prices much better than the other countries reviewed.'

Revelle (1976) has evaluated energy use in rural India. At present, from an energy standpoint, rural India can be considered as a partly closed ecosystem, although rapid population growth is causing fundamental changes. In terms of coal equivalents, the annual energy expenditure corresponds to 0.346 t per capita, compared with 11.15 t per capita in the USA in 1970 (USDC 1972). More than 89 per cent of this energy was provided by the villagers themselves, and less than 11 per cent was from commercial sources, whereas the vast majority of USA energy was from fossil fuels and hydroelectric power.

The future

Numerous authors have discussed the large number of potentially energy-saving measures which could be adopted on farms in developed nations (*see* Wilson and Brigstocke, 1977; 1980). However, although technically significant, their adoption is likely to take place slowly because agriculture is able to absorb a very much greater support-energy input cost before it is forced to consider more dramatic alternatives. Thus, although the biological efficiency of food energy production by animals is low, the high economic cost of the energy inputs, including support energy, can always be met if a high demand results in a correspondingly high premium price for the animal products. Nevertheless, within the different systems of animal production, there would still be an economic incentive to reduce high-cost energy inputs wherever possible, without loss in productivity. Beef and lamb are more likely to be produced from uncroppable land, while poultry and pigmeat might be fed on higher proportions of waste or byproducts of arable farming, or from extraction processes based on very high-yielding forage. In this way the reliance of livestock on high-cost cereals as the main source of feed energy can be reduced (Wilson and Brigstocke, 1978; Wilson, Brigstocke and Cuthbert, 1981). The correlation between high-energy input in the agricultural sector and high gross of national product is likely to persist in developed countries at least until the end of the century. Indeed, Doering (1977) believed that there would be little change in the pattern of the USA food system before 2000 AD. This seems to be the likely energy pattern in most developed countries.

However, for developing countries different criteria apply. In these cases it is not so much a question of reversing the substitution process but of slowing down the substitution rate. Much will depend on the availability and relative costs of labour, draught power and fossil fuels for mechanically driven implements. Where labour is plentiful and not costly, the substitution process may be halted. Where, for political reasons, labour costs are forced up, some degree of partial substitution may take place, but it is unlikely to proceed as far as it has done in the developed world. There can be no general solution or simple answer to the question of optimal agricultural development strategy for developing countries. Each situation is unique and will demand an individual approach.

Similar arguments can be advanced in respect of biological agriculture. Such systems are already practised in many developing countries and will persist or change very slowly. In the developed countries there may be a slight shift towards biological agriculture, and away from so-called orthodox systems, but such changes will again be slow and the driving force will be the economic costs of chemical fertilizers, herbicides and pesticides, together with the constraints imposed by environmental considerations. Again, broad generalizations are dangerous; different nations will adopt various approaches according to the prevailing socioeconomic climate.

At the end of the day, man eats for greed and not so much for need, in the developed countries. The consequence is that where the demand for food is strong enough to pay the attendant costs, then food will tend to be produced by the method that produces most profit for the farmer. The longer-term considerations, which are the concern of the energy conservationist and the environmentalist, will exert an impact on agricultural production techniques only when they do so by legal or by economic constraints.

Moral and philosophical arguments are not very persuasive to the majority of producers, who are forced to regard agriculture as a business rather than a 'way of life'.

References

ASAE (1974). The Cast Report. *Agric. Engr.*, **55**, 19

BEECH, G. A. (1980). Energy use in bread making. *J. Sci. Fd. Agric.*, **31**, 289

BERARDI, G. M. (1978). Organic and conventional wheat production; examination of energy and economics. *Agro-Ecosystems*, **4**, 367

BLACK, J. N. (1971). Energy relations in crop production. *Ann. appl. Biol.*, **67**, 272

BLAXTER, K. L. (1974). The limits to agricultural improvement. *New Scientist*, **61**, 400

BLAXTER, K. L. (1975). The energetics of British agriculture. *Biologist*, **22**, 14

BLAXTER, K. L. (1977). Energy and other constraints on food production. *Proc. Nutr. Soc.*, **36**, 267

BRODY, S. (1945). *Bioenergetics and Growth with Special Reference to the Efficiency Complex and Domestic Animals*, 1964 Edition, p. 954. New York; Hafner. Publ. Co

BROWN, S. J. and BATTY, J. C. (1976). Energy allocation in the Food System; a micro scale view. *Trans. Am. Soc. agric. Engrs*, **19**, 758

CHANDRA, S., EVENSON, J. P. and DE BOER, A. J. (1976). Incorporating energetic measures in an analysis of crop production practices in Sigatoka Valley, Fiji. *Agric. Systems*, **1**, 301

CHAPMAN, D. (1973). An end to chemical farming. *Environment (St. Louis)*, **15**, 12

CLARK, C. and HASWELL, M. (1970). *The Economics of Subsistence Agriculture*, 4th Edition. London; Macmillan

COOK, E. (1971). The flow of energy in an industrial society. *Sci. Amer.*, **225**, 135

COOKE, G. W. (1971). Fertilisers in society. *Proc. Fertil. Soc.*, No. 121

CROUAU, M. (1977). *Comparative Study of Energy Consumption in Biological and Conventional Agriculture.* IFOAM Bulletin No. 20, p. 4

DEKKERS, W. A., LANCE, J. M. and DE WIT, C. T. (1974). Energy production and use in Dutch agriculture. *Neth. J. Agric. Sci.,* **22,** 107

DE WIT, C. T. (1975). Substitution of labour and energy in agriculture and options for growth. *Neth. J. Agric. Sci.,* **23,** 145

DOERING, O. C. (1977). Agriculture and energy use in the year 2000. *Amer. J. Agr. Econ.,* **59,** 1066

GEDDES, W. R. (1954). *The Land Dayaks of Sarawak.* Colonial Res. Study No. 14. London; HMSO

GIFFORD, R. M. (1973). Energy, food and agriculture. In *Annual Report of the Division of Plant Industries, CSIRO,* p. 19

GIFFORD, R. M. (1976). An overview of fuel used for crops and national agricultural systems. *Search,* **7,** 412

GIFFORD, R. M. and MILLINGTON, R. J. (1973). *Energetics of Agriculture and Food Production with special emphasis on the Australian Situation.* Adelaide; UNESCO Symposium on Energy and How We Live

GIFFORD, R. M. and MILLINGTON, R. J. (1975). *Energetics of Agriculture and Food Production.* CSIRO, Bull. 288

GIFFORD, R. M., KALMA, J. D., ASTON, A. R. and MILLINGTON, R. J. (1975). Biophysical constraints in Australian food production; Implications for population policy. *Search,* **6,** 212

HEICHEL, G. H. (1973). *Comparative Efficiency of Energy Use in Crop Production.* Bull. Conn. Agric. exp. Stn **739**

HEICHEL, G. H. (1976). Agricultural production and energy resources. *Amer. Sci.,* **64,** 64

HIRST, E. (1974). Food-related energy requirements. *Science, NY,* **184,** 134

HODGES, R. D. (1978). *Who Needs Inorganic Fertilisers Anyway? The Case for Biological Agriculture.* Review Paper Series, No. 1. Stowmarket, Suffolk; The International Institute of Biological Husbandry

IFIAS (1974). *Energy Analysis Workshop on Methodology and Conventions, Workshop report No. 6, 25–30 August 1974.* Sweden; Guldsmedshyttan

JACQUES, J. K. and BLAXTER, K. L. (1978). The use of support energy in the meat processing industry. *J. Sci. Fd. Agric.,* **29,** 172

JEAVONS, J. (1979). *How to Grow More Vegetables Than You Ever Thought Possible on Less Land Than You Can Imagine.* Palo Alto, California; Ecology Action

KENWARD, M. E. (1976). *Potential Energy.* London; Cambridge University Press

LEACH, G. (1974). The energy costs of food production. In *The Man–Food Equation,* pp. 139–163. Eds F. Steele and A. Bourne. London; Academic Press

LEACH, G. (1976a). *Energy in Food Production.* Guildford; IPC Science and Technology Press

LEACH, G. (1976b). Industrial energy in human food chains. In *Food Production and Consumption. The Efficiency of Human Food Chains and Nutrient Cycles,* p. 371. Eds A. N. Duckham, J. G. W. Jones and E. H. Roberts. Amsterdam; North-Holland

LEACH, G. and SLESSER, M. (1973). *Energy Equivalent of Network Inputs to Food Processing Processes.* Glasgow; Strathclyde University

LEWIS, D. A. and TATCHELL, J. A. (1979). Energy in UK Agriculture. *J. Sci. Fd. Agric.*, **30**, 449

LOCKERETZ, W., KLEPPER, R., COMMONER, B., GERTLER M., FAST, S. and O'LEARY, D. (1976). *Organic and Conventional Crop Production in the Corn Belt; a Comparison of Economic Performance and Energy Use for Selected Farms.* Report CBNS – AE – 7. Washington University, St Louis, Missouri; Centre for the Biology of Natural Systems

LOCKERETZ, W., SHEARER, G., KLEPPER, R. and SWEENEY, S. (1978). Field crop production on organic farms in the Midwest. *J. Soil and Wat. Conservation*, **33**, 130

LONG, C. (1961). *Biochemists' Handbook*, p. 1112. Ed. C. Long. London; E. and F. N. Spon Ltd

MAFF (1968). *A Century of Agricultural Statistics, GB 1866–1966.* London; HMSO

NEWCOMBE, K. (1976). The energetics of vegetable production in Asia, old and new. *Search*, **7**, 423

NOBLE, G. W. (1980). Farm management problems of the energy crisis. *ADAS Quart. Rev.*, **36**, 1

NORMAN, M. J. T. (1978). Energy inputs and outputs of subsistence cropping systems in the tropics. *Agro-ecosystems*, **4**, 355

OLABODE, H. A., STANDING, G. N. and CHAPMAN, P. A. (1977). Total energy to produce food serving as a function of processing and marketing modes. *J. Fd. Sci.*, **42**, 768

OSBOURN, D. F. (1976). Forage conservation and support energy use. In *Energy Use in British Agriculture*, pp. 29–34. Eds D. M. Bather and H. I. Day. Reading University Agriculture Club

OTHMER, D. F. (1970). Man versus material. *Trans. N.Y. Acad. Sci., Ser. II*, **32**, 287

PIMENTEL, D. (1977). Energy budgets in natural and agricultural eco-systems. In *Living Systems as Energy Converters*, p. 299. Eds R. Buvet, M. J. Allen and J-P Massue. Amsterdam; North-Holland

PIMENTEL, D., HURD, L. E., BELLOTTI, A. C., FORSTER, M. J., OKA, I. N., SHOLES, O. D. and WHITMAN, R. J. (1973). Food production and the energy crisis. *Science, NY*, **182**, 443

RAPPAPORT, R. A. (1968). *Pigs for the Ancestors.* New Haven, Conn., USA; Yale University Press

RAPPAPORT, R. A. (1971). The flow of energy in an agricultural society. *Sci. Amer.*, **225**, 117

REVELLE, R. (1976). Energy use in rural India. *Science*, **192**, 969

RUTHENBERG, H. (1971). *Farming Systems in the Tropics.* Oxford; Clarendon Press

RUTTAN, V. W. (1975). Food production and the energy crisis; A comment. *Science, NY*, **187**, 560

SHANNON, M. J. (1977). *Wall Street Journal*, 17 April 1977, p. 1

SLESSER, M. (1973). Energy subsidy as a criterion in food policy planning. *J. Sci. Fd. Agric.*, **24**, 1193

SPEDDING, C. R. W. (1978). Energy flow in agriculture. Paper presented at the *British Association of the Advancement of Science Annual Meeting*, Bath University, Sept. 1978

SPEDDING, C. R. W. and WALSINGHAM, J. M. (1975). Energy use in agricultural systems. *Span*, **18**, 7

STANHILL, G. (1974). Energy and agriculture. A national case study. *Agro-ecosystems*, **1**, 205

STANHILL, G. (1977). An urban agro-ecosystem, the example of 19th Century Paris. *Agro-ecosystems*, **3**, 269

STANSFIELD, J. R. *et al.* (1974). *Fuel in British Agriculture*. NIAE Report No. 13

STEINHART, J. S. and STEINHART, C. E. (1974). Energy use in the US food system. *Science, NY*, **184**, 307

USDA (1975). *Yearbook of Agriculture*. Washington DC; Govt. Printing Office

USDA (1980). *Report and Recommendations on Organic Farming, July 1980.* Washington DC; Govt. Printing Office

USDC (1972). *Bureau of the Census. Statistical Abstract of the US, 1972.* Washington DC; Govt. Printing Office

WERNICK, S. and LOCKERETZ, W. (1977). Motivations and practices of organic farmers. *Compost Sci.*, **18**, 20

WHITE, D. J. (1975). Energy in agricultural systems. *Agric. Engr.*, **30**, 52

WHITE, D. J. (1976). Energy use in agriculture. In *Aspects of Energy Conversion*, p. 141. Eds I. M. Blair, B. D. Jones and A. J. Van Horn. Oxford; Pergamon Press

WHITE, D. J. (1977). Prospects for greater efficiency in the use of different energy sources. *Phil. Trans. R. Soc. Ser. B.*, **281**, 261

WILSON, P. N. and BRIGSTOCKE, T. (1977). Energy and UK agriculture. *Long Range Planning*, **10**, 64

WILSON, P. N. and BRIGSTOCKE, T. (1978). Animal nutrition over 20 years. *Span*, **21**, 66

WILSON, P. N. and BRIGSTOCKE, T. D. A. (1980). Energy usage in British agriculture—A review of future prospects. *Agric. Systems*, **5**, 51

WILSON, P. N., BRIGSTOCKE, T. D. A. and CUTHBERT, N. H. (1981). Some factors affecting the future composition of UK compound animal feed. *Anim. Feed Sci. Technol.* (In press)

28

ELEMENTS OF ANALYSIS OF BIOLOGICAL HUSBANDRY METHODS ON FOUR FARMS IN SOUTH-EASTERN FRANCE

S. BELLON
Institut National Agronomique de Paris-Grignon, 16 rue Claude Bernard, 75005 Paris, France
J.-P. TRANCHANT
Institut Superieur d'Agriculture, Rhone-Alpes, 31 Place Bellecour, 69002 Lyon, France

Organization and approach of the study

During a 2-month course of lectures (January–February 1980) 20 ISARA (Institut Superieur d'Agriculture Rhone-Alpes) undergraduates—among them Jean-Paul Tranchant—surveyed four farms selected for their mixed farming/dairy farming activities. This first stage resulted in a synthesis of all farms surveyed.

The study was continued by Stéphane Bellon, an undergraduate at INAPG (Institut National Agronomique de Paris Grignon), from March to July 1980. Following more surveying of the farmers in order to check data and collect further information, a report was written about each farm. The texts were submitted to the farmers themselves and, after final adjustment, the four reports were synthesized into a complete study. The survey was conducted according to a system developed at the INAPG, adapted to the special cases of biological farms.

In addition to the usual analytical criteria, we have integrated the farmer's objectives, as it is these which guide his actions or practices. The recent history of farming makes it possible to understand how the present situation has been reached, as well as the reasons for, and conditions of, the change to biological farming.

We have reported as fully as possible the production factors (area and soils, labour, soil rotation, animal raising) and their level of utilization, as well as the objectives and plans of the farmer. The history of the farm is seen through changes in the system of farming, in particular since the conversion to biological farming. We have also attempted to look deeper into the farmers' practices (animal feeding and husbandry) and techniques. The results of the practices have been assessed in terms of working time, mineral and organic changes in the soil, and economic aspects.

WORKING TIME

Previous studies in France showed that one reason for changing to biological farming is an attempt to lower operating expenditure through increased utilization of labour, notably family labour. It therefore seemed to be of interest to assess the labour requirements, comparing labour demand in the main farming activities with the supply from permanent or temporary labour, while making allowances for the family's objectives of time-use.

MINERAL AND HUMIC RESULTS

Here our aim has been to observe the consequences of the farmers' practices on soil fertility and the preservation of humus. These results can be set at two levels.

At the farming level, all purchases of mineral and organic products, both crop and animal production, are recorded and balanced against sales and losses. Similar work can be carried out for a rotation on a given plot by assessing various materials added to the soil (organic matter, soil improvers, fertilizers), elements taken from the soil, and leaching losses. Comparison of the two results permits the evaluation of fertility transfer among plots.

ECONOMIC RESULTS

In the same way we have established a primary balance, in French francs, for each farm as the difference between costs and returns of the enterprise. Two additional balances have been established: one for some products of the farm (gross margin); the other for family income.

Findings

The findings on each farm are presented in a report under the headings: history and development of the farm; present operational technique; animal raising and feeding husbandry, grassland and crop husbandry, labour and work schedule; consequences of these practices on mineral and organic fertility; economic results, and conclusion. For the full reports *see* Bellon (1980): here we present a synthesis based on the four samples.

Introduction to the farms

Mixed farming/animal-raising units have *a priori* seemed best representative of biological agriculture, with animal-raising playing an important part.

SUMMARY DESCRIPTION

On farms with 20–60 hectares (50–150 acres) in cropland, the share of arable land is 20–30 hectares (50–75 acres). Grassland represents 45–75 per cent of the total acreage. Cereals are always found, a variable proportion being fed to animals and the rest marketed (0–70 per cent of all cereals grown). Straw is used as litter for animals, but the area of cereals is generally not enough to meet all requirements: only one farmer produces 100 per cent of the straw required; the other three have to buy about 50 per cent of it.

In all cases dairy cows are found, in herds of 8–20 head of various breeds, as well as small marginal activities, such as poultry and hog husbandry.

Another characteristic common to all farms is the presence of considerable family labour (three permanent units at least). Farmers are 40–60 years of age and have been practising biological farming for about 10 years, a period long enough for us to consider the farms as well established in biological methods.

However, substantial differences appear. On farms with the smallest area of arable land (20 ha or 50 acres) there are orchards and small vineyards. On one farm, goats are raised and cattle fattened. Another farmer owns an important enterprise of industrial poultry production which is by no means biological (12 000 pullets twice per year). Finally, it should be noted that the farms have buildings and crops comparable to those of surrounding farms.

HISTORY

Conversion to biological farming began when the farmers more or less rapidly ceased buying chemicals, with the exception of one who still buys slag and groundnut cakes. On the other hand, the supplies come through different channels. The modification of farming systems during conversion will be briefly expounded later. The reasons given by farmers for turning to organic farming are similar to those recorded in other surveys, namely health problems with animals or with the farmers themselves, but also difficulties with crop production (decreasing yields, weeding which has become difficult, pests and diseases).

In all sampled cases, the present operation depends largely on the history of each farm, including developments before the change to organic farming. However, the changes, often due to the family's increased demand on farm income (for instance when a son takes over from his father) go together with the possibility of getting better products through biological channels.

ANIMAL RAISING

Production can be characterized as follows:

1. Milk production near 3000 kg per lactation, which is slightly lower than the average conventional production in this area.
2. Low stocking rate per hectare.
3. Few health problems, particularly as regards fertility: about one calf per dairy cow per year.
4. Artificial insemination used on two of the sampled farms.
5. Only compulsory vaccination practised, as well as the blood test required for detecting brucellosis.
6. Mineral supplements containing trace elements used, produced and marketed by firms specializing in biological products.

All farmers feed their animals on the highest possible amount of green fodder all the year round: hence the general use of annual or catch-crop fodder (turnip, pumpkin, kale, beet, rape, sorghum). Winter feeding consists of concentrates from home-produced grains and legumes. Fodder, too, is legume-based, except in the case of one farm with abundant permanent meadows. The summer ration is essentially composed of pasture grass or is fed green in stables. Owing to irregular yields of legume seeds in cereal–legume mixtures, farmers have to rely on high-quality fodder to meet the nitrogen requirements of animals.

SYSTEM OF FARMING

Rotations

Following the adoption of organic agriculture, farmers have given up the growing of maize as a single crop and introduced a few cereal–legume mixtures, which they harvest for grain and use as a substitute for cakes (0.5–6.0 ha in 1979). In addition they have increased the fodder area at the expense of grain area. At present, most commonly practised rotations are as follows:

1.   Legumes–cereals–cereals (+ covered legumes)
2.   Root crops–cereals–cereals
3.   Root crops–cereals–legumes–cereals (cf. the Norfolk four-course rotation).

Such rotations mean that an important part is given to legumes (50 per cent of the acreage) and these are in most cases sown with cereals as cover crops. We note that little green manure is grown (except in orchards) because of climate and work schedules.

Farming methods

Manure is not always composted and the choice of the composting technique depends mainly on the labour available. Following their conversion to organic farming, farmers have adopted bigger tractors, although the rest of the equipment is about the same as before conversion. The only machine specific to organic farming is a kind of subsoiler which can dig to about 30 cm. All farmers practise ploughing, with plenty of surface cultivation: for example, before ploughing they control weeds by allowing germination; after late ploughing the soil is crumbled; in spring there is weeding and ploughing-in of organic or mineral fertilizer, sometimes sowing legumes and, in conformity with doctrine, 'stimulating microbial life'.

Results

LABOUR AND WORK SCHEDULE

Family labour is essential, especially at haymaking and straw harvesting time. Furthermore, straw—and sometimes hay—has to be bought in from outside the farm. Other intensive work relates to specific crops: fruit harvesting, potato lifting, hoeing of root crops. Because of high workloads, all farmers plan to reduce their working time.

RESULTS CONCERNING MINERALS

Table 28.1   MINERALS USED ON ALL FOUR FARMS

Sample no.	Compound or Element				
	N	P_2O_5	K_2O	CaO	MgO
1	$\ll 0$	$\gg 0$	$\ll 0$	< 0	> 0
2	> 0	$\gg 0$	$\gg 0$	< 0	< 0
3	> 0	$\gg 0$	> 0	> 0	> 0
4	< 0	> 0	$\ll 0$	$\ll 0$	< 0

$\ll 0$: more than 40 units per hectare decrease
< 0: 40–10 units per hectare decrease
> 0: 10–40 units per hectare increase
$\gg 0$: more than 40 units per hectare increase
} after changing to biological methods

Nitrogen

Table 28.1 shows that sample 3 has a positive balance for all minerals. This is attributable to a considerable purchase of mineral elements through feeds for the industrial poultry operation. Likewise, a positive balance of nitrogen in sample 2 is caused by the purchase of cakes. When a balance of nitrogen appears positive on a farm, therefore, it is because the farmer buys commercial feeds. When the balance is negative, farmers are relying on symbiotic fixation and the mineralization of humus.

Phosphorus (as P_2O_5)

The balance of phosphorus is always positive, because all fertilizers bought by farmers contain some, and phosphorus suffers little leaching.

Potassium (as K_2O)

Two important deficiencies are to be noted. The only supply of potassium on these farms comes from the buying of straw or hay and there is a considerable consumption by root crops such as potatoes. One farmer reaches a balance by buying slag.

Lime

Three cases of deficiency are observed, one of them in very calcareous soil. The importance of these, however, is to be interpreted with caution, as the exact amount of calcium liberated by the soils is unknown and leaching may have been overestimated.

Magnesium

Positive balances are seen on two farms supplied with commercial magnesium compounds. However, the product has been applied for 2–3 years only, since the time when it first appeared on the market.

Mineral analysis shows evidence of fertility transfer. On one farm the transfer benefits orchards (economically very important on this farm). On another the benefit has gone to pasture. On samples where nitrogen is purchased, it is applied only to cereals in spring. No general rule is in force as regards the 'direction' of this transfer, which depends on the place of the given crop in the system.

RESULTS CONCERNING HUMUS

As a whole, the rate of organic matter in soils is maintained at a satisfactory level: 2–3 per cent according to our calculations. The presence of an industrial poultry operation makes it possible to reach a slightly higher content on one sample. However, this content is still satisfactory because of the clay and calcium contents of the farm soils.

These figures are confirmed by soil analysis, although, with calcareous soils, analysis shows figures slightly higher than our own.

ECONOMIC RESULTS

The proportion of animal-raising generating income ranges from 45 to 90 per cent. However, the situation differs according to the farms sampled:

Sample 1.   All cereals go to feed the herd, which produces 90 per cent of the gross income; the remaining 10 per cent comes from 1 hectare in fruit trees.

Sample 2.   Cash grains and orchard make up 55 per cent of the gross income.

Sample 3.   Cash grains and potatoes represent 30 per cent of the gross income.

Sample 4.   Processed cash grains, raised to a premium value through biological channels, constitute 30 per cent of the gross income.

In addition, this economic picture shows evidence of a low expenditure on veterinary products and fees. On the other hand, fertilizer costs are sometimes high. Even though less fertilizer is purchased for the farm (15–40 nitrogen units of fertilization by hectare by those farmers who use it), the cost of these inputs specific to biological farming may be balanced by the need of more cultivation; fuel costs are high in half the cases.

When the sampled farm happened to be definitely a family business, we thought it necessary to take into account all income accruing to the whole family (family allowances, retirement pension). With three of the four samples, this indirect contribution represents an important share of the family income (25–45 per cent).

Finally, the study of the economic side permits a look into the organic farmers' marketing channels.

The share of products marketed through biological channels ranges from 25–50 per cent. This raises two problems: first, there is no organized market for biologically produced meat or milk. Only two farmers process their products into cheese, butter or fresh cream. Meanwhile, byproducts are put to use through a hog operation, the animals being sold to regular customers. Another case of increasing milk value is achieved through retail sale, particularly on one farm (30 per cent of dairy production) situated close to a town. The remaining milk is sold to a cooperative.

There is generally no problem in marketing cereals through biological channels. With wheat, the increase in value ranges from 20–100 per cent of prices on the conventional market (or 90 French francs a quintal: 24 F a bushel), according to whether it is marketed through a specialized organization or through individual buyers. Here again, processing and selling grains through cooperatives or shops that sell products with the biological label will allow an appreciable increase in value of production.

Other crops (potatoes, fruit) are sold as much as possible with the biological label through organizations and specialized shops; whatever is left goes to wholesale dealers, without the label.

Conclusion

It has been our aim in this survey to point out the difference between organic agriculture as described as a doctrine in many books, and as actually practised by a few farmers. Our preliminary study suggests that, should organic farming develop, several problems may arise regarding the viability of such farms.

Technically, many difficulties remain to be solved and should not be hidden behind economic results. This concerns particularly the control of cereal–legume mixtures, as well as legume-based fodder, and therefore the meeting of food requirements of animals. The survey of mineral results shows evidence of considerable imbalance as regards mineral nutrition of plants, and their immediate needs cannot be satisfied.

Because the farming rests on animal raising, variations in cereal production are tolerated for the time being, especially as all farmers produce crops with an important gross margin, all of which are marketed through biological channels (though also subject to climatic variations). Alternatively, they process most of their production.

One in two farms shows evidence of potassium and magnesium deficiencies. While we cannot say when these deficiencies will appear, we have reservations about their ability to maintain soil fertility.

Farmers are practically dependent on outside sources for straw. At a time when better use of agricultural byproducts is being considered, notably with energy recovery in view, organic farmers might be deprived of one of their basic resources.

Are the types of organic fertilizers sold by commercial enterprises adequate in view of crop requirements, and need farmers pay such high prices for them? Spring application to cereals of soluble–natural–nitrogen fertilizers is one answer to this first question. Nevertheless, the cost of the unit of fertility is much higher than that of chemical fertilizers. With regard to other compounds (P_2O_5 and CaO), farmers have found, in conventional outlets, fertilizers or enriching agents whose composition can be compared with other products sold by biological firms, and at much lower prices.

With all samples, the services of family labour present on the farm are called upon, occasionally to meet intensive work (hay and straw and fruit harvest) or to process some of the products (packing), and at all times for all permanent residents on the farm: one son who has settled, the farmer's wife and, sometimes, old people.

The labour supply might decrease and jeopardize present production techniques for three major reasons:

1. Work conditions demanded by children, particularly when a son takes over;
2. The ageing of the farmer and his wife, and their attempt to reduce working time;
3. The decease of old people working on the farm.

It is true that this labour is increased in value through processing and marketing on the farm, as well as retail selling in grocers' shops—with or without the biological label. Nevertheless, looking for new markets is not exclusive to biological farmers: many conventional farmers now plan to practise direct sale and processing on the farm.

Finally, organic farming is no longer alone in filling the quality gap. In 1979, government-approved organizations issued a label with a view to guaranteeing certain rules of fruit production. They control nutritive and taste value of fruit through the use of 'objective' quality criteria. There are other attempts from conventional fruit farmers and market gardeners within orthodox farming outside biological and integrated control. These attempts may direct many farmers towards a way similar to biological farming. It remains to be seen whether consumers will sense the difference of labels and will always be prepared to pay the necessarily higher prices for biological products.

Acknowledgments

The authors are grateful to B. Fabre, Y. Gautronneau, L. Genton, Y. Le Pape, and M. Sebillotte for their help with this paper.

Reference

BELLON, S. (1980). *Logique, ou Biologique.* Course dissertation. Institut National Agronomique de Paris-Grignon

29

ANALYSIS OF VEGETABLES PRODUCED BY ORTHODOX AND BIOLOGICAL METHODS; SOME PRELIMINARY RESULTS

D. LAIRON, P. RIBAUD, J. LEONARDI, H. LAFONT, G. GAUDIN
and M. REYNIER
INSERM, U130, 10 Ave. Viton, 13009 Marseille, France

The research group GRAB (Group de Recherche en Agriculture Bio-logique), was founded in the South of France in 1979, and is composed of scientific researchers, agronomists, farmers and consumers. The aims are to promote research into the scientific basis of different agronomic methods covered by the term 'biological agriculture', to compare those methods with techniques of 'orthodox agriculture', and to disseminate information and results both in the scientific community and in rural and urban groups.

The GRAB manages an experimental field where tests are done on a scientific and practical basis. A series of tests has been started involving six different agronomic methods: 1, control; 2, 'orthodox husbandry' (NPK fertilizers, chemical pest control); 3, brushwood compost; 4, household refuse compost; 5, Lemaire compost; 6, farmyard compost. Methods 3–6 involve pest control without chemicals. The first crops are melons and tomatoes. Yields, fruit size, pest attacks and fruit chemical composition are being assessed.

In addition to the studies performed under controlled conditions on the GRAB experimental field, we are comparing the quality of differently

Table 29.1 ANALYSIS OF BIOLOGICALLY GROWN
AND CONVENTIONALLY GROWN VEGETABLES
(POTATOES, LEEKS, KALE, COS LETTUCES,
TURNIPS)

Component	Mean % difference biological/orthodox
Dry matter	+ 26
Phosphorus	− 6
Potassium	+ 13
Calcium	+ 56
Magnesium	+ 49
Iron	+ 290
Copper	+ 34
Manganese	+ 28
Total amino acids	+ 12
Essential amino acids (excluding trytophan)	+ 35
Nitrates	− 69

raised products bought by consumers. This point is of some importance for nutrition and health; little work has been done on it and few data are available in the literature. The present study compares vegetables sold in orthodox markets with those purchased in cooperative stores which distribute biologically grown produce. Potatoes, leeks, kale, cos lettuces and turnips were analyzed using conventional assay methods in the U-130 INSERM laboratory. Ten samples from biological and orthodox sources were studied in each case.

Results are expressed in terms of fresh matter, and are summarized in *Table 29.1*. A statistical analysis was performed, using Student's t test at the 95 per cent probability level, on 45 different measurements done on two groups of 10 samples each. Twenty-seven of these proved significant. Twenty-five significant differences were found favouring biologically produced vegetables: only two significant differences were found in favour of orthodox vegetables. Eighteen differences were not significant. These data agree well with some previously obtained by other workers, and give data on some additional elements. They indicate a generally higher nutritional content of biologically grown products. Further studies are being pursued.

VI
POSTSCRIPT

30

FEEDING THE WORLD

T. L. V. ULBRICHT*
Agricultural Research Council, London, UK

The International Institute of Biological Husbandry is to be congratulated on having organized such a constructive and useful meeting. A great deal of common ground indeed exists among the participants, based on a shared concern for developing an agriculture that will be sustainable in an uncertain future.

For a very long time on this earth—ever since settled agriculture began—we have had biological agriculture. It has been the economic base of great civilizations which sometimes have lasted for several thousand years. Only in very recent times, as a result of a great industrial development based to a significant degree on the availability of cheap energy, have we witnessed the growth of what some workers call 'conventional agriculture'. A certain kind of exploitation of scientific knowledge made that possible. Perhaps 'conventional agriculture' in its extreme form will turn out to be quite a temporary phenomenon—a little blip on the timescale of the earth's history. Perhaps we shall return to a biological agriculture, enriched by scientific knowledge, but still based on the fundamental natural cycle of nutrients.

Long before the Second World War, when an energy crisis was not even the figment of anyone's imagination, when no one worried about the environment and ecology was an obscure science, when the use of artificial fertilizers was a fraction of what it is today and before herbicides and synthetic pesticides were invented, and when the little processed food then consumed contained far fewer chemical additives, there were already organic farmers. They already stood against a trend which became really strong only after the war, a trend which has led to an agriculture heavily dependent on inputs of support energy, chemicals and machinery. Crop yields in the UK have doubled, but at the cost of a 16-fold increase in draught power and a 20-fold increase in the use of nitrogen fertilizer. As pointed out by Spedding, Walsingham and Wagner, as agriculture changed, there has been more and more substitution of nonbiological for biological processes. What was a cyclic process based on renewable resources has become more and more like a linear industrial process with inputs, outputs, and 'waste'.

---

*This paper expresses the personal views of the author, not those of his organization.

A number of factors have made us aware that such an approach is not really satisfactory. It is too narrow—it neglects the impact of this kind of farming on the environment, both in the sense of wildlife and landscape, and also in the even wider sense of people, rural communities, and ultimately, society as a whole. There is actually nothing inherently unscientific about trying to think in terms of the whole; it is simply that our minds are lazy, and it is easier to think in terms of parts. Our tendency is to tackle problems as though they were self-contained and isolated. So, if a field becomes infected with a disease, it is sprayed with a pesticide—which really assumes that the field consists only of crop, pest and pesticide, and that there is no world outside. But why did the field become infected? That would mean thinking about the farming system as a whole, which is more difficult.

It was genuine and legitimate questions about the environment that, for many, first led to the realization that we must take a wider view. In fact, the natural environment is extraordinarily resilient, and pollution from agriculture has been less damaging than from man's other activities. What we see, however, is that the narrow approach seems to be leading to a treadmill. Pests become resistant, so more pesticide has to be used: then resistance increase to the point when the pesticide cannot be used any more, and another has to be found. Can this process go on indefinitely? Can we not find something more stable—more sustainable?

It is interesting to reflect on what happened when the situation I have described really did become serious, namely with the resistance of red spider mite in the glasshouse industry. A method of biological control was developed—in one of my organization's Institutes, I am happy to say—and has become generally adopted. Perhaps some of you do not know, so I will tell you that the first step in the system is to infect the crop with the red spider mite! You will understand that, psychologically, this made the system rather difficult to sell at first. But the predator must have something to feed on . . . there has to be a balance.

The success of this approach has led to others, and much research on biological and integrated control is in progress. Lisansky in his chapter gives some very interesting examples, including control of skeleton weed, *Chondrilla juncea* by the rust fungus *Puccinia chondrilla*. The scope of biological control may be wider than is generally believed. It always necessitates the attempt to understand the system as a whole and the dynamic relationships within it. It is difficult, because we are still very ignorant. So we need more basic ecological studies, and for some years the ARC has been funding a major research project on cereal ecology. We are also engaged in monitoring for certain insect pests, partly to try to understand the factors that control their numbers, and partly so that no spraying is carried out unless it is really necessary.

The second factor that has led us to wonder whether we can continue indefinitely in the 'old way' (although it is a relatively new way, actually) is the cost of energy. For a time, energy was abundant and ridiculously cheap. Now that it is rather expensive we have come to realize how profligate we have been in its use, and that however successful the search for so-far undiscovered deposits may be, fossil fuels must be limited. Agriculture and its support industries use a significant amount of energy: typically in a developed country it is about 15 per cent or so of the national energy expenditure. We may comfort ourselves with the thought that, if energy were in short supply, food production would receive top priority; but how efficient is the use of fossil fuel energy in agriculture?

Wilson and Brigstocke point out to us that simple, extensive systems of agricultural production are very efficient energetically and use only a fraction of the energy inputs of sophisticated systems. Can we reduce energy inputs and maintain high outputs? The two biggest energy inputs in conventional agriculture are tractor fuel, and artificial nitrogen fertilizers. Therefore, if only from a cost and energy point of view, there is good reason to be interested in the recycling of nitrogen, and to consider animal effluents not as waste, but as a resource.

It is relevant to consider here the fact that many organic farmers, although not all, use inputs from conventional agriculture—either manure, or animal feed. In many cases that is how they compensate for the nutrients that leave the farm in the form of grain, vegetables, meat, milk and so on. For some, the ideal would be that all the waste and excreta produced by the number of people fed by a farm, should return to it. In their paper, Martin and Keable enumerate the many problems that will have to be solved to be able to return to a natural and sustainable cycle of this kind on a larger scale. Plaskett has shown that levels of organic carbon cannot be maintained by, for example, ploughing in straw, or by using the output of collectable manure. A possible solution to this problem would be to increase the production of organic carbon by agriculture itself. He considers the example of a perennial weed, part of which could be used as animal fodder (yielding manure among other outputs) and part a material for anaerobic digestion, the residue from which would be used as an organic soil conditioner. His calculation suggests that one hectare of such a plantation could produce the humus needs for one hectare of arable land.

If for a moment we envisage a time when nitrogen fertilizers are in short supply or become very expensive indeed, then a limitation to the expansion of biological farming will be that such enterprises are almost always mixed farms. On a large scale this would severely restrict the acreage that could be devoted to arable crops—yet arable farming is the most efficient, both in terms of energy and in producing food for humans. Again we return to the question of reconstituting the natural cycle on a larger scale, and to the problems of the bulkiness of organic materials, the need for better processes for treating raw sewage sludge and municipal waste, the lack of an infrastructure for collecting different waste materials, and so on. Also, the possibility of a biological arable farming system, with green manuring by leguminous crops, needs to be investigated.

Perhaps this is the right place to make a general point. The accusation that no research relevant to biological agriculture has been done, really is not justified. In the first place, all the very considerable research in the basic sciences, such as soil chemistry and microbiology, plant and animal physiology, biochemistry, genetics, etc., are relevant to all types of agriculture. But there has also been much relevant applied work. As I have just been writing about recycling nitrogen, I think naturally of all the ARC's work on nitrogen fixation. This is both at a quite fundamental level (the biochemistry of the process and its genetic control) as well as at a directly applied level—the breeding of clover and other leguminous forages, the study of legume–*Rhizobium* interactions which has led to the possibility of inoculating pastures with productive strains of rhizobia, the agronomy of mixed grass–clover swards etc. There again, however, it is necessary to look at the whole system, because herbicides and pesticides are generally harmful to rhizobia.

The third factor which has had an impact on how people look at agriculture is the concern of a sector of the population about the quality of food produced by modern intensive farming. (This was one of the reasons why the Netherlands Government set up a committee to study what they called 'alternative agriculture'*.) Hodges has reviewed this difficult subject for us. The evidence is conflicting and I think he rightly concludes that it is insufficient as far as long-term experiments are concerned. The basic hypothesis is that the year-by-year addition of rather pure NPK fertilizer is likely to alter the mineral balance of the products, compared with what would be obtained if all minerals were recycled. The preliminary results reported here by Lairon *et al.* show that biologically grown vegetables contain less water, much less nitrate and more trace minerals, especially iron, than vegetables grown by 'orthodox' methods. But even if further analyses show consistent quantitative differences, how are we to interpret them? I agree therefore that what we need is, in effect, a bioassay. We cannot do it on humans but we can at least study the effects of food produced in different ways on animals, on their growth, health and reproductive capacity.

Turning to the wider problem of feeding the world, let us first try to be clear about the actual situation now, and what it is likely to be during the next two decades. We are told that millions are starving, that a large proportion of the world's population is undernourished, and that the situation is likely to get worse. We are also told that world food production has been increasing faster than the world's population, that vast areas of land are not being utilized and that the world could support a population many times larger than the present. Which statement is true?

Strange as it may seem, both are true. There is a belt of starvation stretching right across Africa, from the Sahel in the West to Somalia in the East, and an estimated 700 million people throughout the world are seriously undernourished. Over the world as a whole, food production has increased slightly faster than population, but such aggregated figures hide the really important facts. There are now only four countries in the world which are net exporters of grain (the US, Canada, and to a lesser extent, Australia and Argentina) and almost every Marxist country in the world (even Cuba) is importing food from the capitalist West. Most significant of all is the situation in the Lesser Developed Countries (LDCs). Whereas in aggregate they were net exporters of grain in the 1950s, in 1970 they imported 42 million tonnes of grain, and 80 million tonnes in 1979. The total increase in food production (from yield and acreage increases) in those countries in recent years has been 2.8 per cent per year, but demand for

*An English translation of the committee's report has been published in *Agriculture and Environment* (1980), Vol. 5, Nos. 1–2.

food (from combined growth of population and income) increased by 3.5 per cent per year. Almost two-thirds of all developing countries showed a fall in self-sufficiency from 1961–63 to 1970–72. Even if grain production increased by 2.5 per cent yearly in the food-deficit LDCs (it grew by only 1.7 per cent per year in 1967–74) then, by 1985, the International Food Policy Research Institute estimate that there will be a deficit of 100 million tonnes of grain in those countries. The World Bank's 1979 World Development Report shows that, of 37 LDCs, the index of food production per head has declined for 28 countries, is unchanged in one and has increased in only eight.

It used to be believed that a major reason why so many people are undernourished is that their diet is deficient in protein. We now realize that, except for quite limited numbers (subsisting on very low-protein staples such as cassava) this is not true. If people could simply eat more of what they are eating now, they would be adequately fed. (They may *appear* to be protein-deficient because, on a very low calorie diet, the protein they do eat is burnt for energy.) Unfortunately, they cannot afford to eat more—they are too poor. The food-deficit LDCs import what they can (they are generally among the world's poorest countries) and receive aid, but in those countries, and in others (for example India) where the local situation is much better, food does not reach many people who need it.

Disaster on a large scale can be kept at bay by shipping grain from North America to the LDCs. There is, however, an actual physical limit to what the world's ships, and more particularly port facilities and grain stores, can cope with.

It is true that world food production could be very greatly increased; theoretically the earth could support a very much larger population. However, such estimates neglect to consider what would be involved in bringing about such increases. Calculation shows that using irrigation to double the arable land from 1.5 to 3 billion hectares would require 3000 billion litres of fuel per year; this use of fuel alone would exhaust all our known reserves in little over 20 years. Or, consider the fact that to double the average rice yield (to 5.8 tonnes/hectare, which is far below what is already possible) would require additional inputs (especially nitrogen fertilizers) representing a ninefold increase in the use of fossil-fuel energy. But the poorest countries have no oil and no fertilizer industry, and it is the poorest farmers, who can least afford extra inputs, that most need the extra food.

We can now see more clearly the problems involved in 'feeding the world'. What is necessary is to increase food production in the food-deficit LDCs, the great majority of whose peoples are poor farmers with very small holdings. It is obvious that conventional agriculture cannot be adopted by these countries when you consider that the energy used by agriculture in the US—production, processing and distribution—is equivalent to the total energy consumption of India. Yet at one time it seemed that almost everyone thought simply in terms of transferring a western-style input-intensive agriculture to these countries. The success of the 'Green Revolution' in the Punjab was misleading. The size of holdings there is much larger than in the rest of India, and the agriculture was already prosperous and technologically well advanced. The farmers actually had the money, or could get the credit, to buy the fertilizer and pesticides that the new wheat varieties had to have. It is a situation not at all typical of the LDCs.

It was good to have quite a number of papers at this Conference on agriculture in LDCs, by Abidogun, Ochwoh and Zake, Ojomo, Wilson and Kung, Dalzell and Moss. Although the latter emphasizes the diversity of tropical soils and environments, it is generally true that organic matter decomposes very quickly in tropical soils and fertility is much more difficult to build up and maintain than in a temperate climate. The low level of organic carbon means that nutrients are not retained, but are readily leached out. Often, dried dung has to be burnt for fuel, for lack of any practical alternative. It was encouraging that the papers from tropical countries at this Conference recognized those difficulties; they appreciated the need to conserve organic carbon and the kind of agricultural systems which might help to do this, such as 'alley cropping'. The comments by Spedding, Walsingham and Wagner on small-scale holdings are also very relevant. Locally adapted systems need to be developed based on crops, animals and rotations really suited to those conditions and manageable by the farmer. Moss has emphasized the diverse and beneficial use of trees (including leguminous species) for fuel, for food (nuts etc.) and for nonfood products (such as gum arabic) as well as shelter, in various types of agroforestry to maintain the organic content of the soil. Much better utilization of leguminous crops (including shrubs and trees) may be possible, aided by inoculation with appropriate strains of *Rhizobium*. Lopez-Real reported on the selection of *Rhizobium* strains for high nitrogen-fixing ability and for the presence of a hydrogenase which catalyzes the oxidation of hydrogen (produced by the nitrogenase) with the generation of ATP, leading to higher yields of, for example, soybeans. In some cases, the best use of limited supplies of artificial fertilizer may be to use low levels to help establish the growth of legumes during the first weeks before nodulation occurs. It is also worth remembering that, up till now, very little effort has been put into breeding legumes compared with that put into breeding grain crops. Indeed, one of the unforeseen consequences of the Green Revolution in India has been that the acreage of traditional pulse crops has markedly declined. This is particularly unfortunate as the amino acid composition of the pulses beautifully complements that of both wheat and rice, and so provided a sound and healthy diet for the largely vegetarian population.

It is my impression that it is increasingly realized by those who run aid programmes, by the World Bank, and by the International Institutes funded by the Consultative Group, that their initial approach to agricultural development in the LDCs was too much conditioned by western experience, and too narrow. Increasingly now it is the local farming *system* that is being studied with a view to understanding what technology is appropriate to it. It is appreciated, too, how efficient—in terms of utilizing the inputs available—small labour-intensive holdings often are. Thus, rather than being swept away, the traditional systems are seen as systems to build on and adapt. Those who have studied and practised biological agriculture will be naturally sympathetic to such an approach, and could play a valuable part in helping the LDCs. I know that it is one of the things that the IIBH always hoped it would be able to do.

In many developed countries, however, farming enterprises are still looked at in much too narrow a light. One reads in the British press, for example, that the EEC Common Agricultural Policy is wasteful and unfair because it is subsidizing many small and 'inefficient' dairy farmers in Germany and other countries. As I understand it, though, it seems that the German Government has asked itself some questions which seem to me rather sensible. 'These small farms are inefficient when looked at simply in terms of agricultural productivity. But if they disappeared, what would happen to that land? Would we let it revert to unsightly scrub? Would we plant new forests? What would happen to the schools, shops, local hospitals and other facilities if these marginal rural areas became depopulated? How would these people earn their living? Where would they go? In the cities there is high unemployment, acommodation is scarce and expensive . . . . Are these people not much more content where they are? Do we want to increase the already existing discontent in our cities? In terms of overall costs to the State, in terms of the preservation of the countryside, as well as in terms of human welfare, is it not better that they stay where they are?' Do we not, at least in part, subsidize hill-farmers in the UK for rather similar reasons? Should we not perhaps acknowledge that more openly and indeed be confident that it is right?

For similar reasons, I agree with what is said in the opening paper of our Conference, 'An Agriculture for the Future', by Hodges. That is, that assessments of the 'efficiency' of organic farms have left a great deal out of account. Their energy inputs are lower, they pollute the environment less, they are more labour-intensive and therefore help maintain the life of rural communities, and they preserve our precious soil from erosion.

It is interesting that in the Netherlands, where the Mansholt plan to increase the size of agricultural holdings and to intensify agriculture was enthusiastically adopted, the Government now has quite a different view. It considers that this process has gone quite far enough. It is worried about the natural environment, about maintaining rural communities, about pesticide residues in food, about energy . . . all the things this Conference has been discussing. In the long term experiment it has set up on one of the new polders, it is comparing not only a conventional with a 'pure organic' system, which has been attempted before, but is also trying to develop a third system that will use elements of both, but which is intended to be less dependent on energy and also minimally polluting.

As I have already said, scientific research has contributed a great deal to our understanding of natural processes important in biological agriculture. As now practised, in developed countries, it is a system significantly different from that which existed before, let us say, the time of Liebig, who initiated the study of the nutrient requirements of plants. Consider only how our knowledge of microbiological processes taking place in manure, about which we were informed by Gray and Biddlestone, has transformed the handling of this resource. I am sure that there will be further significant scientific inputs into biological agriculture, and there was an intriguing hint of new possibilities in the paper by Lovett and Levitt on allelochemicals. But there does exist a problem about plant breeding. The new high-yielding crop varieties have been bred and tested solely with the present conventional system in mind. Even the selection of leguminous crops for high yield has invariably taken place in plots given nitrogen fertilizer. It is unlikely that such varieties will be best suited to biological agriculture, and one of the recommendations in the USDA report on organic farming is for appropriate research on that. For example, emphasis should be given to the selection of host-plant genotypes that may enhance nitrogen fixation. Big differences do exist in intraspecific variability for nitrogen fixation.

It has been encouraging at this Conference to find so much common interest among such a diversity of participants. We have learnt and still have a great deal to learn from each other. An undogmatic approach and a willingness to experiment and put to the test of experience will lead to the development of a better agriculture. I hope I will be forgiven if I express some disagreement with one thing said in the paper by Wilson and Brigstocke: 'Moral and philosophical arguments are not very persuasive to the majority of producers who are bound to regard agriculture as a business rather than a 'way of life' '. That many do so regard it I do not deny. But are they *bound* to? I believe that for many farmers—not only organic ones—farming *is* a 'way of life'. I would be surprised if most participants in the Conference did not agree with me that that is a healthy thing. Perhaps too, they would share my personal wish that many more people in our world could have the experience that earning one's living can be a meaningful part of a way of living.

INDEX

Acetylene, plant growth and, 69
Acioa baterri, 195, 197
Actinomycetes,
  in composting, 102
  in nitrogen fixation, 67
Aeschynomene virginica, 122
Africa,
  crop yields, 58
  minimum tillage in, 227
  mulch farming in, 227
  soil moisture, 41
Agriculture *see also under* types of
    agriculture,
  biology in, 251
  chemical development, 6
  future, 1
Agrobacterium radiobacter, 123
Agroforestry in tropics, 64
Alchornia sp., 195
Alcohol, as fuel from maize, 214
Aldicarb, 81, 86
Aldrin, 82, 84
Algae,
  atmospheric nitrogen fixation, 164
  blue green, 67, 141, 211
  hydrogen production by, 152
  in sewage utilization, 140
  nitrogen fixation by, 60, 67, 141, 155,
    162
Alkaloids,
  plant uptake, 175
  secretion of, 174
Allelochemicals, 169–180
  future role of, 176
  indirect effect, 177
  in weed/crop associations, 170
  nature of, 169
Allelopathy,
  bacteria in, 171
  nature of, 169
Alley cropping, 198, 338
Allolobophora caliginosa, 26
Allotments, 256
Ammonia loss in composting, 103, 108
Anabaema, 155

Anabaena cylindrica, 152
Animal fodder, intercropping and, 215
Animals *see also* Livestock,
  feeding, 323, 335
  fertility, effect of fodder on, 292
  invertebrate, influence of agriculture
    on *see also under* specific species,
    79–97
  on smallholdings, 259
Animal slurries,
  ARCUB system, 107
  as fertilizer, 137
  effect on invertebrate animals in soil, 90
  Licom method, 107
  methane production from, 136, 137
  separators, 148
  transport of, 136
  treatment of, 106, 245
Animal waste, annual production in UK,
  148
Anthonata macrophylla, 195, 197
Anthropods, 2
Antibiotics, 155
Aphids, 174, 296
  control of, 133
Apples, disease in, 296
ARCUB process of animal slurries, 107
Aristida oligantha, 178
Ark project, 283
Aromatic substances, liberation of, 172
Arthropods, effect of straw burning, 94
Atrazine, 87, 88
Australia, agriculture in, energy aspects,
  314
Azolla, 141, 210
  plant nutrients from, 211
Azolola filiculoides, nitrogen fixation by,
  190
Azospirrillum ligoferum, 68
Azotobacter sp., 162, 166
Azotobacter paspali, 68
Azotobacter vinlandii, 166

Bacillus popillae, 123
Bacillus thuringiensis, 123

339